FISH ECOLOGY

FISH ECOLOGY

Editor

Kord Jung

scitus
academics

Fish Ecology

Edited by **Kord Jung**

This Book is Printed on Acid-Free Paper. ♾

ISBN: 9781681179506

Library of Congress Control Number: 2017949088

SCITUS Academics LLC,

616, Corporate Way, Suite 2, 4766,

Valley Cottage, NY 10989

www.scitusacademics.com

Printed in United States of America.

Preface

Fish are a vital component of marine habitats. They are complexly related to other organisms – through the food web and through other mechanisms. Fish play a far more important role as contributors of nutrients to marine ecosystems than previously thought. Fish are very diverse with many adaptations to the different habitats provided by the sea and the continental waters. They can be found from freezing temperatures to the tropics, and have a wide range of life histories. The maximum diversity is found in tropical littoral areas, probably related both to the stability of the oceanographic conditions and the complexity of the habitat. Diversity is maintained through predation and competition. Besides the ecological importance of fish as components of the ecosystem and as part of the energy transfer in the food chain, man in fisheries ranging from a small- to an industrial- scale heavily exploits fish. Fish have adaptations to the fluid medium in which they inhabit, such as the hydrodynamic shape of all active swimming species. The ecosystems that support fisheries, together with other economic activities, are subject to a number of alterations of significant relevance to their functioning and resilience and to the goods and services they can provide. The ecosystem approach for fisheries management is a widely accepted concept and various international instruments require its application. However, there are various interpretations of the ecosystem approach and its application almost always brings about confrontations and resistance among managers, proponents, and stakeholders.

This volume 'Fish Ecology' deals with the latest researches in the fields of fish ecology, limnology and eco-conservation. The purpose of this book is to introduce the ecology of fishes by describing the inter-relationships between fishes and the aquatic habitats they occupy. It covers the main themes of ecology, including habitat use, species interactions, migration, feeding, population dynamics and reproduction in relation to the major habitats occupied by fishes. It moves on to address freshwater, estuarine, and marine ecosystems and identifies factors that affect the distribution and abundance of fishes. It is aimed not only for the cause of research but also for the scientists and administrators involved in the planning and implementation of policies related to fish ecology.

Table of Contents

CHAPTER 1

Ecological Risk and Distribution of Polychlorinated Biphenyls in Fish

T. Brázová[1], V. Hanzelová[1] and P. Šalamún[1]

[1] *Institute of Parasitology, Slovak Academy of Sciences, Košice, Slovakia*

ABSTRACT

The distribution and concentrations of polychrolinated biphenyls (PCBs) were determined in nine freshwater fish species and the parasite Acanthocephalus lucii in Zemplínska šírava, a heavily polluted water reservoir in Slovakia. The study performed at two different time points five years apart (2004 and 2009) revealed excessive PCB contamination of the fish muscle tissue and significant interspecies as well as tissue-specific differences in PCB uptake by fish. Total PCBs broadly correlated with the trophic position of individual fish species within a food chain ($P < 0.01$). The concentrations were particularly high in predatory fish species, perch, pike and pike-perch (108.0, 90.1, 113.0 $mg.kg^{-1}$ lipid wt, respectively), but comparable PCB values were also found in non-predatory detritivorous freshwater bream (128.0 $mg.kg^{-1}$ lipid wt). Tissue-specific differences in PCB concentrations were observed: liver > adipose tissue > muscles > hard roe > bones > brain. Maximum concentrations of PCBs were recorded in the liver of northern pike (214.0 $mg.kg^{-1}$ lipid wt) and the hepatopancreas of freshwater bream (163.0 $mg.kg^{-1}$ lipid wt). Individual congeners were not distributed homogeneously within the investigated organs and the adipose tissue. PCB 153 was present in the higher concentrations than other congeners in all fish organs as well as the adipose tissue. Acanthocephalans absorbed significantly higher concentrations of PCBs ($P<0.001$) than the muscles, liver, kidney, brain and adipose tissue of their host. About 20-fold lower PCB levels was detected in the liver and almost three times in the muscles of infected perch. Data on PCB accumulation in perch infected with acanthocephalans demonstrated a decline of PCB values in all organs as well as the adipose tissue compared to uninfected fish. The study has shown that the fish species, its feeding habit and specific conditions of the habitat are mutually interrelated factors that are responsible for significant variations in fish body burdens.

Keywords: polychlorinated biphenyls, pollution, fish, parasites, water reservoir Zemplínska šírava

1. INTRODUCTION

The ability of persistent organic pollutants (POPs) to spread through long distances and remain in the environment has resulted in their presence worldwide. They have a tendency to accumulate within the food chain and, as a result, pose a high risk to human and animal health. The risks caused by industrial POPs are best illustrated by polychlorinated biphenyls (PCBs), extensively used in a wide variety of applications from the early 20th century. PCBs became ubiquitous contaminants of various biotic and abiotic environments worldwide due to their massive and uncontrolled use in industry and agriculture [1, 2].

PCBs are highly persistent compounds in the environment, especially in aquatic sediments, which act as a stable reservoir from which PCBs can continue to be released over a long period of time, because of their low solubility in water and low volatility [3]. The involvement of PCBs in the food chain occurs through the incorporation of suspended particles in phytoplankton and zooplankton, at the base of the food chain. Bottom feeders and other aquatic organisms ingest, accumulate and pass PCBs upward in the food chain [4, 5]. Fish near the top of the aquatic food web have a relatively long life span and concentrate high amounts of PCBs [6]. The concentration of poorly metabolized chemicals accumulated in fish can thus reflect the degree of pollution in an aquatic system [7, 8].

Biomonitoring is a vital and rapidly growing field that uses several biological groups, such as phytoplankton, macrophytes, invertebrates and fish, as bioindicators [9]. This ecological methodology is increasingly used to assess the level of pollution in aquatic environments [10–13]. Numerous studies have also documented the close relationship between aquatic pollution and parasitism [e.g. 13, 14]. Certain parasitic organisms have the ability to concentrate high quantities of pollutants in their tissues and organs and thus can provide information about the chemical state of the environment [15, 16]. These studies have mainly focused on intestinal parasites, mostly acanthocephalans, as indicators of heavy-metal pollution

[14, 17, 18], but data on organic pollutants, including PCBs and their bioaccumulation in parasitic organisms, are still scarce.

The Zemplínska šírava water reservoir (eastern Slovakia; Fig. 1) is one of the most PCB-contaminated sites in Europe [19, 20]. Large amounts of PCB compounds have been released to the reservoir from a chemical factory in the nearby town of Strážske without treatment (decontamination) or preventive measures, leading to heavy contamination of soil, superficial and underground water and subsequently food chains in this area [21]. The PCBs in the reservoir and the inflowing Šíravský canal thus belong to the so-called old environmental burdens, and their amount varies from tens to hundreds of milligrams per kilogram of sediment dry weight. At least 40, 000 tons of PCB-contaminated sediments are assumed to be still present in the reservoir, effluent canal and Laborec River [22]. Studies over the past 20 years have indicated that elevated concentrations of PCBs in the reservoir pose a potential hazard to human and aquatic health [19, 23].

2. MATERIAL AND METHODS

2.1. Study Area

The Zemplínska šírava reservoir (48°47′32″N; 22°0′42″E), one of the largest artificial water-storage sites in Slovakia, is located in the eastern part of the country (Fig. 1).

It was built between 1961 and 1965 for regulating flood overflows from the Laborec River and partly for irrigation purposes in the area. The reservoir is 11 km long, ca. 3.5 km wide and covers an area of 33 km^2. PCB compounds from the chemical factory in Strážske were released into the Strážsky (effluent) canal to the Laborec River and from there by the Šíravský (influent) canal to the Zemplínska šírava reservoir (Fig. 1). The reservoir has abundant ichthyofauna, represented predominantly by dense populations of common carp, *Cyprinus carpio* L.; freshwater bream, *Abramis brama* L.; pike-perch, *Sander lucioperca* L.; northern pike, *Esox lucius* L.; European perch, *Perca fluviatilis* L.; Wels catfish, *Silurus glanis* L. and goldfish, *Carassius auratus* L. The PCBs in fish were monitored twice, first in 2004 and then in 2009, at eight sites in the reservoir (Fig. 1).

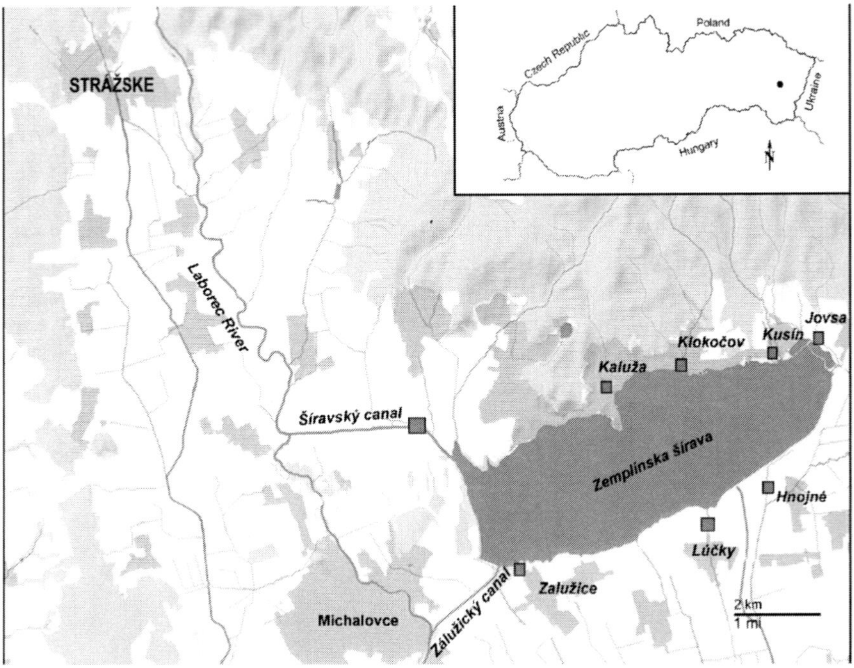

FIGURE 1. Location of the sampling sites in the Zemplínska šírava reservoir.

2.2. Fish and Parasite Collection

A total of 50 fish of nine species belonging to five families (Anguillidae, Cyprinidae, Esocidae, Percidae and Siluridae) were collected. The distribution and concentrations of PCBs were determined in predatory (*P. fluviatilis*, *E. lucius*, *S. lucioperca* and *S. glanis*) and non-predatory (*C. carpio*; the European eel, *Anguilla anguilla*; *Ab. brama*; *C. auratus* and the roach, *Rutilus rutilus*) species. The fishes were transported to the laboratory alive, weighed, measured and divided into feeding guilds [24, 25] (Table 1).

Only muscle tissues (19 samples) were collected in 2004. A more detailed study in 2009 focused on the tissue-specific distribution of PCBs to assess the temporal variations of PCBs in the reservoir. Samples of the liver (hepatopancreas in cyprinids), kidney, adipose tissue, hard roe, bone and brain were collected from 31 fish belonging to seven species [26].

TABLE 1. Number of fish and tissue/organ samples examined, food items and trophic levels. *Abbreviations*: Musc, muscle; Liv, liver; Kidn, kidney; Hroe, hard roe; AT, adipose tissue (Data from [26, 27]).

Fish species	*N*		Major food item	Trophic level*	Number of samples						
	2004	2009			Musc	Liv	Kidn	Hroe	AT	Bone	Brain
Predators											
European perch	0	16	Zoobenthos, nekton	3.2–4.4	6	5	5	–	3	–	5
Northern pike	1	2	Nekton	3.8–4.5	2	2	2	1	1	1	2
Pike-perch	2	3	Nekton	4.3	3	3	3	–	3	–	2
Wels catfish	0	3	Nekton	4.3–4.4	2	2	2	1	1	1	2
Non-predators											
Common carp	2	1	Zoobenthos	2.1–3.1	1	1	1	–	1	1	1
European eel	7	0	Zoobenthos	2.3–3.5							
Freshwater bream	5	1	Zoobenthos	2.9–3.1	1	1	–	1	1	1	1
Goldfish	1	5	Zoobenthos	2.0	5	5	5	1	–	2	5
Roach	1	0	Zoobenthos, plants	2.3–3.4	1	–	–	–	–	–	–

[i] - *N*, number of fish examined.

[ii] - * Trophic level was assessed as in [25].

Only perch were infected with parasites and so received special attention. Twelve of 16 perch in a single catch were small, and therefore the tissue/organ samples from these fish were pooled by the weight of the fish: up to 10 g (*n*=7) and 14–30 g (*n*=5). The four remaining fish weighed 60–130 g and were analysed separately. A total of 24 samples of muscles, livers, kidneys, adipose tissues and brains were examined for the presence of PCBs (Table 1). Screening of the digestive tracts for parasites using a stereomicroscope found only one helminth species, the acanthocephalan *Acanthocephalus lucii*. The two largest perch (>120 g) contained 50 and 19 acanthocephalans. The remaining fish were either free of acanthocephalans or their infections were low, mostly 1–2 acanthocephalans per fish in the 10–30 g group. Perch weighing 60 g contained a maximum of five parasites. Perch harbouring such low parasite burdens were considered as uninfected for the comparison of PCB accumulation in infected versus uninfected perch. The parasites were washed in saline, frozen and subsequently examined for PCBs. A total of 95 tissue/organ samples of all fish species (Table 1) and two samples of acanthocephalan parasites were analysed spectrophotometrically [27]. The scientific and common names of the fish match those in the FishBase database [25].

2.3. Analytical Procedure

The extraction and clean-up of the samples followed the methods described by Himberg et al. and Fisher and Ballschmiter [28, 29], with slight modifications. Briefly, the fish and parasite samples were homogenized in anhydrous sodium sulfate and extracted with a mixture of petroleum ether (90%) and acetone (10%) using a separation funnel. The extracts were concentrated in a rotary evaporator and then purified using a Florisil chromatographic column according to STN EN 12393-2 (2001). The final extracts were analysed on a 6890 gas chromatograph (Hewlett-Packard) equipped with and electron capture detector. The HP-5 capillary column was 30 m in length, with an i.d. of 0.25. and having a film thickness of 0.25 μm. The chromatograph was operated at an injector temperature of 250°C and a detector temperature of 300°C and used helium as the carrier gas at a flow rate of 1.4 ml min^{-1}. The following oven temperature program was used: start temperature 80°C for 1 min, 80–180°C at 30°C min^{-1} and maintained for 1 min, 185–205 °C at 6°C min^{-1} and maintained for 15 min and 205–290 °C at 20°C min^{-1} and maintained for 7.5 min. Individual PCB congeners were identified by their comparison with the retention times of known standards and qualified by comparing the peak areas to the appropriate peaks in the standard mixture (PCB Mix C-SCA-06). The extracts were injected under a splitless mode. The recovery rates of the PCBs in the spiked samples were 80–95%, whereas the detection limits were 1 μg kg^{-1} based on wet weight. Six PCB indicator congeners – PCB 28, 52, 101, 138, 153 and 180 – were analysed. All PCB concentrations in the biological samples are given in mg kg^{-1} lipid weight (lipid wt).

2.4. Statistical Analysis

The data obtained did not meet the requirements for parametric statistical tests (normality). So they were analysed non-parametrically or were log-transformed [ln $(x + 1)$] to satisfy the assumptions (normal distribution and homogeneity of variances) of analyses of variance (ANOVAs). Differences in PCB load between years and between fish tissues/organs and parasites were determined by *t*-tests. The effects of trophic strategy (predatory and non-predatory fish) and tissue/organ type on PCB levels were tested with two-way ANOVAs. Main-effect ANOVAs were used instead of factorial ANOVAs because no significant interactions were confirmed between the factors. The data were consequently divided into two groups, predatory and non-predatory fish, and were processed separately. The concentrations are expressed as mean ± SD (standard deviation). Differences in mean concentrations (mg kg^{-1} lipid wt) of the congeners and total PCBs in the fish tissues/organs (muscle, liver, kidney, brain, adipose tissue and bone for predatory/non-predatory fish) were tested with a non-parametric Kruskal-Wallis ANOVA with post-hoc multiple comparison. The analyses were performed in Statistica for Windows, version 9.0 [30].

3. RESULTS AND DISCUSSION

3.1. PCB Distribution Associated with Fish Species and Feeding Habit

The total PCB concentrations (the sum of the six PCB congeners) in the muscle tissues varied greatly in the nine fish species collected in 2004 and 2009 from the Zemplínska šírava reservoir (Fig. 2).

Total PCB concentrations were generally associated with the trophic position of the species within a food chain. In the predatory species, pike and pike-perch, the concentrations in both years were particularly high, which was expected for the highest trophic level (Table 1). The bottom-feeding and detritivorous non-predatory freshwater bream that obtains nutrients directly from the sediments exposed to PCBs, however, had comparable PCB concentrations. In contrast, the common carp that also feeds from the bottom sediments surprisingly had the lowest PCB concentrations, along with the European eel with a benthic feeding habit and a high body-fat content (Fig. 2).

PCBs clearly tended to bioaccumulate in the muscle tissues of some species (freshwater bream, pike and pike-perch) during the five years (2004–2009) between analyses (Fig. 2). PCB concentrations increased significantly (*t*=-

2.43, df=13, $P<0.05$) in these species and more than two-fold in freshwater bream. The opposite, but not significant, trend was seen in two cyprinids (common carp and goldfish), in which the PCB concentrations were lower in 2009 than in 2004 (Fig. 2).

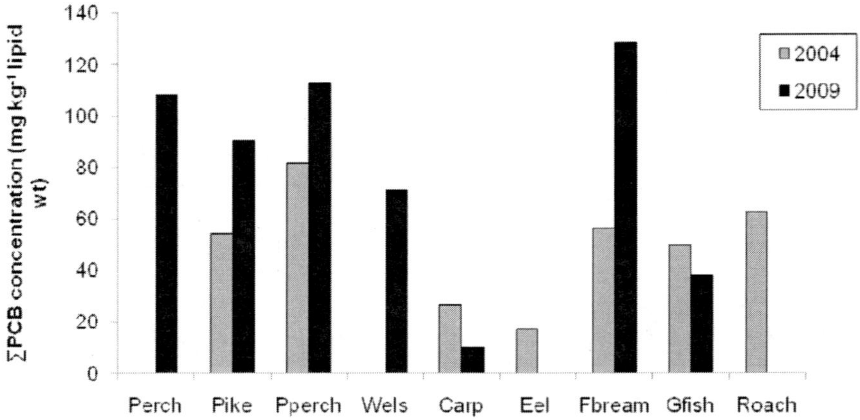

FIGURE 2. Total PCB concentrations (∑PCB) in the muscles of the various fish species from the Zemplínska šírava reservoir in 2004 and 2009. Perch and Wels catfish for 2009 and eels for 2004 were not analysed. *Abbreviations*: Carp, common carp; Eel, European eel; Fbream, freshwater bream; Gfish, goldfish; Perch, European perch; Pike, Northern pike; Pperch, pike-perch; Wels, Wels catfish [26].

3.2. PCB Distribution Associated with Fish Tissues/Organs

The concentrations of the PCB congeners in the tissues/organs of both predatory and non-predatory fish were high (Table 2).

The liver was the major target organ for PCBs in the predatory fish (Table 2), and it received the largest proportion (24%) of the overall contaminant burden. The PCB concentrations (lipid wt) in the tissues/organs decreased in the order: liver > muscles > adipose tissue > hard roe > kidney > bones > brain, with congener 153 markedly dominant in all tissues/organs of the predatory fish. The concentrations of all PCB congeners were significantly lower in the brain and in the liver and muscles, which had the highest concentrations, except for congener 28 in the muscles.

Hard roe had the highest PCB accumulation capacity in the non-predatory fish, followed by adipose tissue, bones, liver, muscles, kidney and brain. Similarly to the predators, the non-predators had high mean concentrations of congener 153 in all tissues/organs. Despite the considerable variation in

PCB concentrations in the various tissues/organs, the observed differences were not significant in the non-predatory fish (*P*>0.05; Table 2).

The two-way main-effect ANOVA indicated that the feeding habits of the fish [$F(6, 77)=9.10$, $P<0.00001$] and particularly the individual fish tissues/organs [$F(36, 340.89)=1.77$, $P<0.01$] affected PCB bioaccumulation. PCB concentrations detected in the muscles and liver were nearly two-fold higher in the predators, which prefer a diet of live fish, compared to the non-predatory fish (Table 2). The concentrations markedly exceeded the acceptable PCB limits established for fish by the Food Codex [31] in the muscle tissues for both groups of fish.

The tissue/organ analysis performed in 2009 showed that fish accumulated PCBs selectively. The liver (hepatopancreas in cyprinids) had the highest PCB burden, with maximum concentrations recorded in northern pike (214.0 mg kg^{-1} lipid wt) and freshwater bream (163.0 mg kg^{-1} lipid wt) (Table 3).

TABLE 2. Mean (± SD) concentrations (mg kg^{-1} lipid wt) of congeners and total PCBs in fish tissues/organs (Data from [27]).

	Muscle	**Liver**	**Kidney**	**Brain**	**Hard roe**	**Adipose tissue**	**Bones**	**Limit**[3]
Predators[1]								
PCB 28	2.60^{a,b}± 0.82	3.63^{b}± 1.62	2.95^{a,b}± 1.97	1.17^{a}± 1.24	1.91 ± 0.290	2.15^{a,b}± 1.28	0.79 ± 0.13	0.2
PCB 52	8.63^{b} ± 5.91	8.23^{b}± 6.07	4.04^{a,b}± 2.76	1.49^{a}± 1.30	11.0 ± 7.44	5.51^{ab} ± 3.37	3.62 ± 3.70	0.2
PCB 101	13.4^{b} ± 6.45	14.3^{b}± 11.2	7.73^{a,b}± 8.72	2.09^{a}± 1.45	9.43 ± 1.77	13.2^{b} ± 9.80	1.95 ± 2.12	0.2
PCB 138	24.7^{b} ± 13.3	29.9^{b}± 28.4	14.7^{a,b}± 17.1	4.35^{a}± 5.74	18.6 ± 9.26	23.5^{b} ± 18.7	5.64 ± 2.31	0.3
PCB 153	29.5^{b} ± 14.8	37.2^{b}± 34.1	20.6^{a,b}± 24.1	7.96^{a}± 9.64	23.4 ± 8.54	26.6^{a,b}± 20.9	5.44 ± 5.77	0.3
PCB 180	20.5^{b} ± 11.7	26.4^{b}± 25.0	11.3^{a,b}± 13.2	2.80^{a}± 2.75	15.0 ± 6.03	22.1^{b} ± 21.4	3.41 ± 1.46	0.2
ΣPCB	**99.3^{b}± 49.9**	**120^{b}± 104**	**61.4^{ab}± 63.0**	**19.9^{a}± 20.1**	**79.4 ± 18.5**	**93.1^{a,b}± 71.5**	**20.9 ± 15.5**	
Non-Predators[2]								
PCB 28	1.31 ± 0.93	1.14 ± 1.02	1.33 ± 1.13	0.84 ± 0.93	1.52 ± 0.110	0.86 ± 0.55	2.05 ± 1.89	0.2
PCB	4.31 ±	4.03 ±	3.52 ±	1.51 ±	4.53 ±	2.28 ±	4.29 ±	0.2

52	2.71	2.94	2.99	2.18	0.24	2.04	5.31	
PCB 101	5.82 ± 4.15	4.87 ± 4.84	4.56 ± 3.82	1.60 ± 1.98	8.68 ± 6.30	6.73 ± 7.93	8.81 ± 12.2	0.2
PCB 138	8.17 ± 12.1	10.6 ± 15.4	1.51 ± 0.82	1.73 ± 1.51	23.2 ± 19.7	20.2 ± 25.2	3.33 ± 4.84	0.3
PCB 153	16.0 ± 15.1	15.4 ± 18.2	9.25 ± 7.99	3.67 ± 4.26	28.7 ± 24.5	24.3 ± 30.5	20.8 ± 33.1	0.3
PCB 180	11.3 ± 11.9	12.0± 15.0	6.90 ± 5.55	1.93 ± 2.32	21.6 ± 19.8	18.4 ± 23.1	14.6 ± 23.9	0.2
ΣPCB	**46.9 ± 42.2**	**48.1 ± 54.7**	**26.1 ± 20.4**	**11.3 ± 11.6**	**88.2 ± 69.9**	**72.7 ± 89.4**	**53.8 ± 80.2**	

[i] - Significant differences ($P<0.05$) between tissues/organs for mean congener concentrations are indicated by different letters. Means followed by the same letters in the same row are not significantly different (post-hoc multiple comparison, $P=0.05$).

[ii] - [1] Means without letters (hard roe and bones) are not included in the comparisons, because these samples were too small.

[iii] - [2] No significant differences were confirmed for the non-predatory group.

[iv] - [3] Acceptable limits established for food [31].

The storage capacity of the muscles was slightly lower, but in some cases, PCB concentrations were higher in the muscles than in the liver, as in the European perch (108.0 and 80.5 mg kg^{-1} lipid wt in muscles and liver, respectively) and goldfish (38.0 and 32.6 mg kg^{-1} lipid wt, respectively). PCBs accumulated to a lesser degree in the kidney and brain than in the muscle tissue and liver (Table 3), with a significant difference only for the brain ($P<0.05$).

Most organic pollutants in aquatic ecosystems tend to accumulate in organisms, in which they can reach levels hundreds or thousands of times higher than the levels in water. Fish are at the top of the aquatic food web and concentrate high amounts of contaminants, including PCBs [6]. Ecotoxicological studies that have used fish as bioindicators have highlighted that the bioaccumulation of organic pollutants is affected substantially by biological factors, such as species, age, size, physiological condition and also parasitism, which was revealed recently [32, 33]. The position of fishes in a food chain and their lipid contents represent other important predictors of organochlorine accumulation [34, 35].

TABLE 3. Mean concentrations (mg kg^{-1} lipid wt) of PCBs in individual tissues/organs collected in 2009 [26].

Fish species	Muscle	Liver	Kidney	Brain
Predators				
European perch	108.0	80.5	75.9	27.3
Northern pike	90.1	214.0	14.3	13.6
Pike-perch	113.0	128.0	110.0	31.1
Wels catfish	71.1	71.1	13.1	3.74
Non-predators				
Common carp	10.2	10.2	3.80	1.44
Freshwater bream	128.0	163.0	-	1.73
Goldfish	38.0	32.6	30.5	15.2

Fish are exposed to PCBs both via direct contact with the water and via food uptake by the membranes of the gastrointestinal tract [36]. Most of the PCBs, which have low solubility in water and fish, are likely derived from food rather than the ambient environment. Predatory fish, by eating other contaminated fish, accumulate higher amounts of organochlorine pollutants than fish with other feeding habits. Some authors [37], however, have arrived at opposite conclusions, and some studies have reported only slight differences between predators and bottom feeders [38, 39]. The results obtained in the present study are not entirely unequivocal; PCB concentrations were highest in the predatory perch, pike and pike-perch, but the non-predatory freshwater bream had comparable concentrations. The feeding behavior of bream is characterized by permanent direct contact with the sediments and a small radius of migration. Bream were thus once considered an important transfer vector for PCBs and one of the best organisms for monitoring freshwater contamination [40]. Our results support this finding.

European eel is commonly classified as a predaceous carnivorous fish, although its position in the trophic scale is lower than that of typical predators such as pike and pike-perch [25]. PCBs are lipophilic compounds prone to high bioaccumulation in eels, which have a high fat content. Accordingly, PCB concentrations in eels are usually higher than in other freshwater fish species [34, 41]. In our study, the muscles of the European eel had an unexpectedly low level of total PCBs (17.1 mg kg^{-1} mean lipid wt), which is the second lowest level in the nine species examined. Due to the interrupted natural flow of groundwater, we assumed that the sedentary eels from the reservoir do not use their energy reserves to migrate upstream, and so may have stored PCBs for long periods in their metabolically inactive adipose tissues. Adipose tissue can retain PCBs for a long time, or PCBs may accumulate in other fat-rich tissues/organs, most probably the liver. We could not confirm the latter hypothesis, as we did not capture any eels in 2009, when the tissue-specific accumulation was investigated.

The closely related cyprinids (carp, freshwater bream, goldfish and roach) with similar ecologies and food preferences [25] differed markedly from each other in PCB content. PCB concentrations were relatively similar in these fishes in 2004, ranging between 27.2 mg kg^{-1} lipid wt in carp and 62.5 mg kg^{-1} lipid wt in roach, but varied widely in 2009. Freshwater bream had extraordinarily high PCB concentrations in 2009 (128.0 mg kg^{-1} lipid wt). Species-specific differences in PCB concentrations were also found in the individual tissues/organs of the cyprinids. PCBs preferentially accumulated in the hepatopancreas in freshwater bream, but the concentrations were equal or nearly equal in the hepatopancreas and muscles of carp and goldfish, respectively (Table 3). These differences could not be attributed to different diets, because these species feed on similar foods (bottom-dwelling invertebrates, plankton in the deeper parts of water bodies, plants and detritus). In addition to biological indices, the environmental conditions prevailing at a particular field site must be considered, because even slight environmental perturbations can induce a high degree of heterogeneity in biochemical markers [42].

Among the potential causative factors responsible for the high variation of PCB levels in our fish, particularly the tissues/organs in the same species (compare pike, pike-perch, carp and bream in 2004 and 2009), the highly varying hydrological conditions in the reservoir may have played a large role. The regular monitoring of the reservoir from 1999 to 2007 [43, 44] revealed large variability in the deposition of sediments at different sites and even the movement of sediments (so-called bottom waves) during seasonal floods, which released PCBs from the sediments and became more available to aquatic organisms. The concentrations of PCB indicator congeners in the bottom sediments increased substantially between 1999 and 2003 (467 and 1633 μg kg^{-1} dry wt, respectively) but were much lower at the same sites in 2005 and 2007 (428 and 692 μg kg^{-1} dry wt, respectively) [43, 44]. The study by Šalgovičová and Zmetáková [20] is consistent with a role for hydrological conditions in PCB accumulation. PCB concentrations in their study showed a decreasing tendency in non-predatory fish species. The levels of PCBs in goldfish reached 133.7 mg kg^{-1} in 2002 but had decreased nine-fold by 2004.

Most information about PCB compounds in aquatic organisms is from the muscles of fish and piscivorous birds [19, 45, 46], and very few data are available for the PCB contents in other organs. More attention has only recently been paid to the tissue-specific distribution of PCBs in fish [47].

The amount of total PCBs in a tissue or organ depends on the lipid content [19, 48]. The most important depot sites for PCBs in the predatory and non-predatory fish examined in the present study, next to the intestinal fat with obviously high lipid content (ca. 50%) [49], were the liver, hard roe and muscles. Other studies [50–52] have also reported higher PCB concentrations in the liver than in other organs, which could be due to its high fat content.

Ovaries are also rich in fat and serve as important depots of contaminants such as PCBs [51]. The lipid and consequently PCB contents in hard roe, unlike in the liver, however, can vary considerably. During maturation of female fish, lipids pass from somatic to gonad tissue, accompanied by an increase in PCBs in the hard roe. In contrast, lipid-rich eggs released during spawning represent a major route of PCB loss in female fish [53]. The high PCB concentrations we found in the hard roe in both fish groups may thus be associated with the spawning period, because the majority of the fish in this study were caught in May, shortly before the peak of the spawn.

As shown by our results, the distribution of lipophilic compounds is not always directly proportional to the lipid content in individual tissues/organs. The brain, for example, with a relatively high fat content [54], appears to be much better protected against the accumulation of PCBs than other tissues/organs. This protection may be due to the efficient blood–brain barrier, as indicated by the relatively low concentrations of all investigated PCB congeners in our study, ranging from 0.84 (PCB 28 in non-predators) to 8.0 (PCB 153 in predators) mg kg^{-1} lipid wt. Several other conditions/factors, including species, age, size, feeding habits, metabolic activity of individual organs or the complex PCB transport in an organism may control PCB accumulation [26, 55].

3.3. Concentrations And Profiles Of The PCB Congeners In Fish

The levels of six indicator congeners, 28, 52, 101, 180, 138 and 153 (in order of increasing concentrations), were high. All the congeners reached their upper acceptable limit in both predators and non-predators (Table 2). The various congeners were not distributed homogeneously within the tissues/organs. PCB 153 was present in higher concentrations than the other congeners in all tissues/organs, representing 31% and 34% of the mean total PCB concentrations in predators and non-predators, respectively (Table 2). PCB 138 and 180 accounted for approximately 25% and 21% of the total amount of PCBs in predators and non-predators, respectively.

Despite the differences in PCB concentrations in individual tissues/organs, the relative proportions of the congeners within the major target tissues/organs, the liver, hard roe, muscle, adipose tissue and kidney, were relatively similar in both fish groups. PCB 153 and 138, the congeners most frequently found in animal samples [56], were also predominant in our assays. Congeners 138, 153 and 180 are less hydrophobic and not as tightly bound to sediments as the more highly chlorinated octa-, nona- and deca-PCBs and are thus more readily available to aquatic organisms [57]. In addition, these congeners with chlorine atoms in positions 2, 4 and 5 in one (PCB 138) or both (PCB 153 and PCB 180) rings may be more resistant to metabolism and elimination from the organism than the congeners with fewer chlorine atoms, such as 28, 52 and 101 [58, 59]. The very high proportion of PCB 138, 153 and 180 found in our tissue/organ samples (76% and 79% of all PCB congeners in predators and non-predators, respectively) may reflect their low rates of biotransformation and resistance to metabolism.

3.4. PCB Concentrations in Perch and Parasites

The accumulation of PCBs was studied in the perch–*A. lucii* host–parasite system. Acanthocephalans attached to the intestinal mucosa accumulated significantly higher concentrations of PCBs (single-means *t*-test, $t < -4.2$, $P < 0.001$) than did the tissues/organs of their hosts (Fig. 3 and Table 4).

PCB 153 predominated in acanthocephalans (as in the fish), followed by PCB 138 and 180 (Table 4). Notably, the concentration of PCB 28 was approximately 10-fold higher in the parasites than in the muscles, liver and adipose tissue of the perch (Table 4).

The numbers of fish tissue/organ samples were too low for reliable statistical analysis (10 samples from infected and 14 samples from uninfected fish), but the results indicated much lower quantities of PCBs in all tissues/organs from the fish infected with acanthocephalans. The concentrations of PCBs were approximately 20- and 3-fold lower in the liver and muscles, respectively, of the infected perch (Fig. 3).

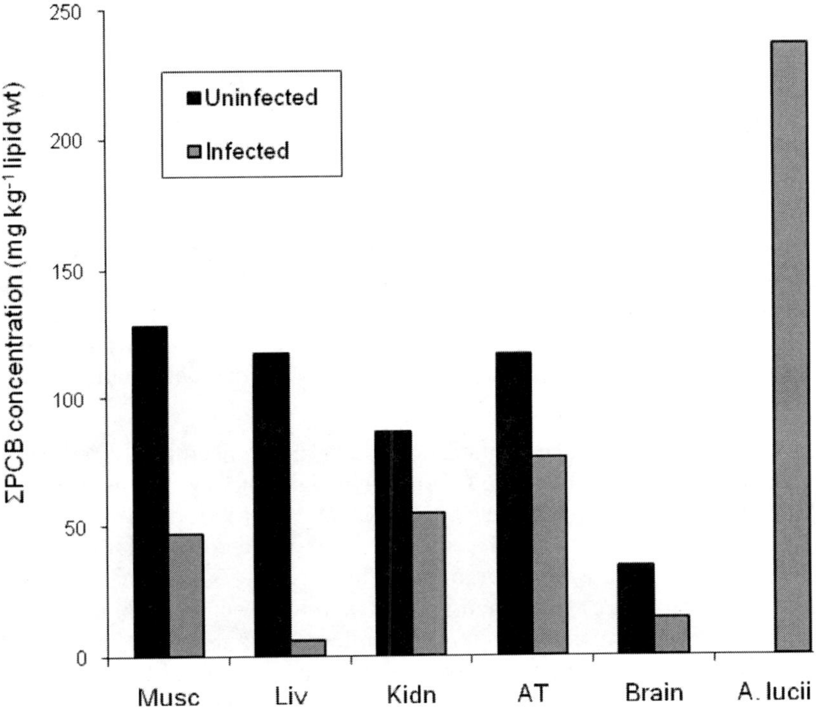

FIGURE 3. PCB concentrations in the tissues/organs of infected and uninfected fish and in *Acanthocephalus lucii* (Parasite). *Abbreviations*: Musc, muscle; Liv, liver; Kidn, kidney; AT, adipose tissue [27].

TABLE 4. Mean and total (∑PCB) concentrations (mg kg^{-1} lipid wt) of the congeners in various tissues/organs of perch (*P. fluviatilis*) and in its intestinal parasite (*A. lucii*) [27].

	Muscle	Liver	Adipose tissue	Kidney	Brain	Parasite
PCB 28	2.9	2.9	2.0	5.0	3.2	20.9
PCB 52	6.3	2.5	5.4	4.1	4.6	7.2
PCB 101	16.8	10.2	20.0	11.7	14.7	36.2
PCB 138	26.9	17.5	22.9	15.9	20.8	61.4
PCB 153	31.4	28.4	24.6	22.1	26.6	63.6
PCB 180	23.8	19.0	21.9	17.1	20.5	47.2
∑PCB	108.1	80.5	96.9	75.9	90.3	237.0

Any kind of pollution represents environmental stress that influences host–parasite interactions and could cause outbreaks of fish diseases [60]. The immune system of fish is severely impaired by exposure to contaminants that consequently leads to an increase of parasitism. Sagerup et al. [61] found a positive correlation between PCB contamination of the habitat and intensity of nematode infection in gulls (*Larus hyperboreus*). Immunosuppression induced by PCBs was confirmed experimentally [62]. European eels infected with the swim-bladder nematode *Anguillicola crassus* were exposed to sublethal concentrations of PCB 126, which completely suppressed the antibody response in the eels. The fish exposed to the PCBs were consequently more easily infected than unexposed fish. A combination of anthropogenic stressors (e.g. PCB pollution) and natural stressors (e.g. parasitic infection) usually has a more detrimental effect on fish than each stressor alone [63].

Parasites are able to accumulate pollutants and consequently reduce their incorporation in the hosts [64, 65], particularly heavy metals. Supporting data for this role of parasites in fish PCB contamination, however, remain scarce. *Pisidium amnicum*, a freshwater clam, infected with larval stages of digeneans and exposed to organic pollutants (2,4,5-trichlorphenol (TCP) and benzo(a)pyrene) contained approximately 12% and 40% less TCP and benzo(a)pyrene, respectively, than uninfected clams [66]. Persson et al. [67], however, reported a slightly lower concentration of PCBs (1.92 ng g^{-1} lipid wt) in the tapeworm *Eubothrium crassum* than in their salmon (*Salmo salar*) hosts (2.65 ng g^{-1} lipid wt), which was attributed to the absence of an alimentary canal in all developmental stages of the tapeworms.

The present study provides the first evidence of the ability of fish parasites (acanthocephalans) to accumulate relatively high levels of PCBs. Our data demonstrated lower PCB concentrations in all tissues/organs of infected relative to uninfected perch under natural conditions. These data indicate that some parasitic organisms may have a positive influence on their hosts in PCB-contaminated environments, as previously demonstrated for heavy-metal pollution [14, 65].

4. CONCLUSIONS

Our results demonstrated a persistent problem with the "old environmental heritage" of PCBs, leading to high risks mainly for predatory fish species in aquatic ecosystems, and for the human populations living near these regions. The commercial production of PCBs has been banned or severely restricted in Slovakia since the 1980s, but the fish in the Zemplínska šírava reservoir continue to accumulate these chemicals from the sediments polluted in the previous decades. Critical PCB levels recorded in fishes and the tendency of bioaccumulation in some species in the reservoir are essential for predicting which food webs are at risk for higher rates of bioaccumulation that endanger the health of upper-trophic predators, including humans living in the area. This study indicated that the different fishes, their dietary intakes and the chemical properties of the pollutants were interrelated factors, all of which were important in the tissue/organ burdens. Fish parasites, due to their good properties of bioaccumulation and position in the food chain, should be carefully considered in any ecotoxicological research.

5. ACKNOWLEDGEMENTS

We thank Dr. Ľudmila Turčeková for her cooperation in collecting fish samples and Dr. William Blackhall for his editorial comments on the text. We gratefully acknowledge the funding of this study by the Slovak Research and Development Agency, project No. LPP-0151-07, Grant Agency of the Ministry of Education of the Slovak Republic, and Slovak Academy of Sciences (VEGA), project No. 2/0080/10. The publication has been realized as part of the project of the Centre of Excellence for Parasitology (Code ITMS: 26220120022) supported by Operational Programme "Research & Development" funded by the European Regional Development Fund (rate 0.9).

REFERENCES

1. Meijer NS, Ockenden AW, Sweetman A, Brevik K, Grimalt OJ, Jones CK. Global distribution and budget of PCBs and HCB in background surface soils: Implications for sources and environmental processes. Environmental Science and Technology 2003; 37: 667–672.
2. Yang F, Wilcox B, Jin S, Aguirre AA, Rougée L, Xu Y, Lu Y. Detection and quantitative analysis of polychlorinated biphenyls in tilapia from Hawaiian waters. Chemosphere 2008; 73: 133–137.
3. U.S. Environmental Protection Agency (1999). Polychlorinated biphenyls (PCBs) update: Impact on fish advisories. Office of Science and technology, Washington, DC. EPA-823-F-99-019.
4. Larsson P, Okla L, Ryding SO, Westoo B. Contaminated sediment as a source of PCBs in a river system. Canadian Journal of Fisheries and Aquatic Sciences 1990; 47: 746–754.
5. German AV, Zakonnov VV. Accumulation of polychlorinated biphenyls in the Sheksninskii pool of the Rybinsk reservoir. Water Research 2003; 30: 524–528.
6. Bervoets L, Blust R. Metal concentrations in water, sediment and gudgeon (*Gobio gobio*) from a pollution gradient: Relationship with fish condition factor. Environmental Pollution 2003; 126: 9–19.

7. Sultana R, Rao DP. Bioaccumulation patterns of zinc, copper, lead and cadmium in grey mullet, *Mugil cephalus* (L.), from harbour waters of Visakhapatnam, India. Bulletin of Environmental Contamination and Toxicology 1998; 60: 949–955.
8. Fang JKH, Wu RSS, Zheng GJ, Au DWT, Lam PKS, Shin PKS. The use of muscle burden in rabbitfish *Siganus oramin* for monitoring of polycyclic aromatic hydrocarbons and polychlorinated biphenyls in Victoria harbour, Hong Kong and potential human health risk. Science of the Total Environment; 407: 4327–4332.
9. Rosenberg DM, Resh VH. Freshwater Biomonitoring and Benthic Microinvertebrates; Chapman and Hall, New York, 1993; 1–488.
10. MacKenzie K, Williams HH, McVicar AH, Siddal R. Parasites as indicators of water quality and the potential use of helminth transmission in marine pollution studies. Advances in Parasitology 1995; 35: 85–144.
11. Sures B. The use of fish parasites as bioindicators of heavy metals in aquatic ecosystems: A review. Aquatic Ecology 2001; 35: 245–255.
12. Sures B. Environmental Parasitology: Relevancy of parasites in monitoring environmental pollution. Trends in Parasitology 2004; 20: 170–177.
13. Vidal-Martínez VM, Pech D, Sures B, Purucker ST, Poulin R. Can parasites really reveal environmental impact? Trends in Parasitology 2010; 26: 44–50.
14. Brázová T, Torres J, Eira C, Hanzelová V, Miklisová D, Šalamún P. Perch and its parasites as heavy metal biomonitors in a freshwater environment: The case study of the Ružín water reservoir, Slovakia. Sensors 2012; 12: 3068–3081.
15. Gelnar M, Šebelová Š, Dušek L, Koubková B, Jurajda P, Zahrádková S. Biodiversity of parasites in freshwater environment in relation to pollution. Parassitologia 1997; 39: 189–199.
16. Jankovská I, Miholová D, Lukešová D, Kalous L, Válek P, Romočuský Š, Vadlejch J, Petrtýl M, Langrová I, Čadková Z. Concentrations of Zn, Mn, Cu and Cd indifferent tissues of perch (*Perca fluviatilis*) and in perch intestinal parasite (*Acanthocephalus lucii*) from the stream near Prague (Czech Republic). Environmental Research 2012; 112: 83–85.
17. Sures B, Siddall R. *Pomphorhynchus laevis* (Palaeacanthocephala) in the intestine of chub (*Leuciscus cephalus*) as an indicator of metal pollution. International Journal of Parasitology 2003; 33: 65–70.
18. Taraschewski H. Host–parasite interactions in Acanthocephala: A morphological approach. Advances in Parasitology 2000; 46: 1–179.
19. Kočan A, Petrik J, Jursa S, Chovancová J, Drobná B. Environmental contamination with polychlorinated biphenyls in the area of their former manufacture in Slovakia. Chemosphere 2001; 43: 595–600.
20. Šalgovičová D, Zmetáková Z. Polychorinated biphenyls in muscle tissue of freshwater fish in East Slovakia. Journal of Food and Nutrition Research 2006; 45: 171–178.
21. Langer P, Tajtáková M, Kočan A, Trnovec T, Šeboková E, Klimes I. From naturally occurring goitrogens to the effects of anthropogenic endocrine disruptors on the thyroid in Slovakia. Bratislavské Lekárske Listy 2003; 104(3): 101–107.
22. Dercová K, Čičmanová J, Lovecká P, Demnerová K, Macková M, Hucko P, Kušnír P. Isolation and identification of PCB-degrading microorganisms from contaminated sediments. International Biodeterioration and Biodegradation 2008; 62: 219–225.
23. Petrik J, Chovancová J, Kočan A Holoubek I. Project TOCOEN. The fate of selected organic pollutants in the environment VIII. PCBs in human adipose tissues from different regions of Slovakia. Toxicological and Environmental Chemistry 1991; 34: 13–18.
24. Baruš V, Oliva O. Mihulovci (Petromyzontes) & ryby (Osteichthyes) (1). Academia Praha; 1995; 1–623.
25. Froese R, Pauly D (2014) FishBase. World Wide Web electronic publication, www.fishbase.org, version (09/2014).
26. Brázová T, Hanzelová V, Miklisová D, Šalgovičová D, Turčeková Ľ. Biomonitoring of polychlorinated biphenyls (PCBs) in heavily polluted aquatic environment in

different fish species. Environmental Monitoring and Assessment 2012; 184: 6553–6561.

27. Brázová T, Hanzelová V, Miklisová D. Bioaccumulation of six PCB indicator congeners in a heavily polluted water reservoir in Eastern Slovakia: Tissue-specific distribution in fish and their parasites. Parasitology Research 2012; 111: 779–786.
28. Himberg K, Julkunen P, Pyysalo H. A method for the quantification of polychlorinated biphenyls. Technical research centre of Finland, Espoo; 1987; 1–27.
29. Fisher R, Ballschmiter K. Congener-specific identification of technical PCB mixtures by capillary gas chromatography. Fresenius' Zeitschrift für Analytische Chemie 1989; 335 (5): 457–463.
30. StatSoft, Inc. (2009) STATISTICA (data analysis software system), version 9.0. www.statsoft.com.
31. Food Codex of the Ministry of Agriculture of the Slovak Republic, Regulation no. 18558/2006–SL.
32. Fisk AT, Hobson KA, Norstrom RJ. Influence of chemical and biological factors on trophic transfer of persistent organic pollutants in the northwater polynya marine food web. Environmental Science and Technology 2001; 35: 732–738.
33. Borgå K, Fisk AT, Hoekstra PF, Muir DCG. Biological and chemical factors of importance in the bioaccumulation and trophic transfer of persistent organochlorine contaminants in arctic marine food webs. Environmental Toxicology and Chemistry 2004; 23: 2367–2385.
34. Roche H, Buet A, Jonot O, Ramade F. Organochlorine residues in European eel (*Anguilla anguilla*), crucian carp (*Carassius carassius*) and catfish (*Ictalurus nebulosus*) from Vaccarès lagoon (French National Nature Reserve of Camargue): Effects on some physiological parameters. Aquatic Toxicology 2000; 48: 443–459.
35. Zhou HY, Wong MH. Screening of organochlorines in freshwater fish collected from the Pearl River delta, People's Republic of China. Archives of Environmental Contamination and Toxicology 2004; 46: 106–113.
36. Burreau S, Zebuhr Y, Broman D, Ishaq R. Biomagnification of polychlorinated biphenyls (PCBs) and polybrominated diphenyl ethers (PBDEs) studied in pike (*Esox lucius*), perch (*Perca fluviatilis*) and roach (*Rutilus rutilus*) from the Baltic Sea. Chemosphere 2004; 55: 1043–1052.
37. Gonzales GJ, & Fresquez PR. Polychlorinated biphenyls (PCBs) in predator and bottom-feeding fish from Abiquiu and Cochiti Reservoirs in North-central New Mexico. Los Alamos National Laboratory report LA-14289. Los Alamos, NM; 2006.
38. Kidwell JM, Phillips LJ, Bichard GF. Comparative analysis of contaminant levels in bottom feeding and predatory fish using the National Contaminant Biomonitoring Program Data. Bulletin of Environmental Contamination and Toxicology 1995; 54: 919–923.
39. Caldas ED, Coelho R, Souza LCKR, Siba SC. Organochlorine pesticides in water, sediment, and fish of Paranoá Lake of Brasilia, Brazil. Bulletin of Environmental Contamination and Toxicology 1999; 62: 199–206.
40. Marth P, Oxynos K, Schmitzer J, Schramm K-W, Kettrup, A. Levels of chlorinated hydrocarbons (CHC1) in breams (*Abramis brama*) from the River Elbe (A contribution to the Federal Environmental Specimen Bank). Chemosphere 1997; 34: 2183–2192.
41. Geeraerts C, Belpaire C. (2010). The effects of contaminants in European eel: A review. Ecotoxicology 2010; 19: 239–266.
42. Roche H, Bogé G. Fish blood parameters as a potential tool for identification of stress caused by environmental factors and chemical intoxication. Marine Environmental Research 1996; 41: 27–43.
43. Hucko, P. PCBs in sediments of the Strazsky Canal, the Laborec River and the Zemplinska Sirava Reservoir. In: Contaminated Sites. Conference Proceedings. Bratislava, 2009.

44. Hucko, P. Trend of contamination of the Laborec River, Strazsky Canal and Zemplinska Sirava Reservoir by polychlorinated biphenyls. In: Proceedings Water days of the region. Michalovce, 2009.
45. Baeyens W, Leermakers M, Elskens M, VanLarebeke N, DeBont R, Vanderperren H, Fontaine, A, Degroodt J–M, Goeyens L, Hanot V, Windal I. PCBs and PCDD/FS in fish and fish products and their impact on the human body burden in Belgium. Archives of Environmental Contamination and Toxicology 2007; 52: 563–571.
46. Sagerup K, Savinov V, Savinova T, Kuklin V, Muir DCG, Gabrielsen GW. Persistent organic pollutants, heavy metals and parasites in glaucous gull (*Larus hyperboreus*) on Spitsbergen. Environmental Pollution 2009; 157: 2282–2290.
47. Monosson E, Ashley JTF, McElroy AE, Woltering D, Elskus AA. PCB congener distribution in muscle, liver and gonad of *Fundulus heteroclitus* from the lower Hudson River Estuary and Newark Bay. Chemosphere 2003; 52: 777–787.
48. Svobodová Z, Lloyd R, Máchová J. Vykusová B. Water quality and fish health. EIFAC Technical Paper. No. 54. Rome, FAO, 1993.
49. Kmínková M, Winterová R, Kučera J. Fatty acids in lipids of carp (*Cyprinus carpio*) tissues. Czech Journal of Food Science 2001; 19: 177–181.
50. Gabrielsen GW, Skaare JU, Polder A, Bakken V. Chlorinated hydrocarbons in glaucous gulls (*Larus hyperboreus*) in the southern part of Svalbard. Science of the Total Environment 1995; 160/161: 337–346.
51. Dabrowska H, Bernard E, Barska I, Radtke K. Inter-tissue distribution and evaluation of potential toxicity of PCBs in Baltic cod (*Gadus morhua* L.) Ecotoxicology and Environmental Safety 2009; 72: 1975–1984.
52. Wang D, Weilin SS, Shannon A, Mellish J-A, Li QX. Tissue distribution of polychlorinated biphenyls and organochlorine pesticides and potential toxicity to Alaskan northern fur seals assessed using PCBs congener specific mode of action schemes. Archives of Environmental Contamination and Toxicology 2010; 58: 478–488.
53. Larsson P, Okla L, Collvin L. Reproductive status and lipid content as factors in PCB, DDT and HCH contamination of a population of pike (*Esox lucius* L.). Environmental Toxicology and Chemistry 1993; 12: 855–861.
54. Bachour G, Failing K, Georgii S, Emadfa I, Brunn H. Species and organ dependence of PCB contamination in fish, foxes, roe deer, and humans. Archives of Environmental Contamination and Toxicology 1998; 35: 666–673.
55. Ashley JT, Secor DH, Zlokovitz E, Wales SQ, Baker JE. Linking habitat use of Hudson River striped bass to the accumulation of polychlorinated biphenyls congener. Environmental Science and Technology 2000; 34: 1023–1029.
56. World Health Organization Regional office for Europe. Air quality guidelines, second ed. Copenhagen, Denmark. Cap 5.10 PCB; 2000.
57. McFarland VA, Clarke JU. Environmental occurrence, abundance, and potential toxicity of polychlorinated biphenyl congeners: Considerations for a congener-specific analysis. Environmental Health Perspectives 1989; 81: 225–239.
58. Jacob D, Boer EH. 8-year study on the elimination of PCBs and other organochlorine compound from eel under nature conditions. Environmental Science and Technology 1994; 28: 2242–2248.
59. Nie XP, Lan CY, An TC, Li KB, Wong MH. Distribution and congener patterns of PCBs in fish from major aquaculture areas in the Pearl River Delta, South China. Human and Ecological Risk Assessment 2006; 12: 363–373.
60. Palm HW, Rückert S. A new approach to visualize ecosystem health by using parasites. Parasitology Research 2009; 105: 539–553.
61. Sagerup K, Henriksen EO, Skorping A, Skaare JU, Gabrielsen GW. Intensity of parasitic nematodes increases with organochlorine levels in the glaucous gull. Journal of Applied Ecology 2000; 37: 532–539.
62. Sures B, Knopf K. Individual and combined effects of Cd and 3,3′,4,4′,5-pentachlorobiphenyl (PCB 126) on the humoral immune response in European eel

(*Anguilla anguilla*) experimentally infected with larvae of *Anguillicola crassus* (Nematoda). Parasitology 2004; 128: 445–454.

63. Arkoosh MR, Casillas E, Clemons E, Kagley AN, Olson R, Reno P, Stein JE. Effect of pollution on fish diseases: Potential impact on salmonid populations. Journal of Aquatic Animal Health 1998; 10: 182–190.

64. Sures B, Siddall R, Taraschewski H. Parasites as accumulation indicators of heavy metal pollution. Parasitology Today 1999; 15: 16–21.

65. Turčeková Ľ, Hanzelová V, Špakulová M. Concentration of heavy metals in perch and its endoparasites in the polluted water reservoir in Eastern Slovakia. Helminthologia 2002; 39: 76–80.

66. Heinonen J, Kukkonen JVK, Holopainen IJ. Toxicokinetics of 2,4,5-trichlorophenol and benzo(a)pyrene in the clam *Pisidium amnicum*: Effects of seasonal temperatures and trematode parasites. Archives of Environmental Contamination and Toxicology 2000; 39: 352–359.

67. Persson M, Larsson P, Stenroth P. Biomagnification and polychlorinated biphenyl congener distribution in an aquatic predator-prey, host-parasite system. Environmental Toxicology and Chemistry 2007; 26: 837–843.

CHAPTER 2

Diversity and Ecological Correlates of Red Fluorescence in Marine Fishes

Nils Anthes[1*†], JenniferTheobald1[1†], TobiasGerlach [1], MelissaG.Meadows[1, 2] and Nico K.Michiels [1 †]

[1]*Animal Evolutionary Ecology Group, Faculty of Sciences, University of Tübingen, Tübingen, Germany*

[2]*Biology Department, Saint Francis University, Loretto, PA, USA*

ABSTRACT

Marine environments at depths below −10 to −25 m are almost devoid of ambient red sunlight because water quickly attenuates long wavelengths. This stenospectral light environment presents unique opportunities for organisms that can transform ambient blue-green light into red light by fluorescence. Numerous marine fish species display intricate patterns of fluorescence. Because color vision is a key component of fish sensory ecology, several putative visual functions of red fluorescence have been proposed but are difficult to test experimentally. Here, we follow a comparative approach to assess the consistency between the phylogenetic distribution of red fluorescence with its presumed functions. We collected and analyzed the largest data set of red fluorescence in fishes to date, consisting of confirmed cases in 272 primarily diurnal fish species from 49 out of 90 surveyed fish families and 12 out of 21 surveyed fish orders, contrasted to 393 fish species with confirmed absence of red fluorescence. Based on *a priori* hypotheses on adaptive function, we compare the prevalence of red fluorescence among pre-defined sets of species based on ecological or biological characteristics while controlling for shared ancestry. When comparing between species, we find no evidence that red fluorescence is more prevalent in deep-water species, contrasting with our recent finding that fluorescence brightness increases with depth within species. There is also no evidence for a role in group-driven communication. Phylogenetic patterns are consistent, however, with three other predictions. First, fluorescence with a rather patchy distribution across the body occurred significantly more often among sit-and-wait predators or otherwise sedentary fish than in more mobile species, consistent with background matching for camouflage. Second, small, predatory fishes tended to show red fluorescent irides disproportionally often consistent with a proposed function in prey localization. Finally, sexually dimorphic species showed fluorescent fins more often, as predicted if relevant in sexual communication. From these findings, we derive predictions for experimental investigations of the presumed functions of red fluorescence.

Keywords: fluorescence, animal coloration, color vision, camouflage, prey detection, sexual signaling, visual contrast

INTRODUCTION

The sea appears blue because water primarily absorbs orange and red light (wavelengths >580 nm) whereas it largely scatters blue light (Jerlov, 1968; Lythgoe, 1979). This generates the strongest, most predictable, and most widespread spectral transition zone in nature. In all reasonably clear aquatic environments the ambient spectrum narrows from broad-spectrum sunlight at the surface to a 400–580 nm range at −20 m and 470–490 m below −100 m (Figure 1). Hence, in terms of color vision, the sunlit euphotic zone of marine environments can be subdivided into two zones. The euryspectral zone in the top few meters has an ambient spectrum wider than the core spectral sensitivity of most fish, which often have tuned their peak color sensitivity to the abundantly available blue-green light (450–550 nm) and thus have difficulties perceiving very short (UV) or very long (red) wavelengths (Munz and McFarland, 1973; Partridge, 1990; Losey et al., 2003; Brandley et al., 2013). In the stenospectral zone starting at around −10 m, the ambient spectrum is narrower than the portion of the light spectrum that most fish can perceive (Meadows et al., 2014 and references therein, Figure 1).

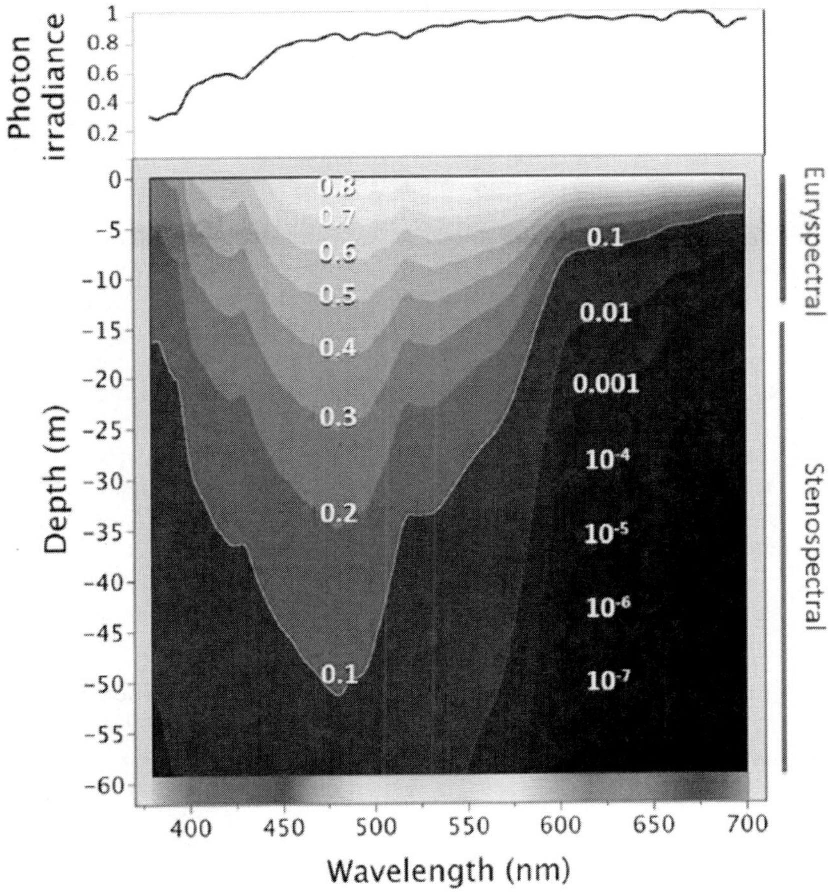

Figure 1. The spectral light environment in tropical marine waters. The top graph shows the ambient spectrum above the water surface (expressed as a proportion relative to the highest irradiance value, Red Sea, March 2013). The lower graph shows how that proportion changes with increasing depth. Each line represents the "iso-brightness" for a given wavelength. Red wavelengths dominate near the surface, but 90% of their irradiance is absorbed within −10 m. For blue light, the same degree of absorbance is only reached at −50 m. Irradiance was measured using a calibrated PhotoResearch PR 670 photospectrometer fitted with a CR-670 cosine receptor between 0 and −25 m depth and is expressed in photons.s−1.m−2.nm−1. These data were used to calculate spectral attenuation coefficients to estimate values down to −60 m (Meadows et al., 2014). Note that the measurements ignore Raman scattering, which is very weak but explains the presence of long wavelengths even in very deep water (Johnsen, 2012, p. 168).

This phenomenon forms the basis for the prevalent view that long wavelengths ("red" to humans) are irrelevant to most marine fishes. The implicit assumption is that animal coloration is exclusively generated by pigments or optical nanostructures that differentially absorb and reflect parts of the incoming light (Endler, 1990). Such subtractive mechanisms cannot reflect wavelengths that are absent in the ambient spectrum, and pigmentation that appears red at the surface will turn gray at depth.

The recent discovery of red fluorescence among marine fishes (Michiels et al., 2008; Sparks et al., 2014), however, indicates that long wavelengths may be far more relevant to fish visual ecology than commonly accepted. Red fluorescence is well-known from algae and corals, where it may enhance photosynthesis (Schlichter and Fricke, 1990), stimulate symbiotic zooxanthellae (Field et al., 2006), provide photoprotection (Salih et al., 2000; Ben-Zvi et al., 2014), or generate visual contrast (Gruber et al., 2008). Its occurrence in fishes suggests additional visual functions (Gerlach et al., 2014; Meadows et al., 2014; Harant et al., 2016; Michiels et al., in submitted, see also Haddock et al., 2005 for an example in deep see siphonophores). Unlike subtractive color mechanisms, fluorescence is an additive mechanisms enabling emission of wavelengths irrespective of their presence in the environment, allowing red coloration even in stenospectral environments.

Fluorescence is obviously not limited to the red range (>580 nm) of the light spectrum (reviewed in Lagorio et al., 2015), and many marine fishes also exhibit green and yellow fluorescence in the 510–580 nm range (Sparks et al., 2014). We focus, however, on long-wavelength fluorescence from 580 nm (orange) to 750 nm (far red) for two reasons. First, in light environments below −10 m red fluorescent emission is more likely to generate strong color contrasts against the blue-green background (Johnsen, 2012, p. 189, see also Haddock and Dunn, 2015, Figure 1). Making use of this free bandwidth would resemble adaptations in vocal communication that focus on frequencies that are rare in the ambient environmental noise (Slabbekoorn and Peet, 2003; Hart et al., 2015). Second, previous research on reef fish vision has focused on the 350–600 nm range of the color spectrum (Marshall et al., 2006, 2015; Brandley et al., 2013). While fluorescent emission in the near red around 600 nm can be assumed to be detectable by many fishes (Kalb et al., 2015), color patterns in the far red range are only scarcely investigated, prompting novel questions about color perception and private signaling (Gerlach et al., 2014, 2016).

To date, red fluorescent spectral emission has only been characterized for a small number of fish species (Michiels et al., 2008; Wucherer and Michiels, 2012, 2014; Meadows et al., 2014; Sparks et al., 2014; Gerlach et al., 2016). Here, we present the most comprehensive dataset of the phylogenetic distribution of red fluorescence among marine fishes to date. For the first time, we also compare species expressing red fluorescence with confirmed cases of its absence. Furthermore, we provide a quantitative overview of emission spectra and their characteristics within and between fish families. Finally, we use comparative analyses to evaluate whether the phylogenetic distribution of red fluorescence is concordant with five non-mutually-exclusive hypotheses on ecological function, as described below.

Hypothesis 1: Short-Distance Visual Functions

Just as long wavelengths from the sun disappear rapidly with depth, red fluorescent emission also attenuates rapidly with distance (Lythgoe, 1979). Hence, in a communication context, it is likely to be functional over very short distances only. While communication distances are poorly known for most fish species, spatial resolution as well as visual ranges and reaction distances toward prey items generally increase with body size in marine fish (e.g., Tamura, 1957; O'Brien, 1979; Schmidt and O'Brien, 1982; Li et al., 1985; Aksnes and Giske, 1993). Therefore, we use body length as a proxy for the distance over which communication or predation typically take place, with small species interacting over shorter distances.

Hypothesis 1 predicts that the prevalence of red fluorescence among marine fish species increases with decreasing body length.

Hypothesis 2: Contrast Enhancement at Depth

Near the surface, red fluorescence is too weak to significantly contribute to color contrast compared to reflective mechanisms (Meadows et al., 2014). In the stenospectral zone, however, red fluorescence is the only non-luminescent mechanism by which red hues can be produced. Hence, if used for visual functions, red fluorescent coloration should predominate in the stenospectral zone compared to the euryspectral zone (cf. Figure 1). Evidence for this comes from within-species comparisons; red fluorescence was brighter at −20 m than at −5 m in 6 out of 8 tested reef fishes (Meadows et al., 2014). Moreover, individual fish boost their fluorescence when exposed to light environments mimicking the low brightness conditions of deeper water (Harant et al., 2016). Here, we assess whether this hypothesis is also supported when examining patterns *between* species.

Hypothesis 2 predicts that the prevalence of fluorescence among marine fish species increases with the maximum depth at which those species occur.

Hypothesis 3: Camouflage through Background Matching

Red fluorescence seems particularly common in cryptobenthic fishes (Michiels et al., 2008). Indeed, Sparks et al. (2014) describe a phylogenetic concentration of green, orange, and red fluorescence in benthic taxa such as eels, lizardfish, blennies, scorpionfish, gobies, and flatfish. In these, red fluorescence could optimize color matching—and thus camouflage—against a substrate on which sessile organisms such as corals, sponges, and/or algae generate a background of irregular, patchy fluorescence (e.g., Alieva et al., 2008; Michiels et al., 2008, Figures 2, 3). Camouflage constitutes a complex interplay between the benefits of being cryptic, the behavioral and perceptive abilities of the observers, the ambient spectrum, and the background against which the cryptic species is observed (Endler, 1981). For a comparative analysis, the difficulty is to define unbiased criteria to score which species may benefit from camouflage and what characterizes a fluorescent pattern as being cryptic.

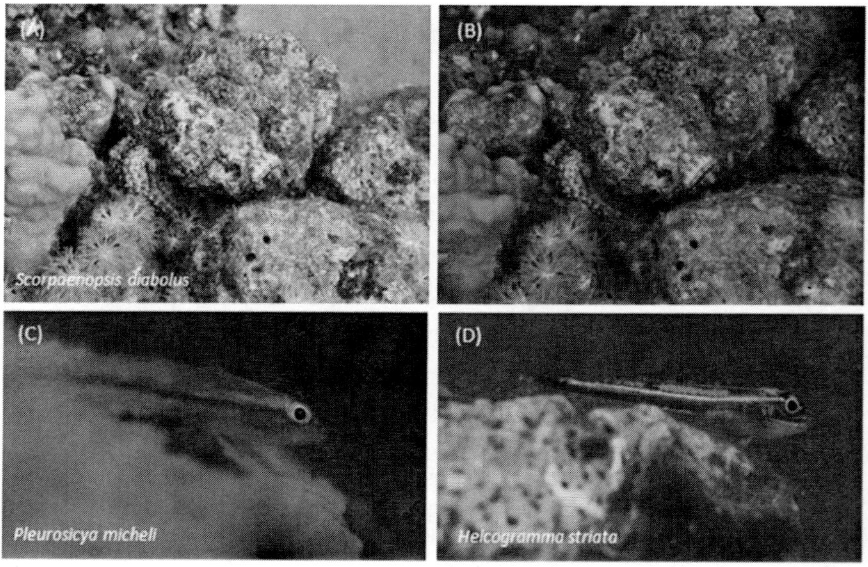

Figure 2. One positive and two negative cases of "patchy fluorescence." The devil scorpionfish (A,B) is a typical motionless sit-and-wait predator with patchy fluorescence that appears similar to that of its background. (C) and (D) exemplify red fluorescent benthic species scored as "negative" for patchy fluorescence. Picture (A) taken without filter, the others with a LEE 164 Flame red filter to enhance long wavelengths, all at ~–20 m with manual white balance (Photos: Nico K. Michiels).

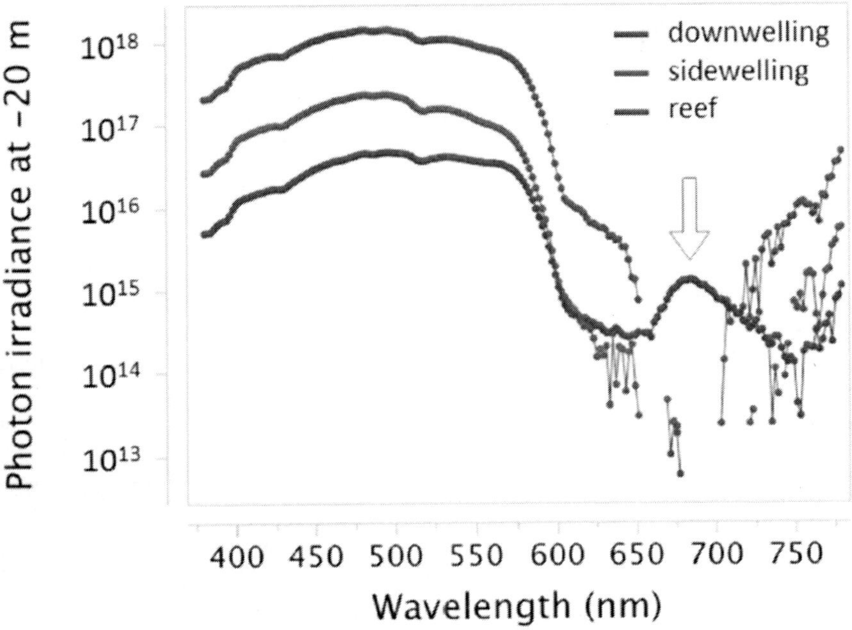

Figure 3. Downwelling light, sidewelling scatter from the open water, and light emitted from a reef in −20 m showing the distinct red fluorescent emission of the reef at this depth (arrow). Measured as photon irradiance (photons.s−1.m−2.nm−1) using a calibrated PhotoResearch PR 670 with CR-670 cosine corrector in El Quseir, Egypt (sunny, midday, March 2013). Raw data shown as connected points. Missing points are wavelengths at which the signal fell below the detection threshold. Note log10 y-axis.

With respect to camouflage benefits, we assume that camouflage by background matching is of relevance to all benthic fishes, and particularly so to rarely moving species as represented by sit-and-wait predators such as scorpionfish. In contrast, we consider free-swimming benthopelagic and pelagic fishes as being generally more conspicuous because of their constant movement in the water column. Free-swimming species are more likely to possess alternative camouflage mechanisms based on different optical principles (e.g., Brady et al., 2013). Hence, we categorized species as (1) free-swimming, (2) benthic and frequently moving, and (3) benthic and mostly motionless as in sit-and-wait predators.

With respect to crypsis, we define those fluorescent color patterns as "cryptic" that resemble the patchiness of fluorescence present on hard substrates such as rocks and reefs (see examples in Michiels et al.,

2008; Sparks et al., 2014). We therefore separated a category of study species showing "patchy" overall body fluorescence from any other body distribution with e.g., large areas of uniform fluorescence or small, well-defined areas such as the iris (Figure 2).

Hypothesis 3 predicts that, among fluorescent fishes, "patchy fluorescence" predominates among benthic fish in general and in motionless foragers in particular.

Hypothesis 4: Prey Detection

Many benthic species including members of the pipefish, goby and triplefin families display striking red fluorescence around their eyes, usually in the irides (Michiels et al., 2008; Sparks et al., 2014). This has raised the idea that red fluorescent irides may induce retro-reflective eyeshine in the eyes of other organisms (prey or predator) facilitating their detection (Bruce, 2009; Meadows et al., 2014; Wucherer and Michiels, 2014). This would be analogous to what has been described for nocturnal flashlight fishes, which produce light through chemiluminescence from a light organ directly below the pupil (Howland et al., 1992). Light emission next to the viewing axis is a crucial feature as this maximizes the retroreflective eyeshine that can be seen in the eyes of other organisms (Jack, 2014). As in *Hypothesis 1*, this is likely to function over short distances only because of attenuation of the fluorescent emission with distance. Hence, such "active photolocation" using fluorescence makes the most sense for small fish that pick individual small prey items that possess eyes (Michiels et al., in submitted). This function of red fluorescence is not expected in species that forage on larger prey with eyes over larger distances and should be absent in species that feed indiscriminately (e.g., filter-feeding, substrate sifting) or visually select organisms that lack eyes (detritivores, corallivores, herbivores).

Hypothesis 4 predicts that red fluorescence is more often near the eye in those fluorescent species that forage on small prey items that have eyes.

Hypothesis 5: Intra-Specific Communication

Fluorescence can be quickly modulated by intra-cellular pigment transport (Wucherer and Michiels, 2012, 2014) and is often associated with signaling structures such as fins (Michiels et al., 2008; Gerlach et al., 2016). This hints at a communication function for fluorescence for courtship, territoriality, or social hierarchies (Sparks et al., 2014). Recent evidence suggests that fluorescent fish perceive (Michiels et al., 2008), and respond to, their own fluorescence. Among wrasses, male-male interactions are mediated by orange-red colors in *Coris julis* (Braun et al., 2014) and by red fluorescence in *Cirrhilabrus solorensis* (Gerlach et al., 2014). Training experiments in the triplefin *Tripterygion delaisi* show that these fish are capable of discriminating between objects in response to weak red fluorescent levels similar to their own (Kalb et al., 2015). Fins are used for signaling by many fishes (overview in Rowland, 1999, see also Ciccotto and Mendelson, 2016a). In contrast to the body, their display angles can be controlled (e.g., pectoral and tail fin), they can be moved even when sitting still (e.g., dorsal fin) and they can be opened or closed to adjust signal size or conceal the signal (e.g., dorsal, caudal, and anal fins).

Hypothesis 5 predicts that red fluorescent fins are more prevalent in fluorescent species that show specific types of intra-specific communication.

To test this hypothesis, we differentiate between two types of intra-specific communication. First, many marine fish live in aggregations of dozens to hundreds of individuals (e.g., Norris and Schilt, 1988). Compared to species that primarily adopt a solitary lifestyle or form stable pairs, grouping generates more continuous signaling among group members. If red fluorescence played a specific role for such interactions, it should be more prevalent in group-living species.

Second, color signals subject to mate choice are often expressed differently in males and females (reviewed in Wyman et al., 2013, see Kraaijeveld et al., 2007 and Baldauf et al., 2011 for exceptions). Hence, if red fluorescent fin displays are favored by sexual selection, we expect them to be more prevalent among species that exhibit sexual dichromatism.

MATERIALS AND METHODS

Data Sources

We assembled information from multiple sources. Our own sampling campaigns combined direct observation and photographic documentation with qualitative and quantitative measurements of fluorescence spectra from live fish, as detailed below. We complemented these with cases documented in the online supplement of Sparks et al. (2014). The latter do not provide a list of negative cases, but we scored all species labeled as showing green but no red fluorescence as negative for red fluorescence. We further included earlier data from our own projects (Michiels et al., 2008; Wucherer and Michiels, 2012; Meadows et al., 2014) and subjected them to categorizations where possible as detailed below. Our survey did not attempt a balanced coverage across the cartilaginous and bony fish phylogeny and explicitly does not aim at reconstructing the evolutionary history of red fluorescence across fish. Instead, while being taxonomically as broad as possible, sampling focused on benthic, mostly shallow water species, automatically generating limited or no coverage in fish clades with largely pelagic representatives such as the Otocephala (including herrings) or the Protacanthopterygii (including salmon and trout).

Spectrometry and Standardized Documentation of Live Specimens

We measured and documented fluorescence from live fish at five localities:

a. **UNIVERSITY OF TÜBINGEN, GERMANY**. We ordered species through the sustainable aquarium trade for spectral measurements in our laboratories in accordance with German animal care legislation (permit ZO 1/12 from the local authority at the Regierungspräsidium Tübingen). These fish served to complement our coverage of fish families and to scrutinize doubtful cases of fluorescence after field observations.

b. **GULF OF AQABA** (northern Red Sea). We collected fish at the Interuniversity Institute for Marine Sciences (IUI) in Eilat, Israel, in March 2012, under the general IUI collection permit (No.: 2012/38470, Israel Nature and National Parks Protection Authority to Roi Holzman).

c. **NORTH-WESTERN RED SEA**. We collected fish from coral reefs in the bays of Sharm Fugani (Mangrove Bay) and Sharm Lassal (Utopia Beach), 15–20 km south of El Quseir, Egypt, in March 2013. Both locations offer protected reefs sloping down to −25 to −30 m. Collection conformed to a 3-year Memorandum of Understanding between the University of Tübingen and the Suez Canal University running 1 Jan 2013–31 Dec 2015.

d. **MEDITERRANEAN SEA**. Collections in June 2013 focused on rocky and sandy environments to −30 m at the Station de Recherches Sous-marines et Océanographiques (Stareso) at Calvi, Corsica, France. We collected and registered fish under the station's general sampling permit.

e. **INDOPACIFIC OCEAN, INDONESIA**. We collected fish at a broad range of coral reef habitats up to −30 m depth at Hoga Island in the Wakatobi archipelago off the SE Sulawesi coast, Indonesia in September 2011. Collection was authorized and registered under a general permit of *Operation Wallacea*.

Most field-sampling focused on diurnal species for which field observations using a red filter mask (Michiels et al., 2008) had already indicated the presence or absence of long-wavelength fluorescence. In the Gulf of Aqaba, collection and measurement occurred blind to the presence of fluorescence.

Fish were collected on SCUBA diving with hand nets after partially anesthetizing individuals using clove oil where required (5% clove oil in 5% ethanol and 90% seawater shaken to emulsify). After transportation to the local laboratory in 100 ml Falcon tubes or 4-L zip-lock bags, fish were maintained in aerated seawater for 1 to 8 h before live measurements, followed by their release on site. Species were identified using the general identification literature (e.g., Debelius, 1998; Allen et al., 2002, 2003, 2014) supplemented by monographs (Clark, 1979; Fricke, 1997; Holleman, 2005), Fishbase (Froese and Pauly, 2014) and expert advice for specific fish groups where necessary. Unidentified species were classified at the family or genus level.

Fluorescence Spectrometry and Spectral Analysis

The fish handling routine, the basic spectrometry setup used in the northwestern Red Sea (Figure 4), and small deviations for the measurements in the Indopacific and the Mediterranean are detailed in Meadows et al. (2014). The same setup was used in the two other locations, with the following modifications: In the Gulf of Aqaba (IUI), we used an Ocean Optics white LED light source (Ocean Optics LLS–Cool White) trimmed to the blue-green range for excitation with a BrightLine HC 533/SP (AHF) short pass filter. The light emitted and/or reflected by the fish was filtered using a stack of four LEE 105 Orange filters in an in-line-filter holder. Measurements at the University of Tübingen used a different bifurcated fiber optics cable (Ocean Optics QR400-7-VIS-BX). Differences between the measuring setups are due to continuous efforts to improve our setup over the years. This precludes direct comparisons of emission intensities and we thus expressed emission intensities on a relative scale as explained below.

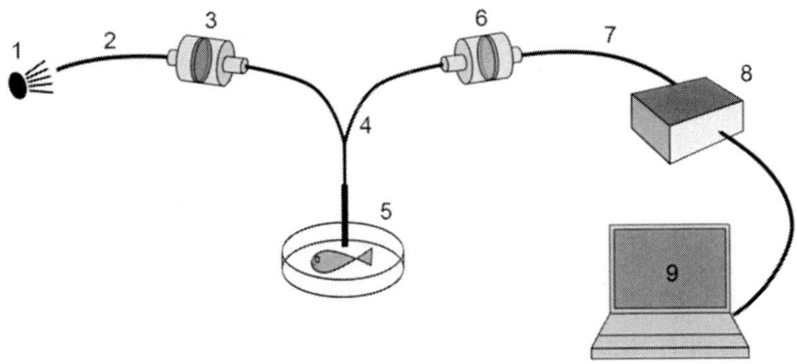

Figure 4. General spectrometry setup. A near-monochromatic green excitation light source (1) was connected (2) via a short-pass or clean-up filter (3) to a bifurcated fiber optics cable (4) that terminated in a hand-held probe (5) close to the measured fish. The emitted light was redirected through a long-pass filter (6–7), analyzed by a high-sensitivity Ocean Optics QE65000 spectrometer (8) and visualized using Ocean Optics SpectraSuite software (9).

Within the excitation range used by Sparks et al. (2014), we used a monochromatic 532 nm light source to excite fluorescent emission that was measured after removing reflected light using a long-pass filter. This single excitation wavelength was chosen, first, to unambiguously differentiate reflection of the excitation light source or short-wavelength fluorescence from red fluorescence with emission peaking at ≥580 nm, and second, to allow comparisons among species without confounding by variation in excitation wavelength. Our unpublished fluorescence measurements across complete spectral gradients indicate that peak emission wavelengths (λ_{max}) are nearly insensitive to variation in excitation wavelength, consistent with a general property of most fluorescent pigments due to molecular structure (Johnsen, 2012).

Fluorescence emission spectra were collected from several body parts: Eye, head, operculum, upper and lower flank, and each fin. To obtain summarizing information on the fluorescence characteristics of each study species, we followed a four-step procedure to condense our original spectral readings. First, we identified distinct fluorescence peaks for each measured individual and body part and characterized their peak emission wavelength (λ_{max}) and emission intensity at λ_{max}. Second, we averaged λ_{max} and intensity per body part and emission peak across all specimens of a given species. Third, across all body parts per species, we grouped measurements with the same single emission peak (near-identical λ_{max}), calculated minimum, mean, and maximum λ_{max}, and selected the highest of the averaged intensity values across all body parts. For each investigated species, we thus obtained a single summarizing measurement (λ_{max} and intensity) for one to four distinct fluorescent emission peaks. Fourth, given that intensity readings are sensitive to deviations in measurement procedure, we grouped intensity values into four quantiles *within* each measurement campaign. Given that the distribution of absolute intensity across measurements showed a close match between measurement campaigns (a) through (d), these datasets were combined prior to assigning intensity categories. Assignment to intensity categories was done separately for campaign (e). Only those four resultant intensity categories are reported in this study.

Fluorescence Photography and Image Analysis

All sampled fish were subjected to standardized fluorescence photography (Meadows et al., 2014). While spectrometric measurements are restricted to point information on fluorescence intensities, we performed an independent scoring of the spatial extension of fish fluorescence based on images recorded in the laboratory as well as in the field. First, we allocated fish to one of four categories to describe the spatial extension of long-wavelength fluorescence:

1. No red fluorescence, or fluorescence restricted to minute spots, often originating from contamination by, e.g., ectoparasites or gut content.
2. Fluorescence covers <10% of the body.
3. Fluorescence covers 10 to 50% of the body.
4. Fluorescence covers >50% of the body.

Second, we scored its presence or absence on specific body parts:

1. **EYE FLUORESCENCE**: Iris or exposed parts of eyeball or eye socket.
2. **PATCHY FLUORESCENCE**: Many dots and patches of variable size and shape across the body.
3. **FIN FLUORESCENCE**: Fluorescence on any fin or set of fins, and irrespective of sex (note that our current sampling did not differentiate sex-specific fluorescence patterns).

In-situ Observations in the Field

Beyond the study sites mentioned above we also conducted field documentation at sites in the Mediterranean Sea [Corsica (~200 dives 2009–2015), Elba (5 dives 2012), Croatia (4 dives 2008)], the Red Sea [Gulf of Aqaba (~30 dives 2011–2012), Marsa Alam (~200 dives 2007–2015)] and the Indo-Pacific Ocean [Lembeh Strait in Indonesia (~30 dives 2013), Raja Ampat in West Papua (~40 dives 2013), Perth in Western Australia (3 dives 2011)]. Fish were scored for the presence of obvious long-wavelength fluorescence perceived as "orange" to "red" by the observer or camera. We only included cases with reliable assessment, requiring conditions that preclude confounding effects by reflective red coloration. Daytime observations were therefore restricted to ≤ −15 m. To facilitate visual or photographic detection, we suppressed the abundant blue-green ambient light with one of several long-pass filters (LEE 105 Orange, LEE 287 Double C.T. Orange, LEE 164 Flame red, LEE 106 Primary red, Nightsea BB62 yellow barrier filter) and documented most species on site using Nikon D300, D700, or D4 digital cameras. To further minimize the likelihood of false positives, observations were cross-checked *in situ* by comparing the putative fluorescent structure with a certified 1.25″ non-fluorescent Spectralon red diffuse reflectance standard (Labsphere) whenever possible, usually in benthic species only. Although we concentrated on species that showed fluorescence under natural illumination, we regularly checked putative cases by highlighting the fluorescence with a blue LED dive light or a blue flashgun. Both were fitted with a short-pass filter to cut out any remaining long-wavelength light (Thorlabs FD2C subtractive dichroic color bandpass filter on Hartenberger Mini Compact blue LED torch; EX-INON Nightsea Excitation filter on Inon Z-240 flashgun). An increase in perceived red emission brightness with increased blue excitation (comparing with and without the blue torch) confirms fluorescence as the origin of the boost in the red signal. Under darker ambient conditions (e.g., shaded substrates, cloud cover), the blue excitation sources could also highlight fluorescence in shallower water (−5 to −10 m). The patterns of fluorescence were categorized for each species as explained above.

Screening Species in the Aquarium Trade

We visually screened all marine species (n = 209) available at one of the largest German wholesale aquarium traders (von Wussow, Pinneberg, Germany) in June 2009. To detect red fluorescence we used blue dive lights (Hartenberger Mini Compact LCD with 7 × 3.5 W 480 nm LEDs) for excitation and three types of LEE filters as barrier filters (LEE 777 Rust, LEE 106 Primary red, LEE 027 Medium Red) with all other light sources in the room switched off. Based on these observations, species were listed as either red fluorescent or non-fluorescent with no further photographic or spectrometric documentation.

Assignment Quality

As an internal control, we compared our observational assessment with spectral measurements of red fluorescence for 91 species for which both types of information were available. Out of 18 species that lacked any fluorescent signal in spectral measurements, two were rated as fluorescent based on observations: In *Coris gaimard*, the spectra were taken on non-fluorescent juveniles, while the adults observed in the field clearly fluoresced in red. In *Nemateleotris decora*, weak fluorescence was visible in the field, but spectral emission peaked in the green-yellow range and thus fell outside the 580 nm cut-off chosen for the current study. Out of the 73 species showing fluorescence in spectral measurements, only the blenny *Atrosalarias fuscus* was categorized as non-fluorescing during field observations. In this species, the fluorescent emission (575 nm) peaked also just outside our cut-off point, and so we rated it as non-red-fluorescent for the current study. Hence, spectral and observational data yielded highly consistent ratings with an error rate of about 3%.

Biological and Ecological Characterization

For all investigated fish, a person blind to the fluorescence rating screened the available literature (primarily Froese and Pauly, 2014, aided by records in comprehensive fish guides, fish ecology books, and individual species papers, see overview in **SUPPLEMENTARY MATERIALS A,B**) to score the following traits (as explained in the Introduction):

(1) Maximum total body length

(2) Maximum recorded depth of occurrence

(3) Substrate association

a. Benthic: Sit-and-wait predators (mostly motionless)

b. Benthic: Active foragers

c. Free-swimming: Benthopelagic or pelagic

(4) Gregariousness

a. Primarily solitary or pair-living

b. Primarily in larger social groups

(5) Sexual dichromatism: Sexually dimorphic coloration present or absent

(6) Primary food items

a. Individually picked microscopic prey with eyes (e.g., microcrustaceans)

b. Individually picked macroscopic prey with eyes (e.g., crustaceans, fish)

c. Eyeless prey or indiscriminate feeding (e.g., filter feeding, sediment sifting, digging, herbivory, corallivory, detritivory, etc).

ANALYSES

Phylogenetic Reconstruction

We used the extensive recent Maximum Likelihood fish phylogeny of Betancur et al. (2013) to generate a family-level tree topology. We pruned this phylogeny to the 71 families also represented in our dataset by removing all non-covered families and maintaining a single terminal node per family. We then manually added the missing 19 families as to reflect established phylogenetic relationships, generating a tree topology that captures the current family-level phylogeny to our best possible knowledge (details in **SUPPLEMENTARY MATERIAL D**). Ancestral character estimation for the presence of fluorescence at the family level used maximum likelihood estimation for discrete characters following Pagel (1994). Given the absence of comparable branch length estimates for the manually added taxa, this analysis rests on branch lengths set to unity (Díaz-Uriarte and Garland, 1998; Garland and Ives, 2000). All tree manipulations and analyses were conducted using the packages APE (Paradis et al., 2004) and phytools (Revell, 2012) for R (R Core Team, 2013).

Statistical Analyses

Our statistical analyses evaluate whether the phylogenetic distribution of fluorescence types can be explained by any of five hypotheses on putative function as outlined in the Introduction. For each hypothesis, we fitted generalized linear mixed effect models (GLMM) using the glmer function in the lme4 package (Bates et al., 2013) for R. The dependent variable was the presence or absence of the defined fluorescence type as specified in the hypotheses as a binary response, modeled using a logit link function. The predictor(s) as given in each hypothesis were included as fixed effect(s). To account for putative confounding effects of body size, we added the covariate maximum body length (log_{10}-transformed to normalize data distribution, z-transformed to improve model convergence) for hypotheses 2 through 5. Initial full models also contained all possible interaction terms among factors in the fixed model component. Taxonomic ranks (genus, family, and order) were included as nested random factors to take into account trait correlations due to shared ancestry (cf. Luiz et al., 2013; Bridge et al., 2016). This is a compromise given that a fully resolved species-level phylogeny is currently not available for the majority of our study species. We then performed stepwise backward model selection based on the Bayesian information criterion (BIC) and hierarchical likelihood ratio tests (as recommended by Zuur et al., 2009). Only fixed factor effects maintained in the final reduced models are reported, with their statistical significance evaluated using Type III Wald chi^2 tests. We further provide an estimate for the total variation explained by the fixed model component expressed as the marginal R^2 (R^2marg .Rmarg.2, Nakagawa and Schielzeth, 2013) as implemented in the piecewiseSEM package (Lefcheck, 2016) for R.

RESULTS

Red Fluorescence Is Phylogenetically Widespread

We could unambiguously assign the presence or absence of red fluorescence in 665 fish species from 90 families and 21 orders (Table 1, **SUPPLEMENTARY MATERIAL A**). Of those, spectral measurements are available for 185 species, documenting red fluorescence in 114 (**SUPPLEMENTARY MATERIAL C**). Observational data for the other 480 species revealed red fluorescence in another 153 species. Hence, we document red fluorescence for 272 species (49 families and 12 orders) in total, representing 41.8% of all species in our database. In all other cases fluorescence was absent or hardly detectable.

Table 1. Sample sizes (number of taxa) and evidence type at different taxonomic levels.

	Investigated	Red fluorescence	No red fluorescence
ALL EVIDENCE COMBINED			
Orders	21	12	19
Families	90	49	72
Genera	277	130	189
Species	665	272	393
VISUAL EVALUATION ONLY			
Orders	21	12	19
Families	87	47	70
Genera	229	99	161
Species	480	153	327
SPECTRAL MEASUREMENTS			
Orders	8	7	5
Families	28	16	19
Genera	94	53	51
Species	185	114	71

Species for which we investigated multiple color forms or individuals from different origins count only as a single entry. Note that numbers for orders, families, and genera within each evidence category do not add up because many taxonomic groups contain representatives both with and without red fluorescence.

Mapping the data onto a family-level phylogenetic tree (Figure 5) shows that red fluorescence is phylogenetically widespread (at least within the teleost fish), lacks a simple association with phylogenetic history, and has likely been repeatedly acquired or lost.

We found substantial family-level variation in the expression of red fluorescence (Figure 5). Even though absolute proportions of fluorescent species are difficult to interpret given that some of our sampling specifically focused on promising candidates for fluorescence, this upward bias applies equally to all target fish families. Out of 13 families with more than 10 sampled species, six showed a predominance of fluorescing species: Gobies (Gobiidae), wrasses (Labridae), triplefins (Tripterygiidae), scorpionfish (Scorpaenidae), pipefish & seahorses (Syngnathidae), and dragonets

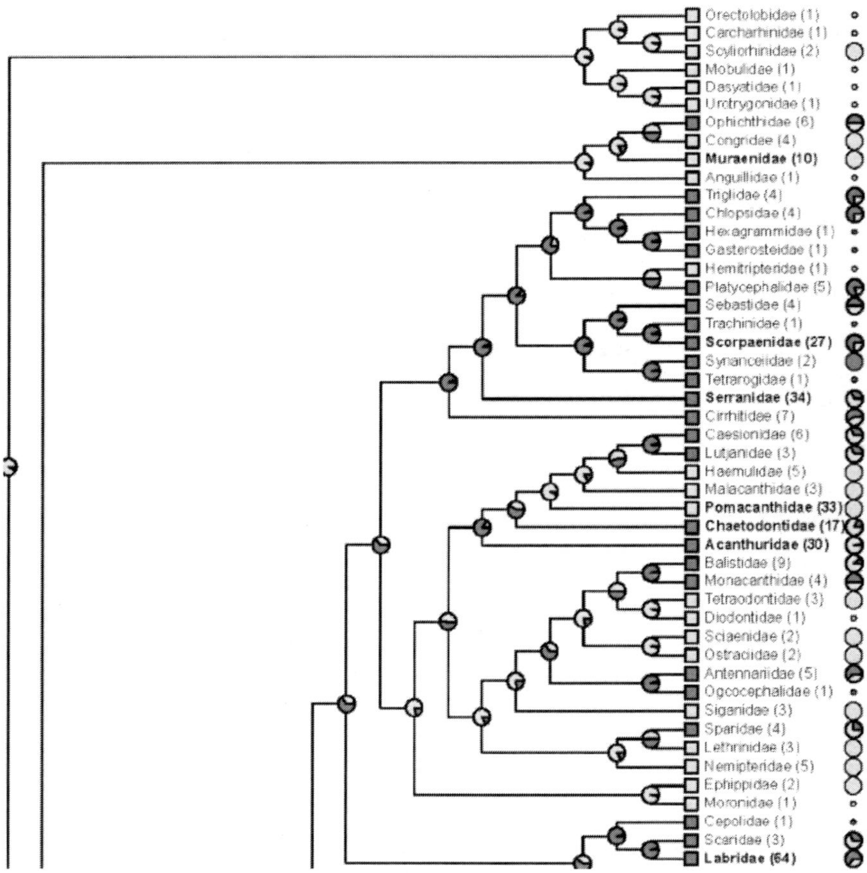

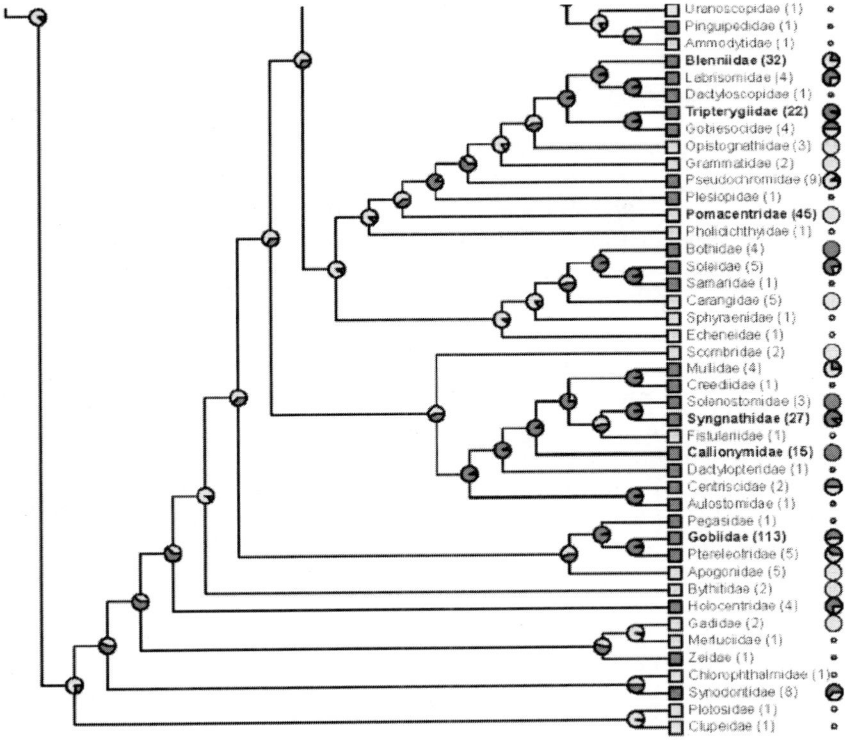

Figure 5. Phylogenetic distribution of red fluorescence among marine fish families. For each family, tip labels show (i) whether fluorescence has been confirmed in at least a single species (red squares vs. blue squares), (ii) the total number of investigated species (in brackets; families with ≥10 species in bold face), and (iii) the proportion of species that exhibit red fluorescence (pie charts; shown for families with ≥ 2 species). At each node, the maximum likelihood for red fluorescence being the ancestral state is indicated by the red pie portion.

(Callionymidae). With the exception of the wrasses, all these families exhibit rather cryptic coloration and a mostly secretive lifestyle. In groupers (Serranidae), blennies (Blenniidae), and butterflyfish (Chaetodontidae), red fluorescent species occurred at intermediate frequencies (Figure 5). In contrast, and despite explicit search for fluorescent cases, red fluorescence was virtually absent from several particularly colorful reef fishes: Damselfish (Pomacentridae), angelfish (Pomacanthidae), surgeonfish (Acanthuridae), and moray eels (Muraenidae) (but conspicuous yellow fluorescence is present in, e.g., moray eels).

Red Fluorescence Occurs in Distinct Types

Distribution of λ_{max} Values

Out of 665 investigated species, 185 could be subjected to spectrometric measurements. Out of these, emission spectra revealed red fluorescence in 114 species (Table 1). Within the investigated 580 to 750 nm range, peak emission wavelengths (λ_{max}) clustered in three distinct groups (Figure 6) that we categorize into "near red" (580–650 nm), "deep red" (650–710 nm), and "far red" (>710 nm). In the near red group, most λ_{max} aggregated between 590 and 630 nm, coinciding well with the abrupt start of light attenuation with increasing depth (Figure 1). The deep red group clusters around the characteristic emission range of ambient fluorescent light produced by chlorophyll *a* at about 680 nm (Figure 3). The far red group contains only few species with λ_{max} at 740 nm, but these come from a diverse array of families with mostly cryptic sit-and-wait predators (see below, Figure 9).

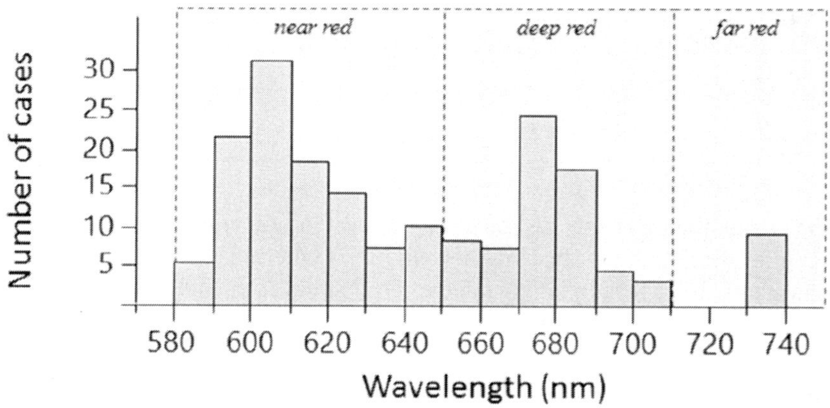

Figure 6. Frequency distributions of peak emission wavelengths (λmax) for 114 fish species. Single species have between one and four emission peaks and can thus be represented multiple times. Hatched red lines delineate our categorization according to peak emission wavelength into near red, deep red, and far red fluorescence.

Single vs. Multiple Peak Emission

Red fluorescence showed a single λ_{max} in 70 out of 114 species (61%) for which fluorescence spectra are available. Twenty-eight species (25%) showed two, 14 species (12%) three, and the remaining two species (1.7%) four emission peaks (the pipefish, *Corythoichthys nigripectus*, and the triplefin, *Enneapterygius mirabilis*). In species with multiple λ_{max}, these peaks were separated by 48.4 nm on average (range: 10.1–105.5 nm,

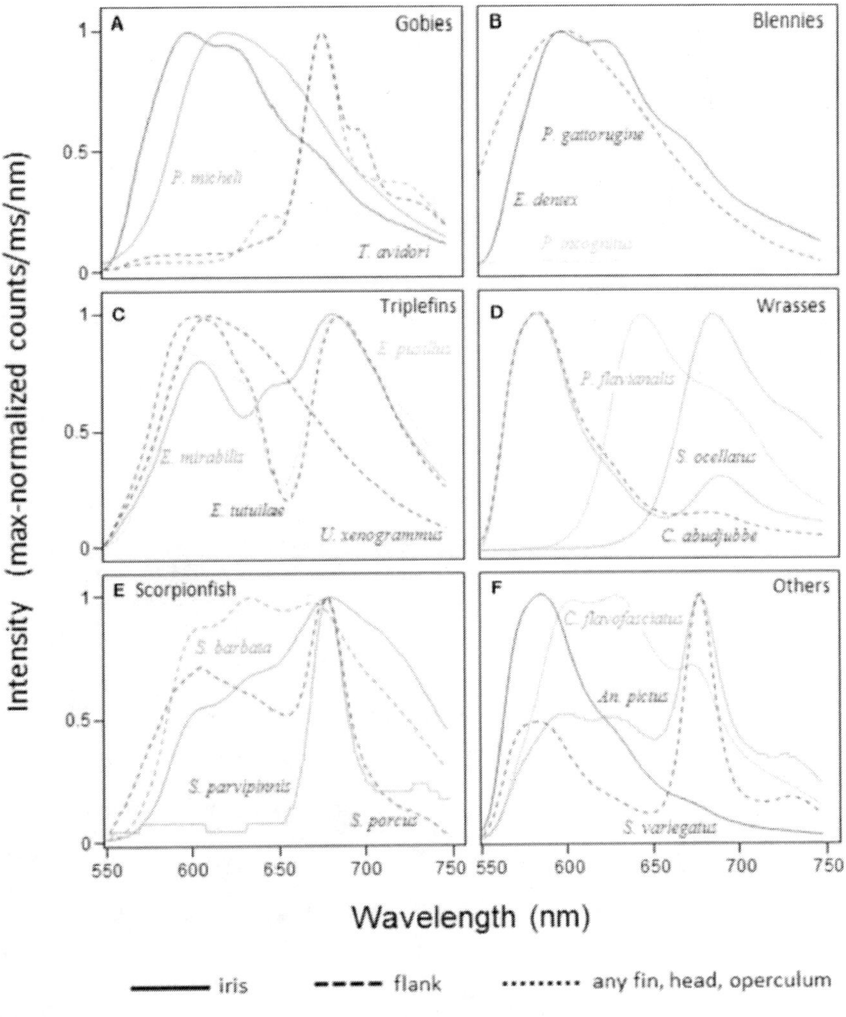

Figure 7. Spectral patterns of red fluorescence. For a selection of families and species, the graphs depict the maximum-normalized shape of fluorescent emission under monochromatic green excitation (see Methods). Within each panel, species are coded by color, body parts by line styles. (A) Gobiidae, (B) Blenniidae, (C) Tripterygiidae, (D) Labridae, (E) Scorpaenidae, (F) Syngnathidae, Antennariidae, and Synodontidae.

Figure 7). Multiple emission peaks were often associated with different body parts, for example with a near red emission in the iris and a deep red emission on the flank in gobies (Figure 7A). Examples of two emission peaks in a single body part are the iris of the blenny *Ecsenius dentex* (Figure 7B), fins of the triplefin, *Enneapterygius pusillus* (Figure 7C), or the flanks of the wrasse, *Cheilinus abudjubbe* (Figure 7D), and the lizardfish, *Synodus variegatus* (Figure 7F). Triple emission peaks as on the flank of the scorpionfish, *Scorpaenopsis barbata* (Figure 7E) or the caudal fin of the pipefish, *Corythoichthys flavofasciatus* (Figure 7F) were rare.

Fluorescent Body Parts and Body Area

The distribution of long-wavelength fluorescence over the fish body shows distinct types (Figures 7, 8). Some species show conspicuous red fluorescence in well-defined, often quite small, uninterrupted areas (e.g., eyes or fin rays), indicative of signaling or prey detection functions (*Hypotheses 4* and *5*) (Figures 8A–E). Others show a scattered, irregular, patchy distribution of red fluorescence across the body that we specifically addressed in *Hypothesis 3* (camouflage function). In these cases, it covers 10 to ≥50% of the body area (Figures 8F,G).

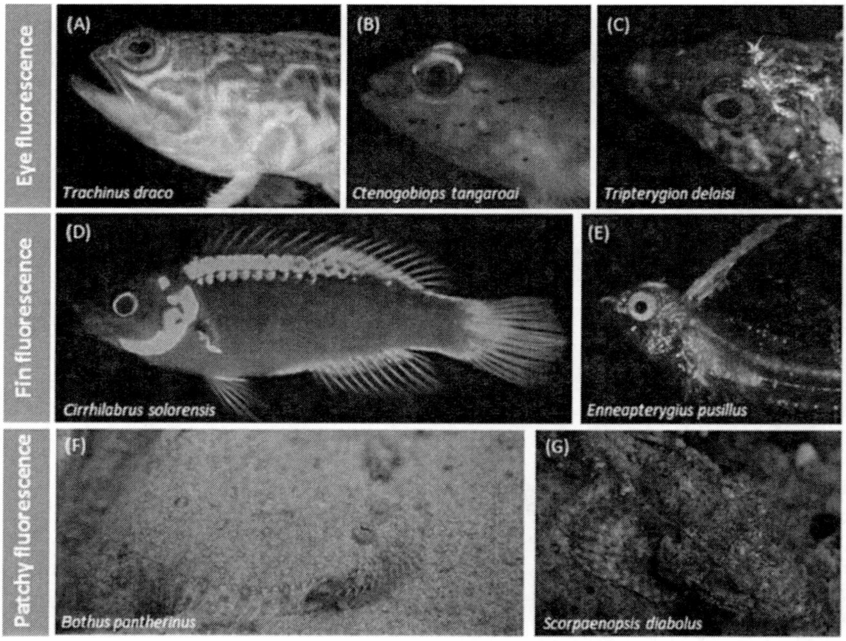

Figure 8. Variation in red fluorescent body patterns. Irides are often the only or clearly most prominent fluorescent structures as in some weevers (A), gobies (B), and triplefins (C). Fin fluorescence may suggest a signaling function as in a fairy wrasse (Gerlach et al., 2014) (D) or triplefin (E). Patchy fluorescence characterizes the irregular distribution of red fluorescence over the body as shown by this flatfish (F) and scorpionfish (G).

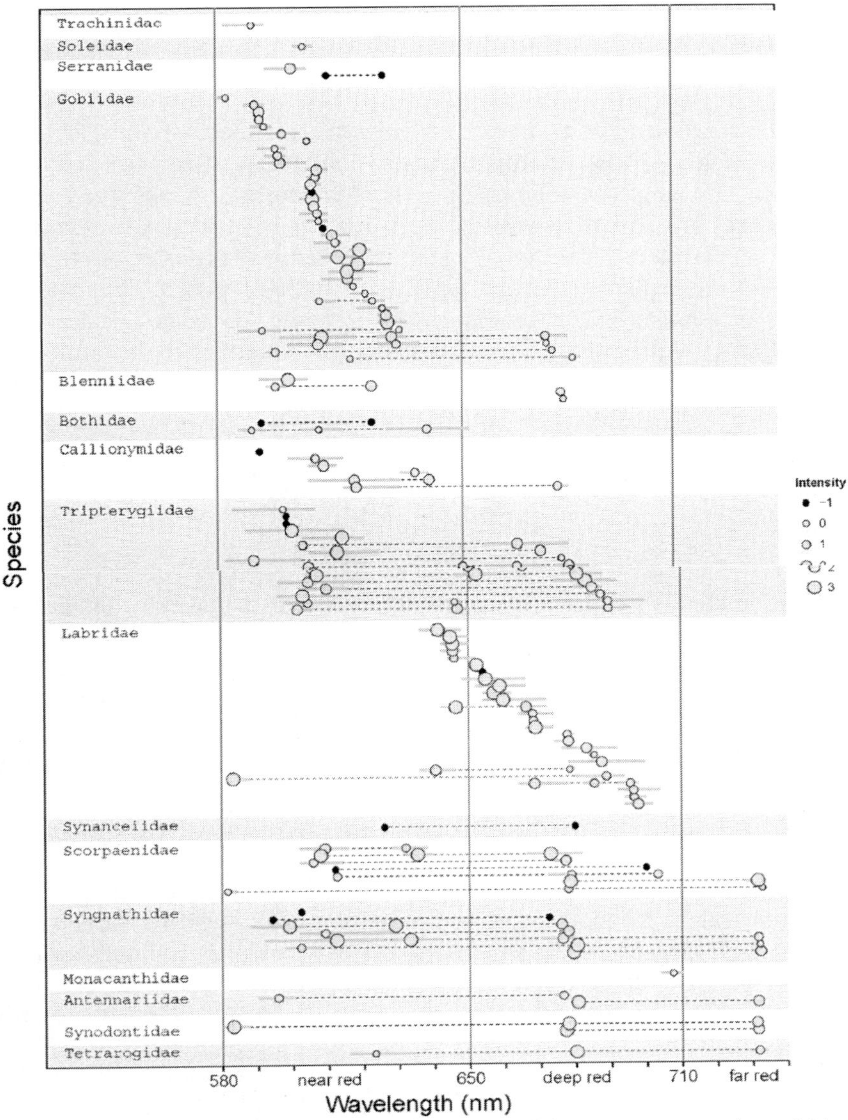

Figure 9. Species- and family-level patterns of fluorescence peak emissions and intensity. The graph indicates the λmax for each emission peak for all 114 measured species, with inter-individual range shown as bars. Multiple λmax values per species are connected by a dotted line. Dot size indicates fluorescence intensity in four intensity quartiles per measurement campaign. Black dots show emission peaks from data sets for which comparable brightness measures were unavailable. Vertical lines separate the data into near red, deep red, and far red fluorescence (cf. Figure 6). Families and species are ordered by average peak emission.

Fluorescence Peaks and Intensities Associate with Fish Families

The distribution of red fluorescent patterns revealed affinities with fish families (Figure 9). For example, gobies (Gobiidae) and wrasses (Labridae) show mostly single-peak emissions. These cluster in the near red range in gobies, but are spread across the deep red range in wrasses. In contrast, triplefins (Tripterygiidae) mostly show two to three peaks in the near and deep red range. Finally, several families, including pipefish (Syngnathidae) and scorpionfish (Scorpaenidae), share triple emission peaks, often including a unique emission peak in the far red. Referring to the family-level phylogeny (Figure 5) it is clear that these patterns are often shared among unrelated families and, hence, probably have evolved independently.

Functional Correlates of Red Fluorescence

Hypothesis 1: Short-Distance Visual Function

We predicted a higher prevalence of red fluorescence in smaller fish species, using body size as a proxy for interaction distance. Smaller species were indeed significantly more likely to express fluorescence than large species (GLMM length effect $\chi^2 = 14.09$, $P = 0.0002$, R^2marg.Rmarg.2 = 0.064, $n = 615$ species, Figure 10A). Given this relationship, we included body length as a covariate into all further models reported below to correct for its potentially confounding effect.

Hypothesis 2: Contrast Enhancement at Depth

We found no support for the hypothesis that red fluorescence should be more prevalent among species whose habitat extends into greater depths. While controlling for body size, we found the incidence of fluorescence to be independent of maximum depth per species (GLMM depth effect $\chi^2 = 1.42$, $P = 0.23$, full model R^2marg.Rmarg.2 = 0.067, $n = 615$ species, Figure 10B).

Hypothesis 3: Camouflage through Background Matching

While taking the effect of body size into account, we found that patchy fluorescence was rarely expressed in free-swimming species, occurred at intermediate frequencies in rather mobile benthic species, and reached a high average incidence of about 70% in motionless sit-and-wait species (GLMM substrate effect $\chi^2 = 10.97$, df = 2, $P = 0.0042$, Figure 11A). This pattern is consistent with *Hypothesis 3* proposing that this type of whole-body red fluorescence may contribute to camouflage.

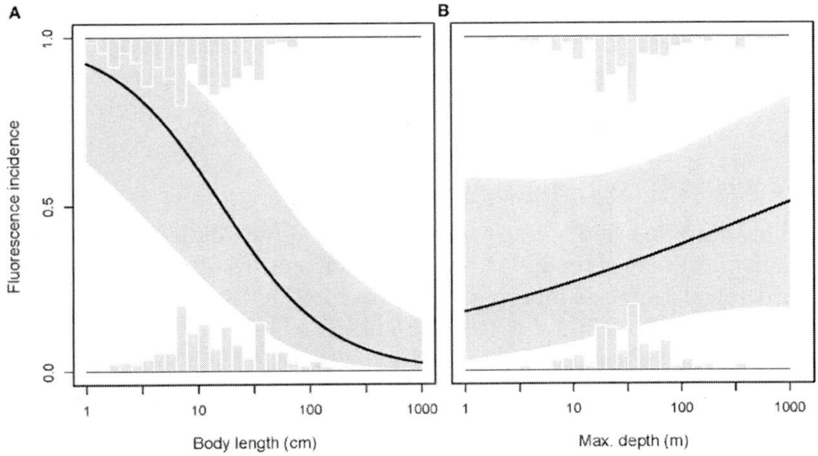

Figure 10. Incidence of red fluorescence in response to maximum body length (A) and maximum depth (B) per species. Marginal histograms display the relative frequency distributions of raw species values. Predicted incidence (black line) and its 95% credibility interval (shaded area) are extracted from binomial models corrected for shared species ancestry. Note log10-scale on x-axes.

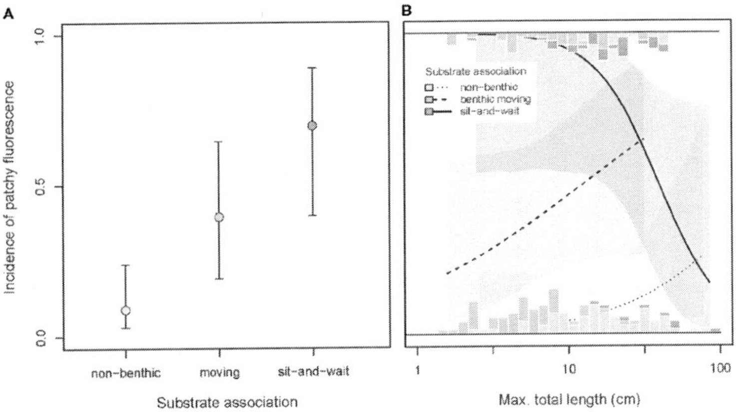

Figure 11. Incidence of patchy whole body fluorescence as predicted by substrate association, maximum body length, and their interaction. (A) Shows the overall predicted incidences of patchy fluorescence (±95% credibility intervals), pooling across body size. (B) Illustrates the interaction between substrate association and body size (note log10 x-scale). Marginal stacked histograms display relative frequency distributions of raw species values. Predicted incidences (lines) and their 95% credibility intervals (shaded areas) are extracted from binomial models taking shared species ancestry into account.

This association showed a significant interaction with body length (GLMM interaction $\chi^2 = 7.37$, df = 2, $P = 0.025$, full model R²marg.Rmarg.2 = 0.264, $n = 187$ species). Among immobile sit-and-wait species, the smaller species were more likely to exhibit patchy fluorescence, while the reverse was true in moving benthic species (Figure 11B).

Hypothesis 4: Prey Detection

Taking variation due to body size into account, we found that the incidence of eye fluorescence tended to increase from species feeding on eyeless food items toward species foraging on microscopic, eyed prey (GLMM prey type effect $\chi^2 = 5.44$, df = 2, $P = 0.066$, Figure 12A). This is consistent with the idea that iris fluorescence may contribute to prey detection when the target exhibits eyes that can directly reflect the incoming light (*Hypothesis 4*).

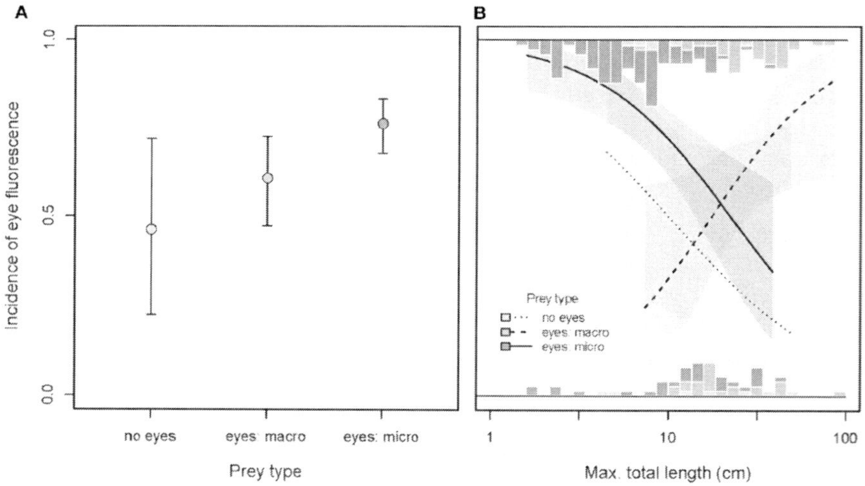

Figure 12. Incidence of eye fluorescence as predicted by prey types, maximum total body length, and their interaction. (A) Shows the overall predicted incidences of eye fluorescence (±95% credibility intervals), pooling across body size. (B) Illustrates the interaction between prey types and body length (note log10 x-scale). Marginal stacked histograms display the relative frequency distributions of raw species values. Predicted incidences (lines) and their 95% credibility intervals (shaded areas) are extracted from binomial models taking shared species ancestry into account.

Body length significantly affected the main prey type effect (GLMM interaction $\chi^2 = 10.76$, df = 2, $P = 0.0046$, full model R^2marg.Rmarg.2= 0.122, $n = 187$ species). Among fish foraging on small, eyed prey or indiscriminately, eye fluorescence was particularly prominent in the smallest species. The reverse was true for fish species foraging on large, eyed prey (Figure 12B).

Hypothesis 5: Intra-Specific Communication

The incidence of red fin fluorescence was significantly higher in sexually dimorphic than in sexually monomorphic species (GLMM dimorphism effect $\chi^2 = 6.21$, df = 1, $P = 0.013$, full model R^2marg.Rmarg.2 = 0.12, $n = 187$ species, Figures 13A,B), consistent with the idea that fluorescence may play a role in sexual communication. In contrast, no difference in the prevalence of fin fluorescence occurred between solitary/pair vs. group-living species (GLMM sociality effect $\chi^2 = 0.57$, df = 1, $P = 0.45$, Figures 13C,D), lending no support for a prime function in group-specific social interactions. Both main effects were statistically independent of variation in body size (no significant interaction, Figures 13 B,D).

DISCUSSION

Our analyses document substantial variation in the spectral characteristics and body topography of red fluorescence within and between 49 families of marine fishes. We analyzed this variation in the context of five different *a priori* hypotheses and found that (1) small fish were more likely to be red fluorescent, (2) maximum depth of occurrence did not predict the presence of red fluorescence, (3) benthic species in general and motionless sit-and-wait predators in particular were more likely to show fluorescent patterns consistent with camouflage, (4) species predating on small, eyed prey were more likely to possess red fluorescent eyes, albeit with marginal significance, and (5) sexually dimorphic species were more likely to show fin fluorescence. The latter could not be demonstrated for group-living species.

Phylogenetic Dynamics

Red fluorescence is a phylogenetically dynamic trait that has been repeatedly acquired and lost, or at least dramatically changed in expression, across the fish phylogeny, extending and confirming previous analyses on a smaller data set (Sparks et al., 2014). The resolution and phylogenetic coverage of our analysis is insufficient to determine whether red

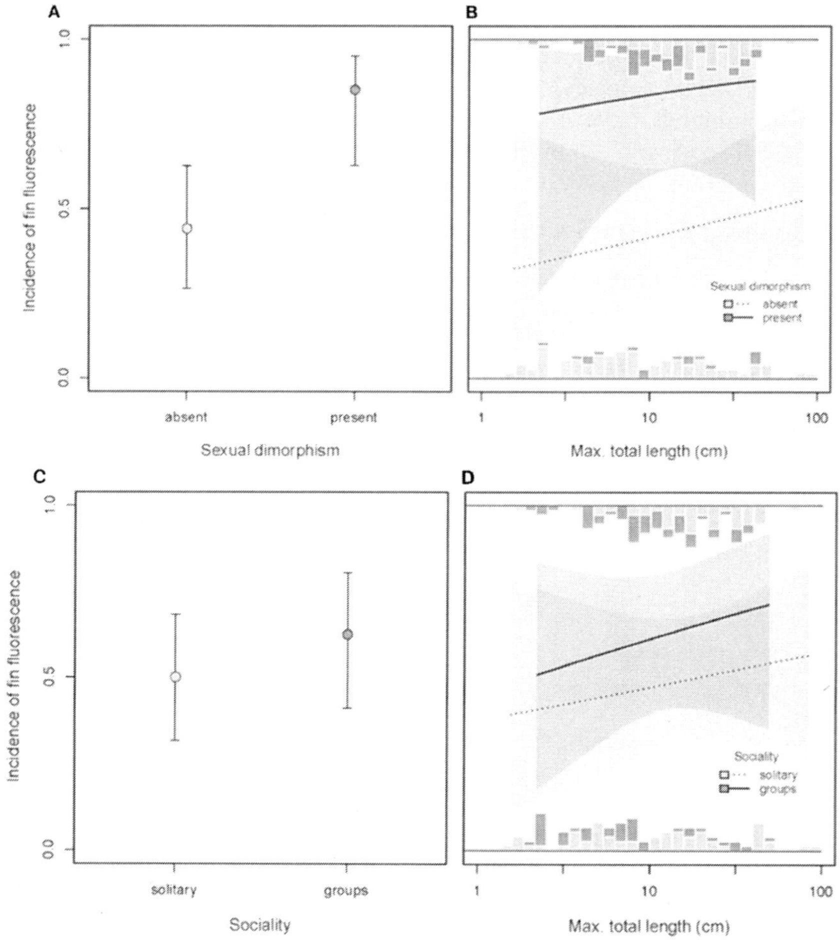

Figure 13. Incidence of fin fluorescence as predicted by sexual dimorphism (A,B) or sociality (C,D) and maximum body length (x-axis). In each row, the left panel illustrates the overall predicted incidences of fin fluorescence (±95% credibility intervals), pooled across body size. The right panel illustrates the interaction between sexual dimorphism or social system and body length (note log10-scale). Stacked marginal histograms display the relative frequency distribution of raw species values. Predicted incidences (lines) and their 95% credibility intervals (shaded areas) are extracted from binomial models taking shared species ancestry into account.

fluorescent pigmentation is ancient within the fish phylogeny, but it clearly appeared early in bony fish evolution. It appears plausible at least that these evolutionary changes are driven by convergent natural selection in independent lineages rather than representing a random, non-functional corollary of other traits. Comparable phylogenetic patterns in color trait evolution have recently been linked to adaptive function also in other fish. For example, the phylogenetically dynamic red and blue body and fin patterns in male darters associate with habitat structure, possibly in the context of predator exposure (Ciccotto and Mendelson, 2016a). In butterflyfish, the evolution of stripe and eyespot patterns was inconsistent with previously assumed predator-avoidance function, but rather co-varied with species ecology (Kelley et al., 2013). Similarly, the evolution of stripe patterns in cichlids (Seehausen and van Alphen, 1999) was phylogenetically highly dynamic and suggested to being driven by ecological specialization beyond phylogenetic affinities.

Interestingly, red fluorescence dominates in cryptic fish groups and seems less prominent in families with conspicuous reflective coloration (see also Sparks et al., 2014)—with the exception of wrasses (Labridae). This suggests that red fluorescence is not primarily used to enhance an already existing reflectance pattern, but rather associates with environmental conditions where red reflectance is not possible, as present in the stenospectral zone.

Variation in Fluorescence Emission Wavelengths

It is striking that the emission peaks (λ_{max}) of fluorescent structures cluster in three distinct ranges in the near, deep, and far red. While this study did not investigate the cellular expression of fluorescence, the pattern suggests the presence of at least three groups of red fluorescent pigments in marine fishes. The actual fluorophores have not yet been characterized, but previous work identified three different fluorescent mechanisms: (i) fluorescent iridophores with fluorescent guanine crystals (Michiels et al., 2008; Wucherer and Michiels, 2014), (ii) fluorescent chromatophores (Wucherer and Michiels, 2012), and (iii) fluorescent scales and fin rays (Michiels et al., 2008). Biochemical analyses are now required to characterize the molecular mechanisms and are likely to reveal cryptic within- and between-family diversity that may be hidden behind similar peak emissions in our study.

Hypothesis 1: Prevalence in Small Fish Confirmed

Our data show that red fluorescence is more common in small species. Assuming that these forage and communicate over short distances, the pattern is consistent with the idea that red fluorescence is associated with short-range vision measured in centimeters rather than meters. Hence, red fluorescence may well be specific for "small world" functions. Such an effect is less likely for yellow or green fluorescence (e.g., Sparks et al.,

2014; Gruber et al., 2016) because the emitted wavelength will travel through water over longer distances. Assuming a function in vision as well, we would predict their prevalence to be independent of body size. For example, the bright yellow tails of some tropical fusiliers (e.g., *Caesio cuning*) and goatfishes (e.g., *Mulloidichthys vanicolensis*) exhibit yellow fluorescence that may enhance the existing reflective signal and serve as a signal for group coherence or as a visual distractor for predators.

Hypothesis 2: Association with Depth Rejected

We could not confirm that species inhabiting greater depths are more likely to express red fluorescence and can think of four non-exclusive explanations. First, light environments poor in long wavelength light are also abundant in shallow water, for example in shaded areas, in turbid water, or at dusk and dawn. Hence, with depth representing just one of several factors favoring red fluorescence, our analysis of gross depth may easily fail to detect this association. Consistent with this idea, the triplefin *T. delaisi* adjusts fluorescence intensity primarily to ambient brightness rather than ambient spectrum (Harant et al., 2016), and may therefore directly respond to diurnal or seasonal changes in overall brightness, irrespective of depth. Second, maximum depths reported in the literature may be rather inaccurate (cf. Bridge et al., 2016) because they are neither systematically assessed nor representative for a species' average depth distribution. Third, our sampling efforts were restricted to depths within reach of regular SCUBA diving, above −30 m. Given substantial within-species, depth-associated variation in fluorescence (Meadows et al., 2014) a general depth effect between species may be obscured. Finally, the benefits of expressing fluorescence may be limited to intermediate depths (e.g., −10 to −100 m) with enough ambient blue-green light to induce fluorescence (unless coupled with a local chemiluminescent source, Douglas et al., 2000). This may be further complicated by the fact that fish may inhabit very different habitats or depths depending on age or season. As for now, however, there is too little information to take such non-linear depth effects into account.

Hypothesis 3: Consistency with Camouflage through Background Color Matching

We statistically confirm an association between a bottom-dwelling, nearly motionless, predatory lifestyle and full-body, patchy fluorescence (Sparks et al., 2014). This supports the suggestion that red fluorescence contributes to camouflage by background matching, where it may complement other camouflaging mechanisms such short-wavelength fluorescence or adjustments in body texture or pigmentation in response to substrate variation. This is particularly likely on complex backgrounds, where sit-and-wait predators are also anatomically very well adapted to blend in with algae, corals, and sponges. The latter generate a background of patchy fluorescence with most emission peaks in the near and deep red. Deep red fluorescence is of particular interest, because chlorophyll produces a distinctive red fluorescent signal around 680 nm. This signal is masked by sunlight in shallow water, but becomes clearly visible at depth (Figure 3). Our measurements show that the deep red fluorescence emission of many cryptic species matches this background emission (Figure 9). Deep red fluorescence for camouflage, however, only makes sense if relevant predators or prey from which a fluorescent target species aims to hide can perceive these rather long wavelengths. Future research therefore needs to assess the spectral sensitivity of candidate species and empirically test whether fluorescence improves camouflage toward potential prey or predators.

Hypothesis 4: Red Fluorescent Eyes More Common in Micro-Predators

Red eye fluorescence tended to be most prevalent in species that forage on small, eyed prey. This pattern is consistent with the idea that fish use reflective and fluorescent structures near their pupils similar to dragonfish (Douglas et al., 1998, 2000) and flashlight fish (Howland et al., 1992). "Active photolocation" using local emission of wavelengths (such as red) that are otherwise rare or absent from the environment may allow small fish to induce a highly contrasting eyeshine in cryptic prey (or predators) over short distances (Michiels et al., in submitted). The association described here confirms and generalizes earlier field observations (Meadows et al., 2014) and experimental laboratory work on the triplefin, *Tripterygion delaisi*, showing active brightness adjustment in its fluorescent irides (Wucherer and Michiels, 2012; Harant et al., 2016).

Hypothesis 5: Red Fluorescence Associated with Sexual Dimorphism, Not Group Living

Gregariousness did not associate with red fin fluorescence. In contrast, and as the strongest pattern among all hypotheses tested, we found red fin fluorescence strongly linked to sexual color dimorphism. This is consistent with the idea that red fluorescent fins could act as an enhancer of sexual signals in mate choice and male-male competition (e.g., in haremic systems), and generalizes earlier findings that orange and red fin displays are important in within-species intra-sexual communication in wrasses (Braun et al., 2014; Gerlach et al., 2014). The degree to which male fluorescent color displays are indeed preferred by females should be tested in experiments similar to those that revealed phylogenetic associations between male coloration and female preference in darters (Ciccotto and Mendelson, 2016b), and then be matched with measurements of spectral sensitivity to assess co-evolution between color pattern and sensory capacities (Pauers et al., 2016).

Can Marine Fish Perceive Red?

Most of the proposed adaptive functions of red fluorescence require that either the emitting species or relevant bystanders can perceive wavelengths beyond 600 nm. Extreme forms of red sensitivity are restricted to deep sea fishes that emit deep red bioluminescence around 700 nm and possess LWS receptors up to λ_{max} = 590 nm in *Aristostomias* (Partridge and Douglas, 1995) and even λ_{max} = 671 nm in *Malacosteus niger* (Douglas et al., 1998). Most marine fish inhabiting the photic zone have a single short wavelength cone and a medium wavelength twin or double cone with λ_{max} in the 500–540 nm range (Losey et al., 2003; Marshall et al., 2006). This arrangement, however, is already sufficient to perceive at least near red fluorescence (600–650 nm). This has for instance been inferred from the degree of overlap between the twin cone sensitivity (λ_{max} = 540 nm) and fluorescence peaking at 606 nm in the benthic goby *Eviota pellucida* (now *E. atriventris*, Greenfield and Suzuki, 2012) (Michiels et al., 2008). Behavioral evidence confirms this assumption in at least three fish species with "regular" LWS receptors (λ_{max}<540 nm). In the triggerfish *Rhinecanthus aculeatus* (LWS λ_{max} = 528 nm), foraging preferences show a bias to red stimuli > 600 nm (Cheney et al., 2013). The triplefin *T. delaisi* (LWS λ_{max} = 530 nm, P.-P. Bitton, unpublished data) recognizes its own red fluorescence (λ_{max} = 600–610 nm) (Kalb et al., 2015), and so does the fairy wrasse *C. solorensis* (LWS λ_{max} = 532 nm, deep red fluorescence at λ_{max} = 650 nm (Gerlach et al., 2014, 2016). Only few shallow-water marine fish possess photoreceptors that are more explicitly tuned to perceive red. For example, the wrasse *Thalassoma duperrey* (LWS λ_{max} = 570 nm) can perceive a red band in its own color pattern under natural light (Barry and Hawryshyn, 1999). Similar long-wavelength photoreceptors have been described for several seahorses and pipefish (LWS λ_{max} = 560 and 580 nm, Mosk et al., 2007) and the goby *Gobiusculus flavescens* (LWS λ_{max} = 553 nm) (Utne-Palm and Bowmaker, 2006). As shown by our current study, all these species belong to families that feature a high incidence of red fluorescence.

A Word of Caution

Our current analyses face two limitations. First, solid quantitative information on ecological and biological traits is only available for a handful of marine fish species. For most of our study species, information on, for example, primary food types, foraging style, sexual dichromatism, or maximum depth range could primarily draw from rather rudimentary statements. This adds substantial noise as typical to large comparative analyses. We carefully avoided confounding biases, but noise in the life history data in particular has potential to substantially lower the statistical power to detect associations between red fluorescence and ecological traits. Second, a fully resolved species-level phylogenetic hypothesis for at least a large proportion of the species under investigation does not exist. Hence, we resorted to simplified analyses, first by investigating trait evolution at the family level, and second by controlling for shared ancestry with a rather superficial correlation matrix, taking the taxonomic levels "order," "family," and "genus" into account. We expect that ongoing advances in overall fish phylogeny will soon enable more fine-tuned phylogenetic analyses on red fluorescence and its association with ecology focused on within-family variation as e.g., in gobies, triplefins, or wrasses.

ACKNOWLEDGMENTS

We would like to thank Jens Pfann (data acquisition), Gregor Schulte (technical assistance), Christoph Hösler and Christopher Schwarzer (raw-data processing routines), Sandra Dangelmayer, Dennis Sprenger, and Klara Wolf (field support), Sabrina Hug (image analysis), Nadine Kalb (trait scoring), Martina Hohloch and Andreas Oelkrug (fish maintenance), as well as Ronald Fricke, Jürgen Herler, Helen Larson, and Rick Winterbottom (fish identification). This project was funded by the German Research Foundation (DFG) Reinhart Koselleck Grant Mi482/13-1 "Red fluorescence in reef fishes: Functions and mechanisms" to NM. We acknowledge support by the Deutsche Forschungsgemeinschaft (DFG) and the Open Access Publishing Fund of the University of Tübingen.

SUPPLEMENTARY MATERIAL A

Systematic list of fish taxa represented in the current comparative dataset. The Table shows our ratings of red fluorescence (0 = absent, 1 = present), substrate association (1 = benthic moving, 2 = benthic sit-and-wait, 3 = non-benthic), prey type (1 = microscopic prey with eyes, 2 = macroscopic prey with eyes, 3 = prey without eyes), sexual color dimorphism (0 = absent, 1 = present), and sociality (1 = solitary, 2 = group-living). Within fluorescent species, we scored (where possible) the presence of patchy fluorescence, of iris fluorescence, and of fin fluorescence (0 = absent, 1 = present). Where spectrometric measurements were available, we add the intensity quartile (ranging from 1 for the lowest intensity quartile to 4 for the highest intensity quartile) and the number of distinct fluorescent emission peaks. Finally, we list the maximum depth of occurrence (in m) and the maximum total body length (in cm) for each species. Empty cells are due to missing data. For details and sources see the Material and Methods section in the main text.

SUPPLEMENTARY MATERIAL B

Major literature sources to complement ecological and biological data to the information available at *fishbase.org*.

SUPPLEMENTARY MATERIAL C

Systematic list summarizing all taxa with spectral data for a detailed characterization of red fluorescence. Each row refers to a single distinct fluorescent emission peak and provides information on peak number, the emission category (1 = near red, 2 = deep red, 3 = far red), the wavelength at peak emission (lambda max), absolute emission intensity (in counts per nm per ms), and the intensity quartile at peak emission.

SUPPLEMENTARY MATERIAL D

Overview about the origin for the phylogenetic position of fish families as represented in Figure 5.

REFERENCES

1. Aksnes, D. L., and Giske, J. (1993). A theoretical model of aquatic visual feeding. *Ecol. Model.* 67, 233–250. doi: 10.1016/0304-3800(93)90007-F
2. Alieva, N. O., Konzen, K. A., Field, S. F., Meleshkevitch, E. A., Hunt, M. E., Beltran-Ramirez, V., et al. (2008). Diversity and evolution of coral fluorescent proteins. *PLoS ONE* 3:e2680. doi: 10.1371/journal.pone.0002680
3. Allen, G. R., Erdmann, M. V., Robertson, D., Anderson, R. C., Baker, N., Lim, K., et al. (2014). *Reef Fishes of the East Indies*. Honolulu, HI: University of Hawaii Press.
4. Allen, G. R., Midgley, S. H., and Allen, M. (2002). *Field Guide to the Freshwater Fishes of Australia*. Perth: Western Australian Museum.
5. Allen, G. R., Steene, R., Humann, P., and DeLoach, N. (2003). *Reef Fish Identification: Tropical Pacific*. Jacksonville, FL: New World Pubns Inc.
6. Baldauf, S. A., Bakker, T. C., Kullmann, H., and Thünken, T. (2011). Female nuptial coloration and its adaptive significance in a mutual mate choice system. *Behav. Ecol.* 22, 478–485. doi: 10.1093/beheco/arq226
7. Barry, K. L., and Hawryshyn, C. W. (1999). Spectral sensitivity of the Hawaiian saddle wrasse, *Thalassoma duperrey*, and implications for visually mediated behaviour on coral reefs. *Environ. Biol. Fish.* 56, 429–442. doi: 10.1023/A:1007556112449
8. Bates, D., Maechler, M., and Bolker, B. (2013). *lme4: Linear Mixed-Effects Models Using S4 Classes. R package version 0.999999-2*. Available online at: http://cran.r-project.org/web/packages/lme4/index.html.

9. Ben-Zvi, O., Eyal, G., and Loya, Y. (2014). Light-dependent fluorescence in the coral *Galaxea fascicularis*. *Hydrobiologia* 759, 15–26. doi: 10.1007/s10750-014-2063-6
10. Betancur, -R. R., Broughton, R. E., Wiley, E. O., Carpenter, K., López, J. A., Li, C., et al. (2013). The tree of life and a new classification of bony fishes. *PLoS Curr.* 5:53ba26640df0ccaee75bb165c8c26288. doi: 10.1371/currents. tol.53ba26640df0ccaee75bb165c8c26288
11. Brady, P. C., Travis, K. A., Maginnis, T., and Cummings, M. E. (2013). Polaro–cryptic mirror of the lookdown as a biological model for open ocean camouflage. *Proc. Natl. Acad. Sci.U.S.A.* 110, 9764–9769. doi: 10.1073/pnas.1222125110
12. Brandley, N. C., Speiser, D. I., and Johnsen, S. (2013). Eavesdropping on visual secrets. *Evol. Ecol.* 27, 1045–1068. doi: 10.1007/s10682-013-9656-9
13. Braun, C., Michiels, N., Siebeck, U., and Sprenger, D. (2014). Signalling function of long wavelength colours during agonistic male-male interactions in the wrasse *Coris julis*. *Mar. Ecol. Prog. Ser.* 504, 277–286. doi: 10.3354/meps10760
14. Bridge, T. C. L., Luiz, O. J., Coleman, R. R., Kane, C. N., and Kosaki, R. K. (2016). Ecological and morphological traits predict depth-generalist fishes on coral reefs. *Proc. R. Soc. Lond. B Biol. Sci.* 283:20152332. doi: 10.1098/rspb.2015.2332
15. Bruce, C. (2009). Fish that see red. *New Sci.* 202, 20. doi: 10.1016/S0262-4079(09)61441-X
16. Cheney, K. L., Newport, C., McClure, E. C., and Marshall, N. J. (2013). Colour vision and response bias in a coral reef fish. *J. Exp. Biol.* 216, 2967–2973. doi: 10.1242/jeb.087932
17. Ciccotto, P. J., and Mendelson, T. C. (2016a). The ecological drivers of nuptial color evolution in darters (Percidae: Etheostomatinae). *Evolution* 70, 745–756. doi: 10.1111/evo.12901
18. Ciccotto, P. J., and Mendelson, T. C. (2016b). Phylogenetic correlation between male nuptial color and behavioral responses to color across a diverse and colorful genus of freshwater fish (*Etheostoma* spp., Teleostei: Percidae). *Ethology* 122, 245–256. doi: 10.1111/eth.12465
19. Clark, E. (1979). Red Sea fishes of the family Tripterygiidae with descriptions of eight new species. *Isr. J. Zool.* 28, 65–113.
20. Debelius, H. (1998). *Red Sea Reef Guide*. Frankfurt: IKAN, 1–321.
21. Díaz-Uriarte, R., and Garland, T. Jr. (1998). Effects of branch length errors on the performance of phylogenetically independent contrasts. *Syst. Biol.* 47, 654–672. doi: 10.1080/106351598260653
22. Douglas, R. H., Mullineaux, C. W., and Partridge, J. C. (2000). Long–wave sensitivity in deep–sea stomiid dragonfish with far–red bioluminescence: evidence for a dietary origin of the chlorophyll–derived retinal photosensitizer of *Malacosteus niger*. *Philos. Trans. R. Soc. Lond. B Biol. Sci.* 355, 1269–1272. doi: 10.1098/rstb.2000.0681
23. Douglas, R., Partridge, J., Dulai, K., Hunt, D., Mullineaux, C., Tauber, A., et al. (1998). Dragon fish see using chlorophyll. *Nature* 393, 423–424. doi: 10.1038/30871
24. Endler, J. A. (1981). An overview of the relationships between mimicry and crypsis. *Biol. J. Linnean Soc.* 16, 25–31. doi: 10.1111/j.1095-8312.1981.tb01840.x
25. Endler, J. A. (1990). On the measurement and classification of colour in studies of animal colour patterns. *Biol. J. Linnean Soc.* 41, 315–352. doi: 10.1111/j.1095-8312.1990.tb00839.x
26. Field, S. F., Bulina, M. Y., Kelmanson, I. V., Bielawski, J. P., and Matz, M. V. (2006). Adaptive evolution of multicolored fluorescent proteins in reef-building corals. *J. Mol. Evol.* 62, 332–339. doi: 10.1007/s00239-005-0129-9
27. Fricke, R. (1997). *Tripterygiid Fisches of the Western and Central Pacific*. Koenigstein: Koeltz Scientific Books.
28. Froese, R., and Pauly, D. (eds.) (2014). *FishBase. World Wide Web Electronic Publication. version (06/2014)*. Available online at: www.fishbase.org

29. Garland, T. Jr., and Ives, A. R. (2000). Using the past to predict the present: confidence intervals for regression equations in phylogenetic comparative methods. *Am. Nat.* 155, 346–364. doi: 10.1086/303327
30. Gerlach, T., Sprenger, D., and Michiels, N. K. (2014). Fairy wrasses perceive and respond to their deep red fluorescent coloration. *Proc. R. Soc. B Biol. Sci.* 281:20140787. doi: 10.1098/rspb.2014.0787
31. Gerlach, T., Theobald, J., Hart, N. S., Collin, S. P., and Michiels, N. K. (2016). Fluorescence characterisation and visual ecology of pseudocheilinid wrasses. *Front. Zool.* 13:13. doi: 10.1186/s12983-016-0145-1
32. Greenfield, D. W., and Suzuki, T. (2012). *Eviota atriventris*, a new goby previously misidentified as *Eviota pellucida* Larson (Teleostei: Gobiidae). *Zootaxa* 3197, 55–62.
33. Gruber, D. F., Kao, H.-T., Janoschka, S., Tsai, J., and Pieribone, V. A. (2008). Patterns of fluorescent protein expression in scleractinian corals. *Biol. Bull.* 215, 143–154. doi: 10.2307/25470695
34. Gruber, D. F., Loew, E. R., Deheyn, D. D., Akkaynak, D., Gaffney, J. P., Smith, W. L., et al. (2016). Biofluorescence in catsharks (Scyliorhinidae): fundamental description and relevance for Elasmobranch visual ecology. *Sci. Rep.* 6:24751. doi: 10.1038/srep24751
35. Haddock, S. H. D., and Dunn, C. W. (2015). Fluorescent proteins function as a prey attractant: experimental evidence from the hydromedusa *Olindias formosus* and other marine organisms. *Biol. Open* 4, 1094–1104. doi: 10.1242/bio.012138
36. Haddock, S. H., Dunn, C. W., Pugh, P. R., and Schnitzler, C. E. (2005). Bioluminescent and red-fluorescent lures in a deep-sea siphonophore. *Science* 309, 263–263. doi: 10.1126/science.1110441
37. Harant, U. K., Michiels, N. K., Anthes, N., and Meadows, M. G. (2016). The consistent difference in red fluorescence in fishes across a 15 m depth gradient is triggered by ambient brightness, not by ambient spectrum. *BMC Res. Notes* 9:107. doi: 10.1186/s13104-016-1911-z
38. Hart, P. J., Hall, R., Ray, W., Beck, A., and Zook, J. (2015). Cicadas impact bird communication in a noisy tropical rainforest. *Behav. Ecol.* 26, 839–842. doi: 10.1093/beheco/arv018
39. Holleman, W. (2005). A review of the triplefin fish genus *Enneapterygius* (Blennioidei: Tripterygiidae) in the western Indian Ocean, with descriptions of four new species. *Smithiana Bull.* 5, 1–28.
40. Howland, H. C., Murphy, C. J., and Mccosker, J. E. (1992). Detection of eyeshine by flashlight fishes of the family Anomalopidae. *Vision Res.* 32, 765–769. doi: 10.1016/0042-6989(92)90191-K
41. Jack, C. B. (2014). *Detecting the Detector: A Widespread Animal Sense?* vixra.org.
42. Jerlov, N. G. (1968). *Optical Oceanography*. New York, NY: Elsevier.
43. Johnsen, S. (2012). *The Optics of Life: A Biologist's Guide to Light in Nature*. Princeton, NJ: Princeton University Press.
44. Kalb, N., Schneider, R., Sprenger, D., and Michiels, N. (2015). The red-fluorescing marine fish *Tripterygion delaisi* can perceive its own red fluorescent color. *Ethology* 121, 1–11. doi: 10.1111/eth.12367
45. Kelley, J. L., Fitzpatrick, J. L., and Merilaita, S. (2013). Spots and stripes: ecology and colour pattern evolution in butterflyfishes. *Proc. R. Soc. B Biol. Sci.* 280:20122730. doi: 10.1098/rspb.2012.2730
46. Kraaijeveld, K., Kraaijeveld-Smit, F. J., and Komdeur, J. (2007). The evolution of mutual ornamentation. *Anim. Behav.* 74, 657–677. doi: 10.1016/ j.anbehav. 2006.12.027
47. Lagorio, M. G., Cordon, G. B., and Iriel, A. (2015). Reviewing the relevance of fluorescence in biological systems. *Photochem. Photobiol. Sci.* 14, 1538–1559. doi: 10.1039/C5PP00122F
48. Lefcheck, J. S. (2016). piecewiseSEM: piecewise structural equation modelling in R for ecology, evolution, and systematics. *Methods Ecol. Evol.* 7, 573–579. doi: 10.1111/2041-210X.12512

49. Li, K. T., Wetterer, J. K., Nelson, G., and Hairston, Jr. (1985). Fish size, visula resolution, and prey selectivity. *Ecology* 66, 1729–1735. doi: 10.2307/2937368

50. Losey, G., McFarland, W., Loew, E., Zamzow, J., Nelson, P., and Marshall, N. (2003). Visual biology of Hawaiian coral reef fishes. *I.* Ocular transmission and visual pigments. *Copeia* 2003, 433–454. doi: 10.1643/01-053

51. Luiz, O. J., Allen, A. P., Robertson, D. R., Floeter, S. R., Kulbicki, M., Vigliola, L., et al. (2013). Adult and larval traits as determinants of geographic range size among tropical reef fishes. *Proc. Natl. Acad. Sci.U.S.A.* 110, 16498–16502. doi: 10.1073/pnas.1304074110

52. Lythgoe, J. N. (1979). *Ecology of Vision*. Oxford: Oxford University Press.

53. Marshall, J., Carleton, K. L., and Cronin, T. (2015). Colour vision in marine organisms. *Curr. Opin. Neurobiol.* 34, 86–94. doi: 10.1016/j.conb.2015.02.002

54. Marshall, J., Vorobiev, M., and Siebeck, U. (2006). "What does a reef fish see when it sees a reef fish?," in *Communication in Fishes, Vol. 2. Visual Communication* eds F. Ladich, S. P. Collin, P. Moller, and B. G. Kapoor (Enfield, CT: Science Publisher Inc), 393–422.

55. Meadows, M. G., Anthes, N., Dangelmayer, S., Alwany, M. A., Gerlach, T., Schulte, G., et al. (2014). Red fluorescence increases with depth in reef fishes, supporting a visual function, not UV protection. *Proc. R. Soc. B Biol. Sci.* 281:20141211. doi: 10.1098/rspb.2014.1211

56. Michiels, N. K., Anthes, N., Hart, N. S., Herler, J., Meixner, A. J., Schleifenbaum, F., et al. (2008). Red fluorescence in reef fish: a novel signalling mechanism? *BMC Ecol.* 8:16. doi: 10.1186/1472-6785-8-16

57. Mosk, V., Thomas, N., Hart, N. S., Partridge, J. C., Beazley, L. D., and Shand, J. (2007). Spectral sensitivities of the seahorses *Hippocampus subelongatus* and *Hippocampus barbouri* and the pipefish *Stigmatopora argus*. *Vis. Neurosci.* 24, 345–354. doi: 10.1017/S0952523807070320

58. Munz, F. W., and McFarland, W. N. (1973). The significance of spectral position in the rhodopsins of tropical marine fishes. *Vision Res.* 13, 1829–1874. doi: 10.1016/0042-6989(73)90060-6

59. Nakagawa, S., and Schielzeth, H. (2013). A general and simple method for obtaining R^2 from generalized linear mixed-effects models. *Methods Ecol. Evol.* 4, 133–142. doi: 10.1111/j.2041-210x.2012.00261.x

60. Norris, K. S., and Schilt, C. R. (1988). Cooperative societies in three- dimensional space: on the origins of aggregations, flocks, and schools, with special reference to dolphins and fish. *Ethol. Sociobiol.* 9, 149–179. doi: 10.1016/0162-3095(88)90019-2

61. O'Brien, W. J. (1979). The predator-prey interaction of planktivorous fish and zooplankton: recent research with planktivorous fish and their zooplankton prey shows the evolutionary thrust and parry of the predator-prey relationship. *Am. Sci.* 67, 572–581.

62. Pagel, M. (1994). Detecting correlated evolution on phylogenies: a general method for the comparative analysis of discrete characters. *Proc. R. Soc. Lond. B Biol. Sci.* 255, 37–45. doi: 10.1098/rspb.1994.0006

63. Paradis, E., Claude, J., and Strimmer, K. (2004). APE: analyses of phylogenetics and evolution in R language. *Bioinformatics* 20, 289–290. doi: 10.1093/bioinformatics/btg412

64. Partridge, J. (1990). "The colour sensitivity and vision of fishes," in *Light and Life in the Sea*, eds P. J. Herring, A. K. Campbell, M. Whitfield, and L. Maddock. (Cambridge: Cambridge University Press), 167–184.

65. Partridge, J. C., and Douglas, R. H. (1995). Far-red sensitivity of dragon fish. *Nature* 375, 21–22. doi: 10.1038/375021a0

66. Pauers, M. J., Kuchenbecker, J. A., Joneson, S. L., and Neitz, J. (2016). Correlated evolution of short wavelength sensitive photoreceptor sensitivity and color pattern in Lake Malawi cichlids. *Front. Ecol. Evol.* 4:12. doi: 10.3389/fevo.2016.00012

67. R Core Team (2013). *R: A Language and Environment for Statistical Computing*. Vienna: R Foundation for Statistical Computing [Online]. Available online at: http://www.R-project.org/.
68. Revell, L. J. (2012). phytools: an R package for phylogenetic comparative biology (and other things). *Methods Ecol. Evol.* 3, 217–223. doi: 10.1111/j.2041-210X.2011.00169.x
69. Rowland, W. (1999). Studying visual cues in fish behavior: a review of ethological techniques. *Environ. Biol. Fish.* 56, 285–305. doi: 10.1023/A:1007517720723
70. Salih, A., Larkum, A., Cox, G., Kühl, M., and Hoegh-Guldberg, O. (2000). Fluorescent pigments in corals are photoprotective. *Nature* 408, 850–853. doi: 10.1038/35048564
71. Schlichter, D., and Fricke, H. (1990). Coral host improves photosynthesis of endosymbiotic algae. *Naturwissenschaften* 77, 447–450. doi: 10.1007/BF01135950
72. Schmidt, D., and O'Brien, W. J. (1982). Planktivorous feeding ecology of Arctic grayling (*Thymallus arcticus*). *Can. J. Fish. Aquat. Sci.* 39, 475–482. doi: 10.1139/f82-065
73. Seehausen, O., and van Alphen, J. (1999). Evolution of colour patterns in East African cichlid fish. *J. Evol. Biol.* 12, 514–534. doi: 10.1046/j.1420-9101.1999.00055.x
74. Slabbekoorn, H., and Peet, M. (2003). Ecology: birds sing at a higher pitch in urban noise. *Nature* 424, 267–267. doi: 10.1038/424267a
75. Sparks, J. S., Schelly, R. C., Smith, W. L., Davis, M. P., Tchernov, D., Pieribone, V. A., et al. (2014). The covert world of fish biofluorescence: a phylogenetically widespread and phenotypically variable phenomenon. *PLoS ONE* 9:e83259. doi: 10.1371/journal.pone.0083259
76. Tamura, T. (1957). A study of visual perception in fish, especially on resolving power and accommodation. *Bull. Jap. Soc. Sci. Fish* 22, 536–557. doi: 10.2331/suisan.22.536
77. Utne-Palm, A. C., and Bowmaker, J. K. (2006). Spectral sensitivity of the two-spotted goby *Gobiusculus flavescens* (Fabricius): a physiological and behavioural study. *J. Exp. Biol.* 209, 2034–2041. doi: 10.1242/jeb.02171
78. Wucherer, M. F., and Michiels, N. K. (2012). A fluorescent chromatophore changes the level of fluorescence in a reef fish. *PLoS ONE* 7:e37913. doi: 10.1371/journal.pone.0037913
79. Wucherer, M. F., and Michiels, N. K. (2014). Regulation of red fluorescent light emission in a cryptic marine fish. *Front. Zool.* 11:1. doi: 10.1186/1742-9994-11-1
80. Wyman, M. J., Stinchcombe, J. R., and Rowe, L. (2013). A multivariate view of the evolution of sexual dimorphism. *J. Evol. Biol.* 26, 2070–2080. doi: 10.1111/jeb.12188
81. Zuur, A. F., Ieno, E. N., Walker, N. J., Saveliev, A. A., and Smith, G. M., (eds.). (2009). *Mixed Effects Models and Extensionis in Ecology with R*. New York, NY: Springer.

CHAPTER 3

Global and Regional Patterns in Riverine Fish Species Richness: A Review

Thierry Oberdorff,[1] Pablo A. Tedesco,[1] Bernard Hugueny,[1] Fabien Leprieur,[2] Olivier Beauchard,[3] Sébastien Brosse,[4] and Hans H. Dürr[5]

[1]UMR BOREA, IRD 207, Département Milieux et Peuplements Aquatiques, Muséum National d'Histoire Naturelle, 43 rue Cuvier, 75231 Paris cedex, France

[2]Laboratoire Ecosystèmes Lagunaires, UMR 5119 (CNRS-IFREMER-UM2-IRD), Université Montpellier 2, Place Eugène Bataillon, 34095 Montpellier Cedex 5, France

[3]Ecosystem Management Research Group, Department of Biology, Faculty of Sciences, University of Antwerp, Universiteitsplein 1, 2610 Antwerpen (Wilrijk), Belgium

[4]Laboratoire d'Ecologie Fonctionnelle, UMR 5545, CNRS-Université Paul Sabatier, 118 route de Narbonne, 31062 Toulouse cedex 4, France

[5]Faculty of Geosciences, Utrecht University, P.O. Box 80115, 3508 TC Utrecht, The Netherlands

ABSTRACT

We integrate the respective role of global and regional factors driving riverine fish species richness patterns, to develop a synthetic model of potential mechanisms and processes generating these patterns. This framework allows species richness to be broken down into different components specific to each spatial extent and to establish links between these components and the processes involved. This framework should help to answer the questions that are currently being asked by society, including the effects of species invasions, habitat loss, or fragmentation and climate change on freshwater biodiversity.

1. INTRODUCTION

The diversity of life, usually referred to as "biodiversity", is not evenly distributed throughout the globe. A considerable proportion is to be found in the tropics, while the poles are only home to a small fraction, and between the two extremes there is a whole diversity gradient. Ecologists, bio-geographers, and paleontologists have studied the reasons for these differences, but the question remains open despite the dozens of hypotheses that have been put forward on the subject [1–5]. The present analysis is limited to one important aspect of biodiversity, species richness, which is defined as the number of species present at a given time in a given place. Species richness gradients can be examined across a variety of spatial extents (extent is the geographic separation between the furthest points) and grains (grain is the area of the sampling unit) [6]. But ecologists, who up to the 90s preferred experimental approaches, mainly focussed on the factors and processes that influence species richness at fine grain sizes and spatial extents (based on published papers in Ecology between 1980 and 1986, cited by May [7]). However, it is now recognized that species richness patterns are directly influenced by processes working at much larger scales; that is, regional or even continental [8–12]. This gave birth to macroecology [13, 14], whose aim is to highlight the statistical properties that emerge from complex ecosystems, in order to identify general patterns at different space-time scales of observation, and particularly at the macroscopic scale. If we follow Brown's ([13, page 6]) definition of macroecology: "*it is a non-experimental, statistical investigation of the relationship between the dynamics and interactions of species populations that have typically been studied on small scales by ecologists and the processes of speciation, extinction, and expansion and contraction of ranges that have been investigated on much larger scales by biogeographers, paleontologists, and macroevolutionists. It is an effort to introduce simultaneously a geographical and a historical perspective in order to understand more completely the local abundance, distribution, and diversity of species, and to apply an ecological perspective in order to gain insights into the history and composition of regional and continental biotas.*" In fact, determining which factors and processes are responsible for the variation in species richness patterns is a crucial issue for conservation planning in the face of current and future global and regional anthropogenic impacts [15].

Here, we review patterns and predictors of riverine fish species richness at the drainage basin grain and at global and regional extents. The "freshwater fish" model is particularly well adapted to this type of study since drainage basins are separated from one another by barriers (oceans, or land) that are—for all practical purposes—insurmountable for strictly freshwater fishes, and thus form a kind of insular habitats. Like remote islands, drainage basins are not under equilibrium conditions, as they receive new colonists so rarely that immigration and speciation often occur on similar timescales. This absence of migration between river basins over large temporal scales implies that extinction and speciation processes are specific of each drainage basin [16]. Thus, river basins are, to some extent, independent entities that could be used in comparative analysis to explore the factors that shape overall fish community richness between them. Incidentally, a considerable amount of exploitable data is now available that enables the use of comparative approaches to test the main ecological hypotheses currently under consideration. In this chapter we will use this natural experiment framework to review and discuss the relative role of regional and continental features in determining river drainage basin diversity patterns.

Unless otherwise specified, the term "river drainage basin" will refer to rivers flowing into the ocean (including all their tributaries). For rivers that are part of a bigger drainage basin, the term "tributary" will be used. In this paper we will focus on two grains sizes (i.e., river drainage basin or tributary drainage basin) at two different extents (i.e., global to regional). The term species richness (or species diversity) describes here the total number of species encountered within a river basin or within a tributary.

2. GLOBAL APPROACH TO RIVERINE FISH SPECIES RICHNESS

At the intercontinental scale, three major hypotheses that sum up the majority of different hypotheses proposed (see [3] for a review) have already been tested to explain the variability of riverine fish species richness.

The first, the area hypothesis [17, 18] refers to the existence of a positive relationship between the number of species present in a given area and the size of this area. This relationship has been described by a power function in the form $S = CA^Z$ (where is the number of species, A is the (surface) area, and C and Z are constants to be fitted) [19, 20]. It suggests that size (the surface of a river drainage basin in the case of riverine fishes) limits the number of species an area can harbor, and, due to its universal application, almost serves as a law in community ecology [21]. Several nonexclusive explanations have been put forward to explain this species-area relationship (Schoener 2010) but three of them are most often invoked: (1) the size-dependent extinction rate [17, 18], (2) the size-dependent speciation rate [22], and (3) the diversity of the habitat [18]. According to the first explanation the probability of extinction of a species increases with a reduction in the size of the "island", due to a decrease in its population size. The second explanation suggests a positive effect of area on speciation rate by exposing species to greater ecological heterogeneity and/or geographical barriers [5]. The third explanation suggests that the heterogeneity of the habitat and the diversity of available food resources increase with the size of the "island" thus offering a large number of available niches and consequently favouring the coexistence of a large number of species [23].

The second hypothesis, the species-energy hypothesis [24, 25] predicts a positive correlation between species richness and the energy available within the system. This hypothesis has received empirical support from a large number of studies carried out on different communities of animals and plants [24, 26–36]. This being said, there is still a certain ambiguity even in the way the hypothesis is expressed. In fact, energy can influence richness by means of two rather different processes. Wright [24] considers energy to be a factor that determines resources available for a given biological community and thus as a productivity factor per se, whereas Turner et al. [33] and Currie [27], for example, consider energy to be a factor that determines the physiological limits of the species. In the former, one would expect a variable such as net primary production to be an important predictor of species richness whereas in the later, variables linked with temperature or available solar energy would predominate [29].

Finally, the third hypothesis, the historical hypothesis [37], attempts to explain differences in richness gradients by the potential for recolonisation of systems and thus by the degree of maturity achieved since the last major climate change or by the degree of stability in past climatic conditions [38, 39]. This last hypothesis, which combines past environmental conditions with geographic contingencies regulating dispersal possibilities, has been relatively neglected compared to the others. Two main reasons can explain this gap: (i) in essence, past conditions are much more difficult to evaluate and accurately measure than present conditions and (ii) current and past conditions are globally highly correlated.

2.1. The Roles of Area and Energy

In the first global studies conducted in this topic Oberdorff et al. [40] and Guégan et al. [41] used data obtained for 292 drainage basins on 5 different continents to identify the factors responsible for variations in riverine fish species richness within the framework of the three above-mentioned hypotheses. The models resulting from these exploratory analyses tend to show that, at this spatial extent, the factors associated with the first two hypotheses (i.e., the area hypothesis and the species-energy hypothesis) predominate. Only taking into consideration three summary factors, that is, the total surface area of the river drainage basin, the mean flow at the river mouth and the net terrestrial primary productivity within the basin, those models explain between 78 and 93% (depending on the statistical model) of the natural variability of the river basin species richness, the mean annual river discharge explaining the greater part of the variance in species richness.

Based on a comprehensive species richness dataset (Figure 1) recently compiled (926 river basins analyzed, see [42, 43] and the Supplementary Appendix available online at doi: 10.1155/2011/967631 for further details on the database), we performed a spatial autoregressive model (SAR, see [44]) accounting for the spatial configuration of drainage basins. The final model explains 77.1% of the total variation in species richness. Results of this new analysis confirm previous findings concerning the effects of area-related and climate-related variables, but also reveal a significant influence of past climatic changes and geographic isolation of drainage basins on species richness patterns (see Table 1 and Figure 2). These historical effects have also been revealed in previous regional analyses (see [45–48] although on a different spatial grain) but, regarding freshwater fish, this is the first time that the effect of past climatic variability (from glacial periods of the Pleistocene to present day) on species richness patterns is detected at the global scale (but see [39] for an effect of climatic variability on beta diversity).

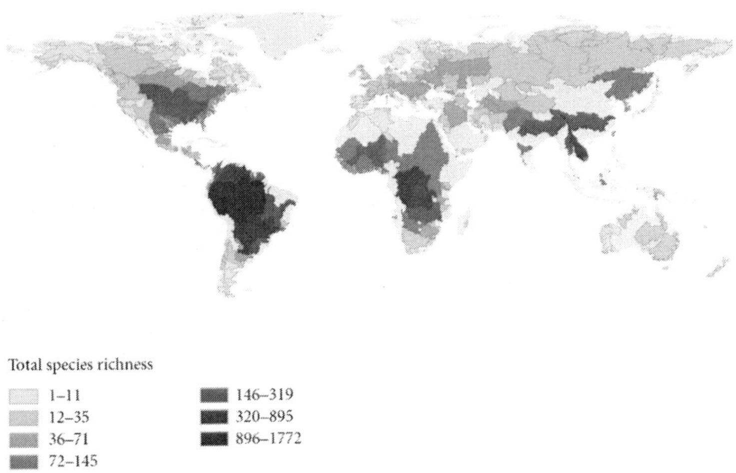

Figure 1: Global freshwater fish species richness patterns at the drainage basin grain.

Table 1: Results from a spatial autoregressive model (SAR) relating species richness to environmental, climatic, and historical variables. Spatial analysis was performed with statistical package [49] and spdep library [50] (see the Supplementary Appendix for further explanations). The spatial structure was implemented by a neighbourhood matrix of the drainage basins (see [46] and the Supplementary Appendix for further explanations) and assuming that the autoregressive process occurs in the error term (i.e., the "spatial error model" described by Dormann et al. [44]). Further methodological details on species richness, environmental variables computing, and modelling procedure are available in the Supplementary Appendix. Habitat heterogeneity was estimated by applying Shannon's diversity index to proportions of biomes (i.e., vegetation types associated with regional variations in climate) within drainage basins. Temperature anomaly represents the Quaternary climate variability measured as the change in mean annual temperature between the present and the Last Glacial Maximum (LGM, circa 21 thousand years ago). Following Oberdorff et al. [51] we also considered whether or not a drainage basin was on a land mass, a peninsula, or an island (LPI; continental mass = 0; peninsula = 1; island = 2). All other variables are fully explained in the Supplementary Appendix. The Moran's I value represents the remaining autocorrelation on the residuals of the model for the first distance class, that is, neighbour drainages (the values for the remaining distance classes are also nonsignificant).

Related hypothesis	Variable	Standardized estimates	Standard error	z value	P-value
Habitat size and diversity	Drainage area	0.548	0.032	17.123	<.0001
	Habitat heterogeneity	0.188	0.031	6.012	<.0001
	Altitudinal range	−0.208	0.194	−1.069	n.s.
	Altitudinal range^2	0.130	0.200	0.649	n.s.
	Runoff	0.784	0.091	8.628	<.0001
	Runoff^2	−0.761	0.098	−7.797	<.0001
Historical climatic stability and geographic isolation	Temperature anomaly	0.559	0.147	3.815	<.0001
	Temperature anomaly^2	−0.37	0.130	−2.857	.0043
	Land-Peninsula-Island (LPI)	−0.257	0.041	−6.349	<.0001
Climate/energy	Actual Evapotranspiration	0.073	0.049	1.493	n.s.
	Precipitation	0.376	0.058	6.464	<.0001
	Temperature	0.778	0.085	9.209	<.0001
	Temperature^2	0.195	0.047	4.176	<.0001
	Precipitation seasonality	0.009	0.040	0.227	n.s.
pseudo R^2		0.771			
AIC		1851.4			
Moran's I		0.0046			n.s.

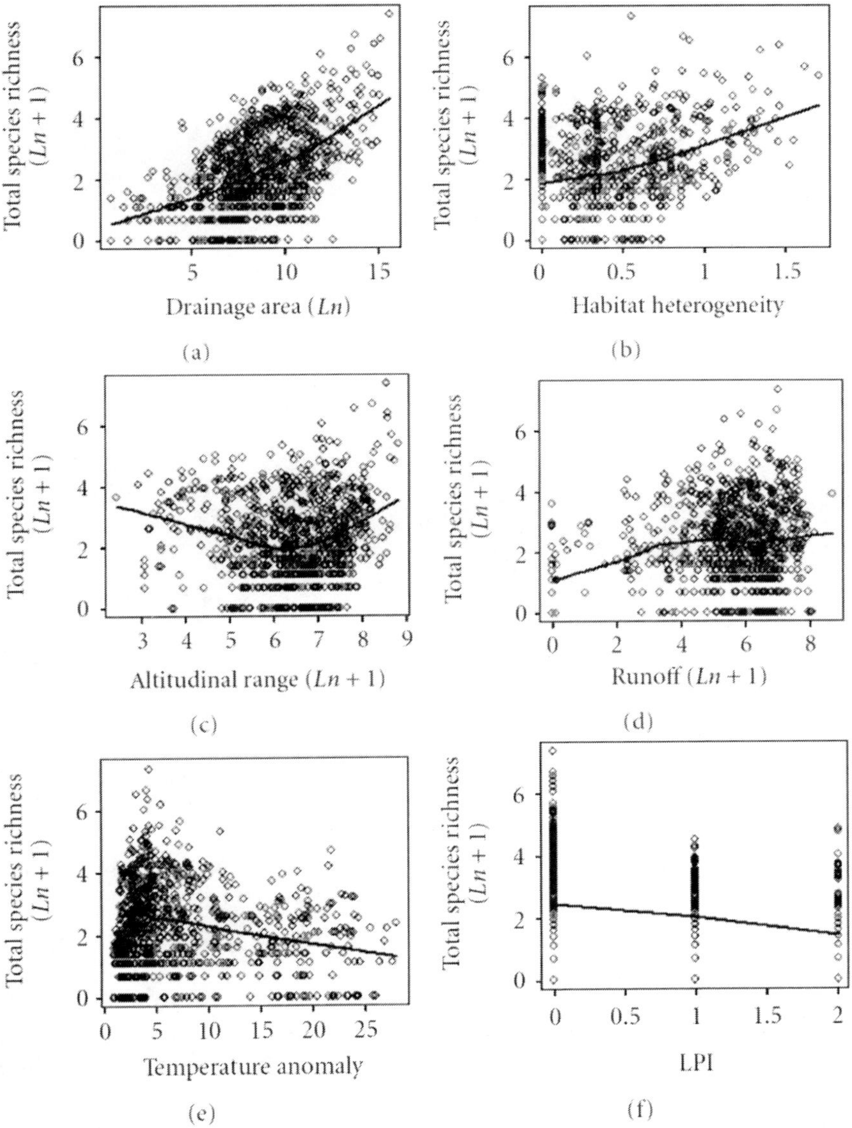
Total species richness (Ln + 1)
Drainage area (Ln)
(a)
Total species richness (Ln + 1)
Habitat heterogeneity
(b)
Total species richness (Ln + 1)
Altitudinal range (Ln + 1)
(c)
Total species richness (Ln + 1)
Runoff (Ln + 1)
(d)
Total species richness (Ln + 1)
Temperature anomaly
(e)
Total species richness (Ln + 1)
LPI
(f)

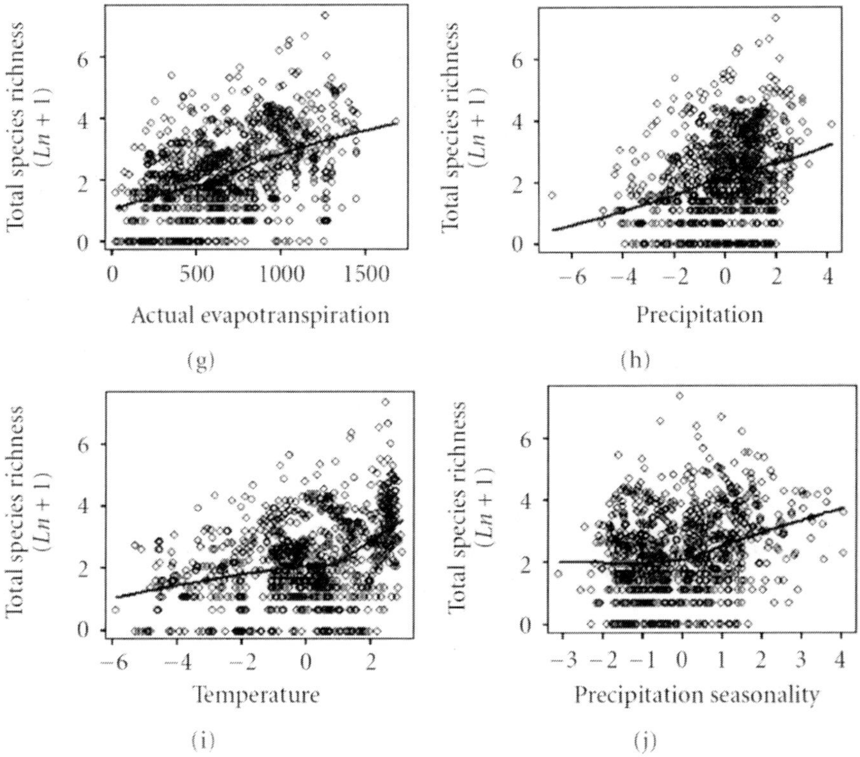

Figure 2: Fish species richness for each river basin as a function of drainage area (a), habitat heterogeneity (b), altitudinal range (c), runoff (d), temperature anomaly (e), LPI (Land-Peninsula-Island) (f), actual evapotranspiration (g), precipitation (h), temperature (i), and precipitation seasonality (j). See Table 1 for variables description.

With respect to the area hypothesis, these results confirm those of several previous studies carried out at the regional scale that identified the size of the river drainage basin and/or the mean flow at the river mouth as important predictors of river basin species richness [46, 52–56]. Furthermore, according to our SAR model, habitat diversity still plays a significant role in explaining richness gradients after accounting for drainage area (Table 1). However these results do not fully answer the questions following from the area hypothesis, that is, are species richness patterns due to area-dependant rates of extinction and/or speciation, or to an increase in habitat diversity, or both?

With respect to the species-energy hypothesis, the results obtained by Oberdorff et al. [40] and Guégan et al. [41] tend to favour the hypothesis of an effect of energy on richness through an increase in available resources for the species. (Net Primary Productivity is an important predictors of species richness.) However, a difficulty in discussing further this last result is that these authors used estimates of terrestrial primary productivity from Lieth's models [57] instead of real aquatic primary productivity (data not available). Even if considering that terrestrial productivity gives a correct estimation of aquatic productivity (as food webs supporting fish are largely based on allochthonous inputs), using estimates of terrestrial primary productivity probably under-estimates true aquatic productivity (see [58] for a review). However, our SAR model also gives support to an indirect effect of energy through species physiological limits (positive effect of variables linked with temperature in the model, see Table 1).

The species-energy theory as developed by Wright et al. [25] posits a positive link between species richness and energy availability [59]. However, in plant and animal communities, a variety of patterns in species richness have been observed over productivity gradients, including positive, negative, and unimodal relationships [60–63]. It is not clear yet why richness shows these (apparently) contradictory relationships with productivity even if some explanations have already been proposed. For example, it has been suggested that all these noted relationships may just be incomplete segments of an overall hump-shaped, unimodal relationship over a broader range of productivity. Nevertheless, evidence for this possibility is currently limited at best [60, 63]. Results from Oberdorff et al. [40], Guégan et al. [41], and our SAR model support the view of a monotonically increase of riverine fish species richness with increasing productivity at the global scale (Figure 2).

At this spatial scale, the only direct historical factor significantly acting on species richness was past climatic variability (see Figure 2 and results of the SAR model in Table 1). It is thus tempting to conclude that history is a minor driver of diversity at the global scale. However we should keep in mind that all the variables used in the SAR model are interrelated to some extent and difficult to separate. This can be visualized in Figure 3, where the explained variance of a linear regression has been partitioned into three different groups of factors related to the area, energy and historical hypotheses. Currie [27], referring to land animals, put forward an explanation for the absence of influence of history on contemporary diversity patterns: that historical factors only influence species richness over relatively short periods, that is, less than the period of time since the last glacial maximum. Nevertheless this explanation seems inappropriate for riverine fishes. In their case, community saturation should be more difficult to reach than for land animals in the sense that their colonization depends on potential connections between river drainage basins. It is thus logical to expect that the influence of historical events should still be detectable in riverine fish communities at the global scale and that the weak influence of this driver most often noticed comes preliminary from difficulties in defining the appropriate variables.

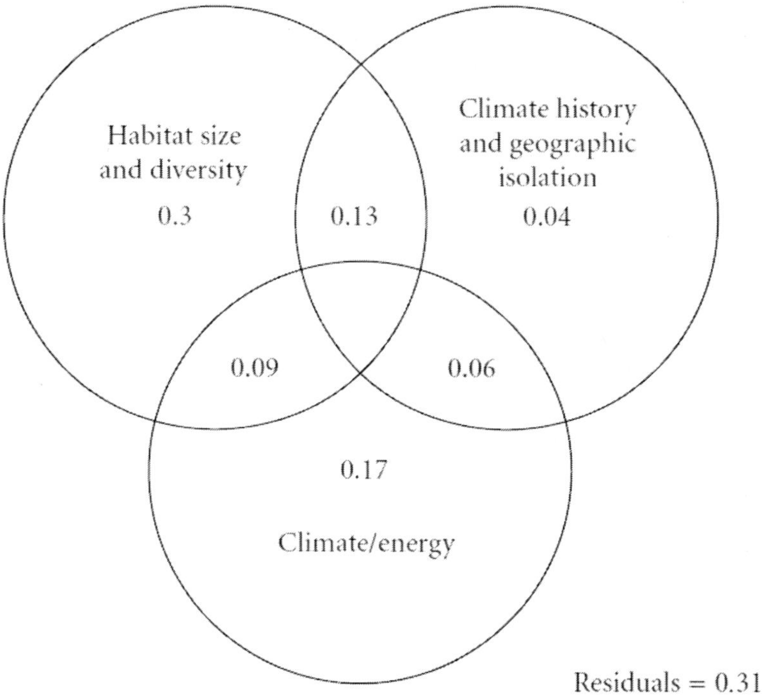

Figure 3: Variance partitioning in explaining species richness gradients between area-related, climate-related, and historical variables. The analysis was performed using the "varpart" function from the vegan R package [64] and grouping variables as in Table 1.

2.2. The Role of History

It is not always simple to separate effects linked to history from those linked to current environmental factors, but comparisons between similar environments in different regions could address variation in speciation and extinction caused by different history [66]. In order to highlight the potential influence of historical factors on species richness, Oberdorff et al. [65] studied rivers on two different continents, North America and Western Europe, which have comparable climatic and environmental characteristics but a rather different history (Figure 4). After having initially identified the main ecological factors responsible for variations in species richness on the two continents (i.e., factors related to river size, productivity, and climate), the second phase of the study integrated in the model factors presumed to reflect historical events (i.e., distance from the larger refugial area and surface area of drainage basin covered by the ice sheet during the last Pleistocene glaciation) in order to examine their relative contribution in explaining species richness gradients.

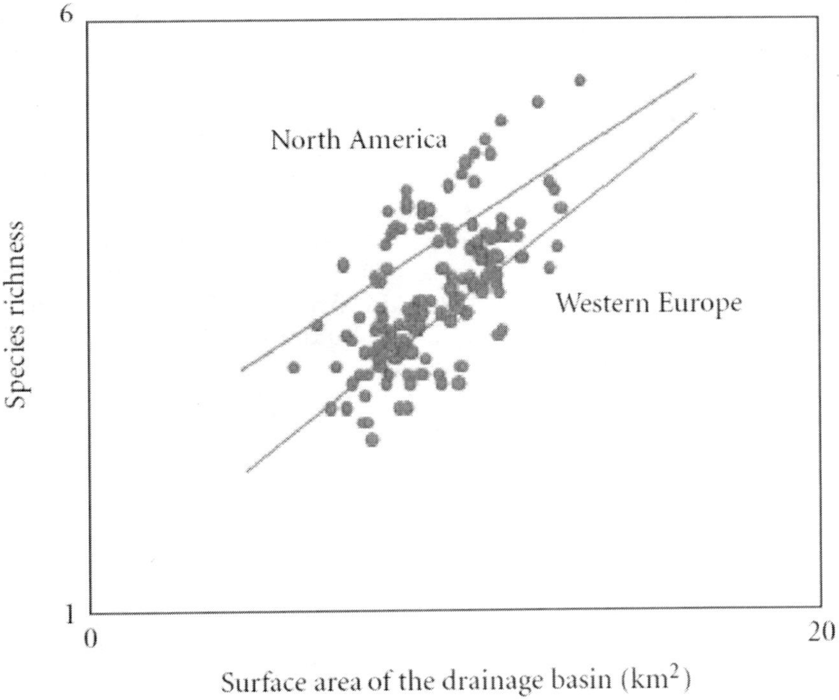

Figure 4: Plot of fish species richness as a function of the surface area of drainage basins for West European and North American rivers. Variables expressed in logarithmic values (Ln): redrawn from Oberdorff et al. [65].

Results showed that ecological factors (particularly the size of the river and to a lesser extent available energy) explain a large part of variations in species richness between the two continents, while one historical factor (distance from the larger refugial area) appears to be more marginal though significant, while the other (surface area of drainage basin covered by the ice sheet during the last Pleistocene glaciation) is invariably rejected. These results thus seem to agree with conclusions reached at the global scale, suggesting a marginal role of history in driving contemporary diversity patterns. This is rather surprising if one accepts the low dispersion capacity generally attributed to fish communities. A preliminary explanation is that the most northern regions of Western Europe and North America are mainly populated with euryhaline species that could have rapidly recolonised rivers via coastal fringes. At the same time, the fact that a "continental" effect is highly significant in the final model leads to think that other historical factors not taken into account in the study are perhaps involved in differences between rivers on the two continents, like, for example, differences in the process of speciation which seem to occur more often in North American refuge zones [67]. In fact, some North American genera such as Notropis have radiated at a rate not encountered in any of the European genus [67]. If speciation rate is assumed to be inversely related to body size [68], a low speciation rate is also suggested by body size distribution of European fish with dominance of medium and large species, conversely to North America were small fish predominate [67, 69, 70]. A complex array of factors is probably involved in this pattern, but speciation events seem to have occurred more frequently in North American refugial areas than in West European ones [67]. Moreover, the data analysed by Oberdorff et al. [65] also show that after river size and net primary productivity have been factored out, North American rivers are still 1.7 times as rich as European ones.

Consistent with this, other recent studies trying to evaluate the role of history in shaping riverine fish diversity patterns at regional and intercontinental scales found a significant influence of history in forging riverine species richness patterns [45–48]. For example, Tedesco et al. [46] have analyzed the effect of rain forest refuges at the last glacial maximum (LGM) on tropical freshwater fish diversity patterns in three different regions, that is, Tropical South America, Central America, and West Africa. At the end of the most recent glacial period (Last Glacial Maximum, LGM; 18,000 years BP), while ice sheets in the Northern Hemisphere extended from the Arctic southward to cover most of North America and central Asia to approximately 45°N latitude, African, and Amazonian rain forests contracted in response to glacial aridity [71]. Following this scenario, in the Northern Hemisphere, high fish species extinctions should have occurred in the rivers totally or partially glaciated, while few extinctions should have occurred in the few refuge zones representing remnants of preglaciation habitats. At the same time, in the tropical zone of the Southern Hemisphere, overall reduced precipitation should have led to high extinction rates in river basins affected by the drought (through a decrease in river basins discharge and active surface area), and few or no extinctions in river basins having kept their characteristic natural (e.g., precipitation patterns and vegetation conditions). Indeed, Tedesco et al. [46] found that both river drainage area and contact (or absence of contact) with LGM rain forest refuges explained the greatest proportions of variance in the geographical pattern of riverine species richness. In the three examined regions, highest richness was found in drainages that were connected to the rain forest refuges (Figure 5). However, they also found that, at the continental scale, South American rivers were more species rich than their African and

Central American counterparts, respectively. Therefore, a historical signal seems to persist even when the regional historical effect (climate at the LGM) has already been accounted for. These results suggest that from the LGM to the present day (a time scale of 18,000 years), extinction processes should have played a predominant role in shaping the current diversity pattern. By contrast, the continental effects could reflect historical contingencies explained by differences in speciation and extinction rates between continents at larger time scales (millions of years). Despite these few studies, the role of historical processes in shaping present-day distribution patterns of diversity is still the subject of considerable debate, stressing the difficulties of testing historical processes based on current species distributions. More refined tests of historical factors involving intercontinental comparisons are needed to better assess the relative importance of ecological and historical processes in shaping contemporary diversity patterns. In this context, endemic species have always been fascinating because they should reflect the roles of speciation, extinction, and dispersal ultimately responsible for their restricted geographic distribution. They are then good candidates for analysing the role of historical processes in present-day distribution patterns of diversity.

3. REGIONAL APPROACH TO RIVERINE FISH SPECIES RICHNESS

Changing the spatial extent we now consider variations in richness in rivers and tributaries of comparable size that flow in the same biogeographical region. In other words, we will now control for the effects highlighted at the intercontinental scale (i.e., energy and history) and focus on factors that intervene at this regional scale. Two main explanations can be put forward to explain differences in species-area relationships at this scale: (1) the area-dependent extinction rates hypothesis and (2) the habitat diversity hypothesis. In order to distinguish effects specific to each hypothesis, Hugueny [54] proposed considering two types of rivers: tributaries, which can be freely colonized by species present throughout the basin and completely isolated river drainage basins. According to the hypothesis of size-dependent extinction rates, tributaries should harbor overall a higher number of species than river drainage basins of similar size, since the immigration rate of the former is higher than zero while the immigration rate of the later is actually zero. Conversely, according to the hypothesis of habitat diversity, the same number of species is to be expected in both categories (tributaries and river drainage basins) as long as they are of comparable size. In order to test the hypothesis of the size-dependent

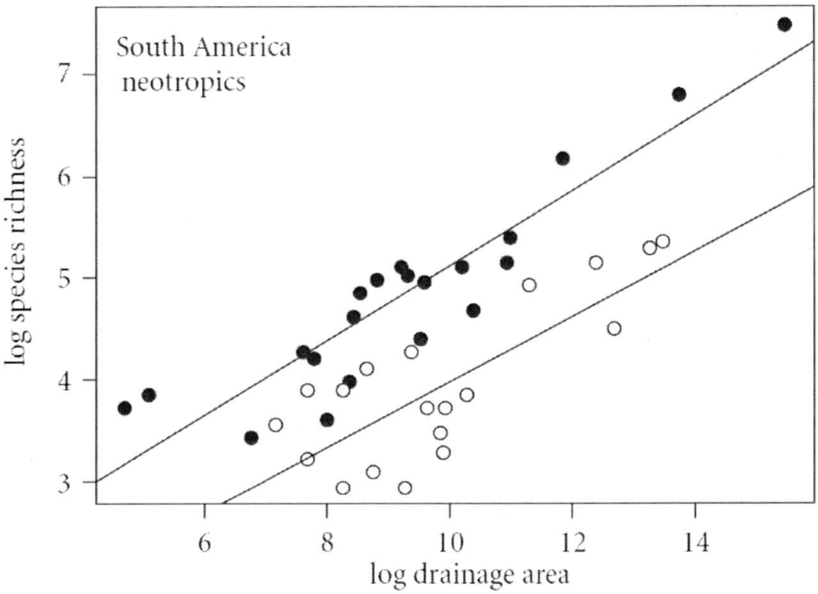
South America
neotropics
log species richness
7
6
5
4
3
6
8
10
12
14
log drainage area

(a)

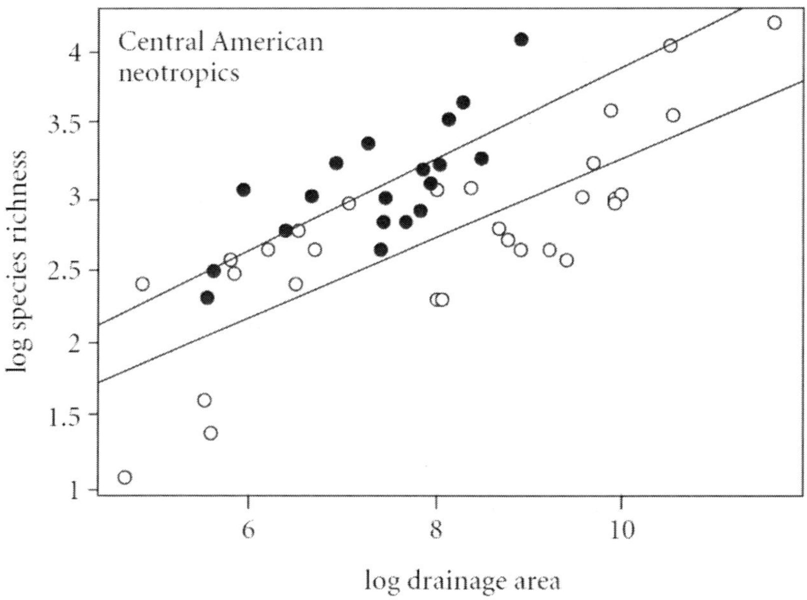
Central American
neotropics
log species richness
4
3.5
3
2.5
2
1.5
1
6
8
10
log drainage area

(b)

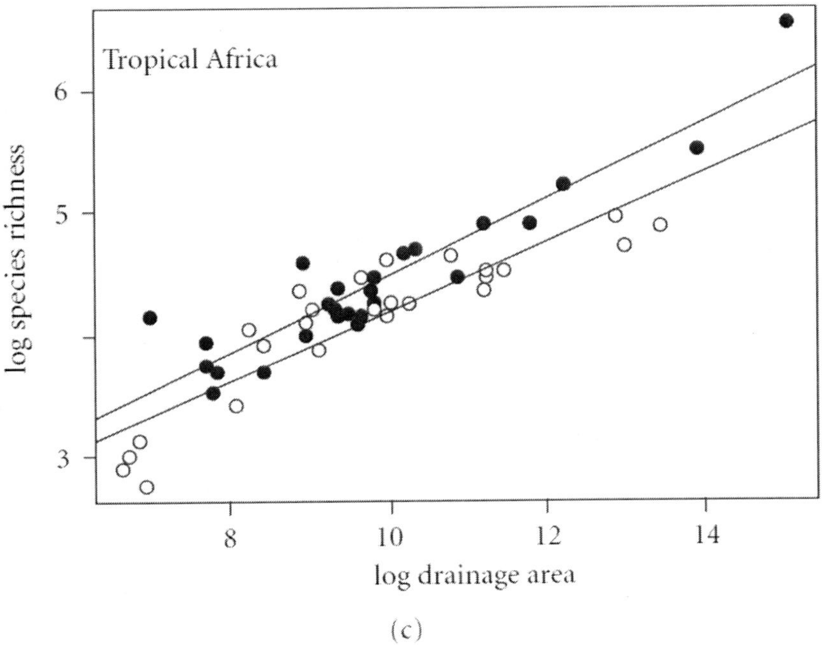

Figure 5: Plot of species richness as a function of total surface area of the drainage for river basins from tropical South America, Central America and tropical Africa. Black circles represent the basins that were in contact with a rain forest refuge zone during the LGM; white circles represent basins completely isolated from rain forest refuges. Lines correspond to simple linear regressions for each group. Variables are expressed in logarithmic values (Ln): redrawn from Tedesco et al. [46].

extinction rate Hugueny [54] and Belkessam et al. [72] compared the species richness of river drainage basins and tributaries of similar size within bio-geographical regions (West Africa and North-west France, resp.). The results of these comparisons showed that species richness of river drainage basins was overall lower than that of tributaries of comparable size (Figure 6). These results thus suggest that regional species richness is regulated in part by processes of extinction and immigration and that a priori, the threshold of species richness is not only determined by habitat diversity. Two other studies analysing colonisation patterns of introduced species in several watersheds of North America supported these previous results [73, 74]. These studies show that fish communities are not saturated and are thus capable of achieving higher species richness if the pool of potential colonisers is artificially increased by the introduction of other species. Another study of freshwater fish communities brought into contact by the opening of the Panama Canal in 1914 also supports this view. Smith et al. [75] analyzed data on communities from Caribbean and Pacific river basins sampled in 1911 and 2002. Numerous species migrated through the canal and have reached similar relative abundances in their new basin. Furthermore, it appears that no species has gone extinct, hence increasing species richness in both basins. The authors concluded that the communities were unsaturated and that community composition and richness were not regulated by interspecific competition at least over ecological time scales (10–100 generations), but rather by regional processes of dispersal.

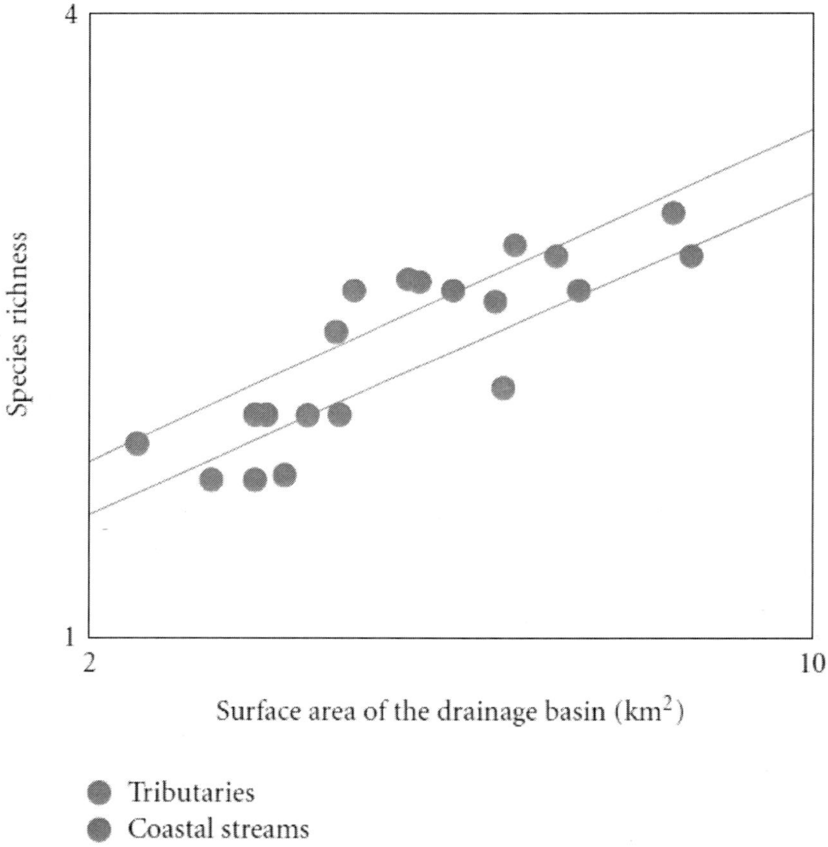

Figure 6: Relationship between species richness and total surface area of the drainage basin for tributaries and coastal rivers of North-Western France: redrawn from Belkessam et al. [72].

4. CONCLUDING REMARKS AND FUTURE DIRECTIONS

At this stage, by synthesizing results presented above, we can develop a preliminary integrated framework that provides a potential mechanistic explanation for riverine fish species richness patterns at the global scale (Figure 7). Although each one of the processes involved in the framework outlined in Figure 7 still needs to be refined, it may be helpful in predicting how current and future changes should alter species richness patterns at large spatial scales. Invasion of exotic species, habitat loss and fragmentation and global climate change held to be the primary causes of endangerment to riverine fish species [76]. Below we focus on how the results detailed in this paper could help us to forecast the consequences of these changes on fish species richness.

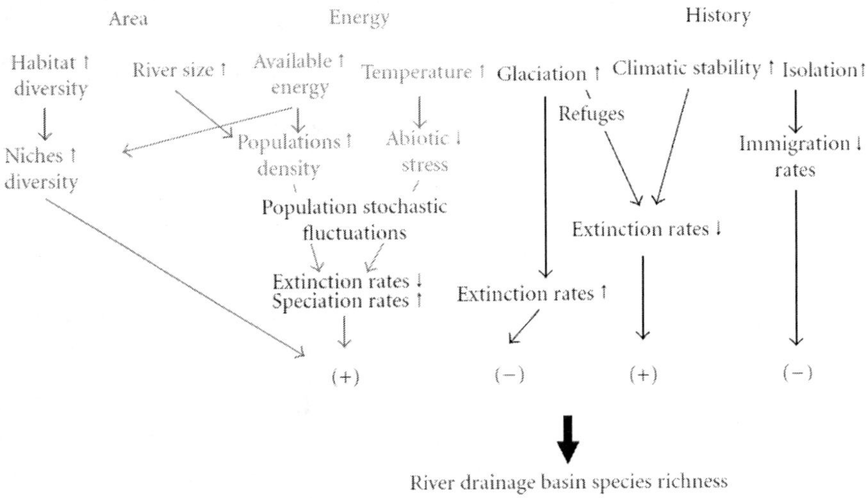

Figure 7: Potential causal pathways between environmental factors and processes affecting positively (+) or negatively (−) the species richness of a river basin.

4.1. Introduction of Exotic Species

Knowledge of the degree of saturation of a community provides an important basis for understanding how the community reacts or will react to the introduction of new species. In this context, understanding to what extent freshwater fish introductions have long-term consequences for biodiversity is crucial to the adoption of sound and effective conservation strategies. As previously said, for strictly freshwater fishes, river basins (flowing into the ocean) can be considered as nonequilibrated islands in which species extinctions (related to historical events) are not fully balanced by colonization from neighbouring river basins [16, 46, 54, 72]. The implication is that river basins are very likely to be unsaturated with species and thus more susceptible to the establishment of nonnative species because ecological space should be less densely packed and interspecific competition should be less intense [75]. An implication is that species introductions might have impacts on fish communities that are smaller than would be expected if the saturation point had already been reached [42, 43, 77, 78].

4.2. Fish Species Facing Habitat Loss and Fragmentation

Habitat loss and fragmentation is one of the greatest threats to biodiversity worldwide, and this certainly holds true for riverine fishes. One of the conclusions of the present paper is that extinction processes are quite important in setting contemporary riverine fish species richness. It is thus almost certain that disturbances, generated for example by dams, weirs, reservoirs for water supply, diversion for irrigation and industrial purposes (creating physical barriers blocking normal migrations and movements of the biota or decreasing habitat availability), flow modification, industrial pollution, and eutrophication, all creating direct or indirect decreases in habitat availability, will endanger or extinguish many freshwater fish species in the future. The global scale models described above, by including factors related to the size of the river basin (i.e., surface area of the drainage basin, river discharge) could be of use to predict patterns of extinction due to this type of disturbance. Nevertheless, we should keep in mind that there is an important time lag between habitat loss and species loss [79, 80]. For example, Morita and Yamamoto [81] have estimated that the probability of extinction of a salmonid species (Salvelinus leucomaensis) in an isolated stream basin of 1 km^2 is equal to 0.8 after 50 years and is equal to 0.1 for an isolated stream basin of 10 km2 for the same time lag. That means that the processes involved are rather slow, even for small rivers. This fact will be further discussed in the next paragraph.

4.3. Future Richness Patterns under Global Climate Changes

Changes in global climate are expected in the twenty first century. At this point it is difficult to make precise predictions about how these changes will affect rivers. The changes may be varied and large, involving diverse characteristics as temperature, hydrology, and water level (through changes in rainfall patterns) and consequently productivity within the systems. The long-term effects of such changes are largely a matter of speculation but by combining results obtained by computer-driven General Circulation Models (GCMs) and global models detailed here, a gross estimation of the potential effects of global warming on river basin species richness could be realised. For example, Xenopoulos et al. [82] and Xenopoulos and Lodge [83] have developed global scenarios of future losses in river discharge from climate change and have applied these results to known relationships between fish species and discharge to build gross scenarios of losses of riverine fish richness at global and regional scales. In rivers with hypothesized reduced discharge (about 30% of the world rivers), these authors predicted fish losses up to 75%. However, this species-discharge-based model only projects the fraction of species "committed to extinction", primarily resulting from decreases in river flow. While such extrapolations are useful for assessing rivers vulnerability to climate change, the lag time between being "committed to extinction" and actually going extinct may range from decades to many millennia suggesting that the realized extinction rates, at a given time, are likely to be lower and perhaps much lower than the projected species "committed to extinction" percentages [84]. New approaches quantifying real extinction rates (i.e., the number of extinctions divided by the time over which extinctions occurred) are now critically needed to start organizing sound-effective remedial conservation actions. In this sense, a promising way is the use of empirical relationships established from historical data (prehuman) between species extinction rates and area of occupancy (extinction-area relationships) to predict true extinction rates (integrating the time lags to extinction) due to future habitat loss. Hugueny et al. [85] provided an empirical extinction-area relationship for riverine fishes by (i) estimating population extinction rates since the fragmentation of paleorivers due to sea level rise at the end of the Pleistocene and (ii) by combining this information with rates estimated using population surveys and fossil records. The use of this empirical extinction-area relationship (EAR) to project riverine fish extirpation rates under future global climate change may be a future fruitful approach.

ACKNOWLEDGMENTS

This paper was supported by the Institut de Recherche pour le Développement (IRD) through the French National Agency for Research (Freshwater Fish Diversity project, ANR-06-BDIV-010) and the 7th Framework European program (BIOFRESH project, Contract no. 226874). The authors would like to precise that the global dataset is available on request from the authors.

REFERENCES

1. M. A. Huston, Biological Diversity: The Coexistence of Species on Changing Landscapes, Cambridge University Press, 1994.
2. E. R. Pianka, "Latitudinal gradients in species diversity: a review of concepts," American Naturalist, vol. 100, pp. 33–46, 1966.
3. K. Rohde, "Latitudinal gradients in species diversity: the search for the primary cause," Oikos, vol. 65, no. 3, pp. 514–527, 1992.
4. M. E. Ritchie and H. Olff, "Spatial scaling laws yield a synthetic theory of biodiversity," Nature, vol. 400, no. 6744, pp. 557–560, 1999.
5. M. L. Rosenzweig, Species Diversity in Space and Time, Cambridge University Press, 1995.
6. R. Field, B. A. Hawkins, H. V. Cornell et al., "Spatial species-richness gradients across scales: a meta-analysis," Journal of Biogeography, vol. 36, no. 1, pp. 132–147, 2009.
7. R. M. May, P. J. Edwards, R. M. May, and N. R. Webb, "The effects of spatial scale on ecological questions and answers," in Proceedings of the 35th Symposium of the British Ecological Society with the Society for Conservation Biology, pp. 1–17, 1994.
8. H. V. Cornell and J. H. Lawton, "Species interactions, local and regional processes, and limits to the richness of ecological communities: a theoretical perspective," Journal of Animal Ecology, vol. 61, no. 1, pp. 1–12, 1992.
9. K. J. Gaston, "Global patterns in biodiversity," Nature, vol. 405, no. 6783, pp. 220–227, 2000.
10. R. D. Holt, "Ecology at the mesoscale: the influence of regional processes on local communities," in Species Diversity in Ecological Communities: Historical and Geographical Perspectives, R. E. Ricklefs and D. Schluter, Eds., pp. 77–88, University of Chicago Press, 1993.
11. J. H. Lawton, "Community ecology in a changing world," in Excellence in Ecology, Oldendorf-Luhe, 2000.
12. R. E. Ricklefs, "Community diversity: relative roles of local and regional processes," Science, vol. 235, pp. 167–171, 1987.
13. J. H. Brown, Macroecology, University of Chicago Press, 1995.
14. J. H. Brown and B. A. Maurer, "Macroecology: the division of food and space among species on continents," Science, vol. 243, no. 4895, pp. 1145–1150, 1989.
15. J. T. Kerr, H. M. Kharouba, and D. J. Currie, "The macroecological contribution to global change solutions," Science, vol. 316, no. 5831, pp. 1581–1584, 2007.
16. B. Hugueny, T. Oberdorff, and P. A. Tedesco, "Community ecology of river fishes: a large-scale perspective," in Community Ecology of Stream Fishes: Concepts, Approaches and Techniques, K. Gido and D. Jackson, Eds., vol. 73, pp. 29–62, American Fisheries Society, 2010.
17. B. H. MacArthur and E. O. Wilson, "An equilibrium theory of insular zoogeography," Evolution, vol. 17, pp. 373–387, 1963.
18. R. H. MacArthur and E. O. Wilson, The Theory of Island Biogeography, Princeton University Press, 1967.
19. F. W. Preston, "The canonical distribution of commonness and rarity—part I," Ecology, vol. 43, pp. 185–215, 1962.
20. F. W. Preston, "The canonical distribution of commonness and rarity—part II," Ecology, vol. 43, pp. 410–432, 1962.
21. J. H. Lawton, "Are there general laws in ecology?" Oikos, vol. 84, no. 2, pp. 177–192, 1999.
22. J. B. Losos and D. Schluter, "Analysis of an evolutionary species-area relationship," Nature, vol. 408, no. 6814, pp. 847–850, 2000.
23. M. Williamson, A. A. Myers, and P. S. Giller, "Relationship of species number to area, distance and other variables," in Analytical Biogeography: An Integrated Approach

to the Study of Animal and Plant Distributions, A. A. Myers and P. S. Giller, Eds., pp. 91–115, Chapman & Hall, London, UK, 1988.

24. D. H. Wright, "Species-energy theory: an extension of species-area theory," Oikos, vol. 41, no. 3, pp. 496–506, 1983.
25. D. H. Wright, D. J. Currie, and B. A. Maurer, "Energy supply and patterns of species richness on local and regional scales," in Species Diversity in Ecological Communities: Historical and Geographical Perspectives, R. E. Ricklefs and D. Schluter, Eds., pp. 66–76, University of Chicago Press, 1993.
26. J. M. Adams and F. I. Woodward, "Patterns in tree species richness as a test of the glacial extinction hypothesis," Nature, vol. 339, no. 6227, pp. 699–701, 1989.
27. D. J. Currie, "Energy and large-scale patterns of animal and plant species richness," American Naturalist, vol. 137, no. 1, pp. 27–49, 1991.
28. D. J. Currie and V. Paquin, "Large-scale biogeographical patterns of species richness of trees," Nature, vol. 329, no. 6137, pp. 326–327, 1987.
29. B. A. Hawkins, R. Field, H. V. Cornell et al., "Energy, water, and broad-scale geographic patterns of species richness," Ecology, vol. 84, no. 12, pp. 3105–3117, 2003.
30. M. Kaspari, S. Donnell, and J. R. Kercher, "Energy, density, and constraints to species richness: ant assemblages along a productivity gradient," American Naturalist, vol. 155, no. 2, pp. 280–293, 2000.
31. J. T. Kerr and L. Packer, "Habitat heterogeneity as a determinant of mammal species richness in high-energy regions," Nature, vol. 385, no. 6613, pp. 252–254, 1997.
32. E. M. O'Brien, "Water-energy dynamics, climate, and prediction of woody plant species richness: an interim general model," Journal of Biogeography, vol. 25, no. 2, pp. 379–398, 1998.
33. J. R. G. Turner, C. M. Gatehouse, and C. A. Corey, "Does solar energy control organic diversity? Butterflies, moths and the British climate," Oikos, vol. 48, no. 2, pp. 195–205, 1987.
34. J. R. G. Turner, J. J. Lennon, and J. A. Lawrenson, "British bird species distributions and the energy theory," Nature, vol. 335, no. 6190, pp. 539–541, 1988.
35. B. J. Van Rensburg, S. L. Chown, and K. J. Gaston, "Species richness, environmental correlates, and spatial scale: a test using South African birds," American Naturalist, vol. 159, no. 5, pp. 566–577, 2002.
36. D. J. Currie, G. G. Mittelbach, H. V. Cornell et al., "Predictions and tests of climate-based hypotheses of broad-scale variation in taxonomic richness," Ecology Letters, vol. 7, no. 12, pp. 1121–1134, 2004.
37. R. H. Whittaker, "Evolution of species diversity in land communities," Evolutionary Biology, vol. 10, pp. 1–67, 1977.
38. M. Dynesius and R. Jansson, "Evolutionary consequences of changes in species' geographical distributions driven by Milankovitch climate oscillations," Proceedings of the National Academy of Sciences of the United States of America, vol. 97, no. 16, pp. 9115–9120, 2000.
39. F. Leprieur, P. A. Tedesco, B. Hugueny et al., "Partitioning global patterns of freshwater fish beta diversity reveals contrasting signatures of past climate changes," Ecology Letters, vol. 14, no. 4, pp. 325–334, 2011.
40. T. Oberdorff, J. F. Guegan, and B. Hugueny, "Global scale patterns of fish species richness in rivers," Ecography, vol. 18, no. 4, pp. 345–352, 1995.
41. J. F. Guégan, S. Lek, and T. Oberdorff, "Energy availability and habitat heterogeneity predict global riverine fish diversity," Nature, vol. 391, no. 6665, pp. 382–384, 1998.
42. S. Blanchet, G. Grenouillet, O. Beauchard et al., "Non-native species disrupt the worldwide patterns of freshwater fish body size: implications for Bergmann's rule," Ecology Letters, vol. 13, no. 4, pp. 421–431, 2010.
43. F. Leprieur, O. Beauchard, S. Blanchet, T. Oberdorff, and S. Brosse, "Fish invasions in the world's river systems: when natural processes are blurred by human activities," PLoS Biology, vol. 6, no. 12, article e322, p. 2940, 2008.

44. C. F. Dormann, J. M. McPherson, M. B. Araujo et al., "Methods to account for spatial autocorrelation in the analysis of species distributional data: a review," Ecography, vol. 30, no. 5, pp. 609–628, 2007.
45. Y. Reyjol, B. Hugueny, D. Pont et al., "Patterns in species richness and endemism of European freshwater fish," Global Ecology and Biogeography, vol. 16, no. 1, pp. 65–75, 2007.
46. P. A. Tedesco, T. Oberdorff, C. A. Lasso, M. Zapata, and B. Hugueny, "Evidence of history in explaining diversity patterns in tropical riverine fish," Journal of Biogeography, vol. 32, no. 11, pp. 1899–1907, 2005.
47. D. Griffiths, "Pattern and process in the ecological biogeography of European freshwater fish," Journal of Animal Ecology, vol. 75, no. 3, pp. 734–751, 2006.
48. D. Griffiths, "Pattern and process in the distribution of North American freshwater fish," Biological Journal of the Linnean Society, vol. 100, no. 1, pp. 46–61, 2010.
49. R Development Core Team R: A language and environment for statistical computing. R Foundation for Statistical Computing, Vienna, Austria, 2009, http://www.R-project.org/.
50. A Birand and D. J. Howard, "The relationship between proportion of endemics and species diversity on islands: expectations from a null model," Ecography, vol. 31, no. 2, pp. 286–288, 2008.
51. T. Oberdorff, S. Lek, and J. F. Guégan, "Patterns of endemism in riverine fish of the Northern Hemisphere," Ecology Letters, vol. 2, no. 2, pp. 75–81, 1999.
52. J. Daget and A. Iltis, "Les poissons de Côte d'Ivoire (eaux douces et eaux saumâtres)," Mémoires de l'Institut Français d'Afrique Noire, vol. 74, pp. 1–385, 1965.
53. J. Daget, "Diversité des faunes de poissons dans les cours d'eau du Portugal," Arquivos do Museu Bocage, vol. 2, pp. 21–26, 1968.
54. B. Hugueny, "West African rivers as biogeographic islands: species richness of fish communities," Oecologia, vol. 79, no. 2, pp. 236–243, 1989.
55. D. A. Livingstone, M. Rowland, and P. E. Bailey, "On the size of African riverine fish faunas," American Zoologist, vol. 22, no. 2, pp. 361–369, 1982.
56. J. J. Sepkoski and M. A. Rex, "Distribution of freshwater mussels: coastal rivers as biogeographic islands," Systematic Zoology, vol. 23, pp. 165–188, 1974.
57. H. Leith, "Modeling the primary productivity of the world," in Primary Productivity of the Biosphere, H. Leith and R. H. Whittaker, Eds., pp. 237–263, Springer, 1975.
58. J. H. Thorp and M. D. Delong, "Dominance of autochthonous autotrophic carbon in food webs of heterotrophic rivers," Oikos, vol. 96, no. 3, pp. 543–550, 2002.
59. D. S. Srivastava and J. H. Lawton, "Why more productive sites have more species: an experimental test of theory using tree-hole communities," American Naturalist, vol. 152, no. 4, pp. 510–529, 1998.
60. G. G. Mittelbach, C. F. Steiner, S. M. Scheiner et al., "What is the observed relationship between species richness and productivity?" Ecology, vol. 82, no. 9, pp. 2381–2396, 2001.
61. M. L. Rosenzweig and Z. Abramsky, "How are diversity and productivity related?" in Species Diversity in Ecological Communities: Historical and Geographical Perspectives, R. E. Ricklefs and D. Schluter, Eds., pp. 52–65, University of Chicago Press, 1993.
62. D. Tilman and S. Pacala, "The maintenance of species richness in plant communities," in Species Diversity in Ecological Communities, Historical and Geographical Perspectives, R. E. Ricklefs and D. Schluter, Eds., pp. 13–25, University of Chicago Press, 1993.
63. R. B. Waide, M. R. Willig, C. F. Steiner et al., "The relationship between productivity and species richness," Annual Review of Ecology and Systematics, vol. 30, pp. 257–300, 1999.
64. J. Oksanen, F. G. Blanchet, and R. Kindt, vegan: Community Ecology Package. R package version 1.17-1, 2010, http://CRAN.R-project.org/package=vegan.
65. T. Oberdorff, B. Hugueny, and J.-F. Guégan, "Is there an influence of historical events on contemporary fish species richness in rivers? Comparisons between Western

Europe and North America," Journal of Biogeography, vol. 24, no. 4, pp. 461–467, 1997.
66. R. E. Ricklefs, "A comprehensive framework for global patterns in biodiversity," Ecology Letters, vol. 7, no. 1, pp. 1–15, 2004.
67. R. Mahon, "Divergent structure in fish taxocenes of north temperate streams," Canadian Journal of Fisheries and Aquatic Sciences, vol. 41, no. 2, pp. 330–350, 1984.
68. J. M. Marzluff and K. P. Dial, "Life history correlates of taxonomic diversity," Ecology, vol. 72, no. 2, pp. 428–439, 1991.
69. R. M. McDowall, "On size and growth in freshwater fish," Ecology of Freshwater Fish, vol. 3, no. 2, pp. 67–79, 1994.
70. P. B. Moyle, B. Herbold, W. J. Matthews, and D. C. Heins, "Life-history patterns and community structure in stream fishes of western North America: comparisons with eastern North America and Europe," in Community and Evolutionary Ecology of North American Stream Fishes, W. J. Matthews and Heins, Eds., pp. 25–32, University of Oklahoma Press, 1987.
71. J. H. Brown and M. V. Lomolino, Biogeography, Sinauer Associates, Inc., 2nd edition, 1998.
72. D. Belkessam, T. Oberdorff, and B. Hugueny, "Unsaturated fish assemblages in rivers of north-western France: potential consequences for species introductions," Bulletin Francais de la Peche et de la Protection des Milieux Aquatiques, no. 344-345, pp. 193–204, 1997.
73. K. B. Gido and J. H. Brown, "Invasion of North American drainages by alien fish species," Freshwater Biology, vol. 42, no. 2, pp. 387–399, 1999.
74. D. Rathert, D. White, J. C. Sifneos, and R. M. Hughes, "Environmental correlates of species richness for native freshwater fish in Oregon, U.S.A," Journal of Biogeography, vol. 26, no. 2, pp. 257–273, 1999.
75. S. A. Smith, G. Bell, and E. Bermingham, "Cross-Cordillera exchange mediated by the Panama Canal increased the species richness of local freshwater fish assemblages," Proceedings of the Royal Society B, vol. 271, no. 1551, pp. 1889–1896, 2004.
76. D. Dudgeon, A. H. Arthington, M. O. Gessner et al., "Freshwater biodiversity: importance, threats, status and conservation challenges," Biological Reviews of the Cambridge Philosophical Society, vol. 81, no. 2, pp. 163–182, 2006.
77. S. Blanchet, F. Leprieur, O. Beauchard, J. Staes, T. Oberdorff, and S. Brosse, "Broad-scale determinants of non-native fish species richness are context-dependent," Proceedings of the Royal Society B, vol. 276, no. 1666, pp. 2385–2394, 2009.
78. F. Leprieur, S. Brosse, E. García-Berthou, T. Oberdorff, J. D. Olden, and C. R. Townsend, "Scientific uncertainty and the assessment of risks posed by non-native freshwater fishes," Fish and Fisheries, vol. 10, no. 1, pp. 88–97, 2009.
79. S. L. Pimm, The Balance of Nature. Ecological Issues in the Conservation of Species and Communities, University of Chicago Press, 1991.
80. S. L. Pimm, G. J. Russell, J. L. Gittleman, and T. M. Brooks, "The future of biodiversity," Science, vol. 269, no. 5222, pp. 347–350, 1995.
81. K. Morita and S. Yamamoto, "Effects of habitat fragmentation by damming on the persistence of stream-dwelling charr populations," Conservation Biology, vol. 16, no. 5, pp. 1318–1323, 2002.
82. M. A. Xenopoulos, D. M. Lodge, J. Alcamo, M. Märker, K. Schulze, and D. P. Van Vuuren, "Scenarios of freshwater fish extinctions from climate change and water withdrawal," Global Change Biology, vol. 11, no. 10, pp. 1557–1564, 2005.
83. M. A. Xenopoulos and D. M. Lodge, "Going with the flow: using species-discharge relationships to forecast losses in fish biodiversity," Ecology, vol. 87, no. 8, pp. 1907–1914, 2006.
84. H. M. Pereira, P. W. Leadley, V. Proença et al., "Scenarios for global biodiversity in the 21st century," Science, vol. 330, no. 6010, pp. 1496–1501, 2010.

85. B. Hugueny, A. Movellan, and J. Belliard, "Habitat fragmentation and extinction rates within freshwater fish communities: a faunal relaxation approach," Global Ecology and Biogeography, vol. 20, no. 3, pp. 449–463, 2011.

CHAPTER 4

Ecology and Biology of The Lake Sturgeon: A Synthesis of Current Knowledge of a Threatened North American *Acipenseridae*

Douglas L. Peterson Æ Paul Vecsei Æ

Fisheries Division, Warnell School of Forestry and Natural Resources, University of Georgia, Athens, GA 30602, USA

ABSTRACT

The lake sturgeon is one of the largest North American freshwater fish and was once common in most inland rivers and lakes of the US and Canadian Midwest. World demand for caviar and sturgeon meat led to a dramatic decline of lake sturgeon populations throughout much of its range. Along with overfishing, lake sturgeon populations have been negatively affected by habitat degradation. Recruitment factors and early life history are poorly understood. Today, renewed interest in lake sturgeon restoration has led to numerous state and federally-funded research activities. Research has focused on identifying and assessing the size structure of remnant stocks, the availability of spawning habitat, and factors affecting reproductive success. Additional studies are needed to improve hatchery techniques, to better understand recruitment mechanisms, and how genetic diversity among and within meta-populations may affect long-term recovery of depleted populations.

Keywords: Biology Ecology Habitat Lake sturgeon Life history

INTRODUCTION

The lake sturgeon (*Acipenser fulvescens*, Acipenseridae) is a large, cartilaginous, benthic fish, endemic to larger mesotrophic and oligotrophic systems of the Central US, Great Lakes, and the Hudson Bay drainages of Canada. Its life history, characterized by long life span, late age-at-maturity, and protracted spawning periodicity, is unique among North America's freshwater fishes (Harkness and Dymond 1961; Scott and Crossman 1973; Becker 1983). Although once abundant throughout their range, severe overfishing in the late 1800s and early 1900s decimated most populations (Auer 1999, 2004; Bogue 2000). Today, few healthy populations remain and many anthropogenic factors continue to hamper most conservation and restoration efforts (Noakes et al. 1999; Auer 2004; Wilson and McKinley 2004). Among these factors, hydroelectric dams that obstruct upstream access to historic spawning grounds and degrade critical downstream habitats, are the most problematic (Auer 1996a; Wilson and McKinley 2004). Currently, the lake sturgeon is listed as extirpated, endangered, threatened or of special concern in 12 states (Leonard et al. 2004; Holey and Trudeau 2005).

The unique appearance and life history of the lake sturgeon has generated considerable interest among many fisheries scientists. Since 1989, five separate international symposia have focussed on the biology and management of lake sturgeons and other imperilled sturgeons. As many fisheries managers have become increasingly concerned with the conservation of endemic species in recent years, renewed interest in lake sturgeon has spurred several new research or restoration initiatives. Although these and many previous studies have improved our knowledge of lake sturgeon life history and population dynamics, the phylogeny of all sturgeon species, including that of lake sturgeon, remains unclear. In this paper, we describe several key taxonomic characters of lake sturgeon as they relate to sturgeon phylogeny. Furthermore, we synthesize existing studies of life history as well as current population status and management efforts.

Taxonomy and systematics

The oldest sturgeon fossils date back to the Upper Cretaceous; however, the earliest members of the group probably evolved in the Lower Jurasic approximately 200 million years ago (Bemis et al. 1997). Although sturgeon phylogeny still remains somewhat uncertain, most taxonomists agree that Acipenseriformes is a monophyletic group, derived from the paleonisciform fishes (Bemis et al. 1997). In recent years however, ichthyologists have developed conflicting theories regarding phylogenetic relationships within the group, particularly within the Acipenser genus. For example, Birstein et al. (2002) questions the monophyly of the group and claims that current members of the genus do not share a single synapomorphic molecular character, which suggests multiple evolutionary lineages. Findeis (1997) proposed a similar argument noting the absence of shared osteological characters within the genus. Regardless, investigators have noted an array of common morphological characters shared by all members of the genus including: a disconnected gill membrane attached at the isthmus; a small, downward-projecting transverse mouth; a long flattened snout that is either conical or narrow; a set of 4 cylindrical or fimbrated barbels; a palatoquadratum connecting the symplecticum; a stylohyale articulating with the posterior section of the symplecticum; a linear arrangement of the palatoquadratum and the upper part of the maxillae; and clustered basihyalia positioned along the median line of the rostrum. (Antoniu-Murgoci 1936a, b, 1942).

Regardless of phylogenetic uncertainties, the sturgeon family (Acipenseridae) is comprised of 27 species distributed among four separate genera. The largest of these, *Acipenser*, contains 17 species, five of which are native to North America (Scott and Crossman 1973; Grande and Bemis 1996; Bemis and Kynard 1997; Findeis 1997). Among *Acipenser* species, only the lake sturgeon completes its lifecycle entirely within freshwater. As such, its evolutionary history and relationship among other members of the genus has been of particular interest to many sturgeon researchers.

Ontogenetic changes in lake sturgeon morphology have prompted many early investigators to suggest that *A. fulvescens* is actually composed of several discrete species (Harkness and Dymond 1961; Priegel and Wirth 1971). During the 19th and early 20th centuries, at least 17 different scientific names have been assigned to the various *Acipenser* populations of the Great Lakes, St. Lawrence River, and Central US (Scott and Crossman 1973). By the 1950s however, ichthyologists had determined that these stocks all belonged to a single species. Following the rules of nomenclature, the oldest scientific designation, *Acipenser fulvescens* (Rafinesque 1817), has since been accepted as the official species name.

MORPHOLOGY

General description

'The stud-like bones are most beautiful objects, being as hard or harder than ivory, with the outer surface indented and marked as though they had been carved by a Japanese artist. When set in silver, selected samples of these shackles form very beautiful ornaments for ladies' dresses, and I certainly would advise my lady readers who are always looking for something new and pretty to try the effect of sturgeon's shackles when worn as ornaments.'

—Frank Buckland (1883), Natural History of British Fishes.

The physical appearance of lake sturgeon (Fig. 1) is similar to that of most other *Acipenser* species; and like all other members of the genus, they are easily recognized by several primitive morphological features that distinguish them from other North American fishes. Perhaps the most noticeable of these is the scaleless body, which is protected by five lateral rows of bony plates or scutes. The heavy-set body is spindle shaped, the greatest body depth occurring slightly anterior to the midsection. The origin of the anal fin is located posterior to that of the dorsal, its tip rarely extending beyond the caudal fulcrate plate. Other morphological features that distinguish lake sturgeon from other North American freshwater fishes include a heavily armoured skull, a spiral valve intestine, and a cellular swim bladder that retains some of the lung-like characteristics of early actinopterigeans (Harkness and Dymond 1961).

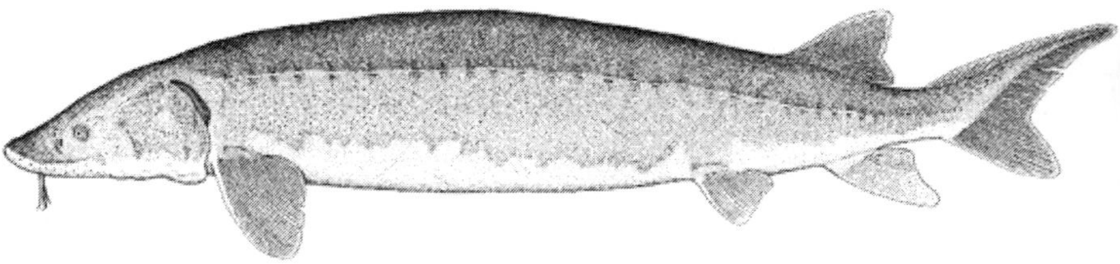

Figure. 1 Basic morphology of the adult lake sturgeon

Basic morphology of the lake sturgeon is similar to that of other Acipenserids. The elongated body in cross-section is pentagonal in young juvenile specimens but becomes progressively more rounded with age (Scott and Crossman 1973). Dorsal, lateral, and ventral scute counts are typically 9–17, 29–42, and 7–12, respectively. Dorsal fin rays number 35–45; anal fin rays 25–30 (Vladykov and Greeley 1963; Scott and Crossman 1973). The slightly upturned rostrum is disproportionately large in juveniles, often exceeding post-orbital distance in juveniles < 50 cm; however, this proportion is gradually reversed with age (Vladykov and Greeley 1963). The large, transverse mouth typically measures approximately 66–93% of the interorbital width (Vladykov and Greeley 1963). The top lip is continuous; the bottom lip interrupted (Fig. 2). Mouth shape and size in proportion to head width is most similar to that of shortnose sturgeon (*A. breviostrum*) (Vladykov and Greeley 1963; Hochleitner and Vecsei 2004). As in the elasmobranches, the sturgeon jaw is detached from the skull, allowing the mouth to project downward during feeding (Vecsei and Peterson 2004). Lake sturgeon barbels are situated closer to the tip of the snout than to the origin of the mouth—an important diagnostic character distinguishing the species from acipenserids. Gill rakers are short and typically number 25–40 (Vladykov and Greeley 1963). The thick-walled, gizzard-like stomach is connected to a spiral-valve intestine, a primitive alimentary arrangement shared by many Acipenserids adapted to a diet of benthic crustaceans and molluscs (Harkness and Dymond 1961).

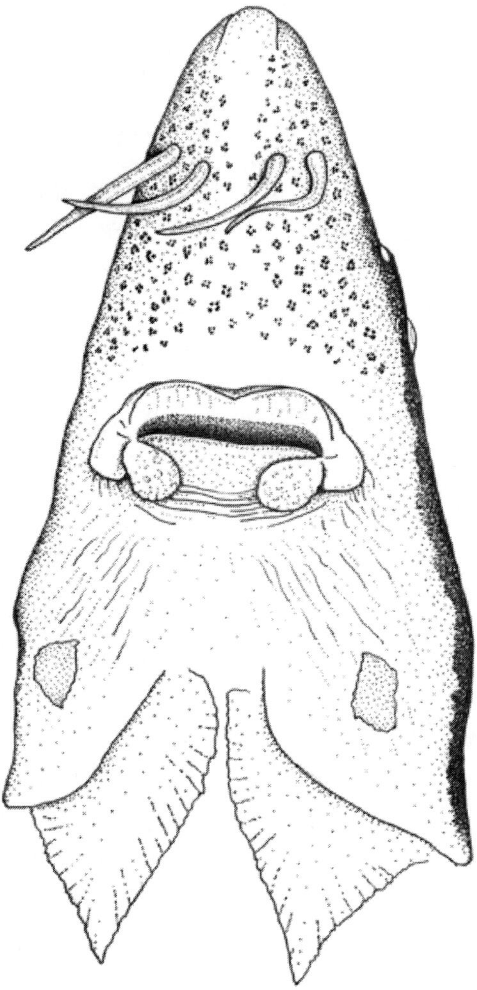

Figure. 2 Ventral view of lake sturgeon head, showing distribution of sensory pits and relative position of barbels and mouth

Body armouring is extensive on juveniles but becomes progressively reduced with age (Priegel and Wirth 1971; Scott and Crossman 1973; Vecsei and Peterson 2004). In juveniles < 100 cm, the laterodorsal and lateroventral surfaces are protected by a layer of tightly-spaced denticles evenly distributed between the five principal rows of scutes. Sharp, apical hooks are particularly prominent on the scutes of juveniles, but these gradually disappear with age until the scutes themselves are almost completely resorbed later in adulthood (Fig. 3). In contrast, most anadromous sturgeons retain ossified scutes that continue to grow throughout their entire lifecycle. Hence, the process of scute resorbtion in adult lake sturgeon probably illustrates an important trade-off in the functional morphology of body armouring within the genus. In freshwater environments devoid of sharks and other large biting predators, the protective advantage of body amour diminishes with increasing body size, yet the energetic costs of overcoming frictional drag caused by the rough armoured surfaces increases exponentially. By adulthood, lake sturgeon have simply outgrown all potential aquatic predators and hence, their need for body armouring.

The skull of all acipenserids, including the lake sturgeon, is heavily armoured by a series of contiguous bony plates that are most apparent in juveniles and sub-adults (Fig. 4). Variation and complexity in the ossification of the *Acipenser* skull roof has been noted by several researchers (e.g. Jollie 1980); however, only the skull structure of shortnose sturgeon has been well studied (Hilton and Bemis 1999). Although the lake sturgeon skull is comparatively less variable, Jollie (1980) noted considerable intraspecific and ontogenetic variation.

While the enodcranial elements of Acipenseridae may be useful in understanding some phylogenetic relationships within the group, they are generally too complex and variable for use in species identification. Fortunately, the number and arrangement of dermal plates on the dorsum and ventrum of the posterior trunk are much more species-specific and less subject to individual or ontogenetic variation. Although Vladykov and Greeley (1963) report that predorsal plates may be used for taxanomic identification, the postdorsal and preanal plates are most commonly used for this purpose. In lake sturgeon the postdorsal plates are typically seen as 1–2 unpaired elements (Vladykov and Greeley 1963; Peterson et al. 2003); however, the second predorsal may appear as a paired element in some individuals. The relatively large preanal plates always occur in single file and number 1–2 (Vecsei and Peterson 2004). These ossifications may be considered definitive in all cases except on very old individuals where they may be completely resorbed.

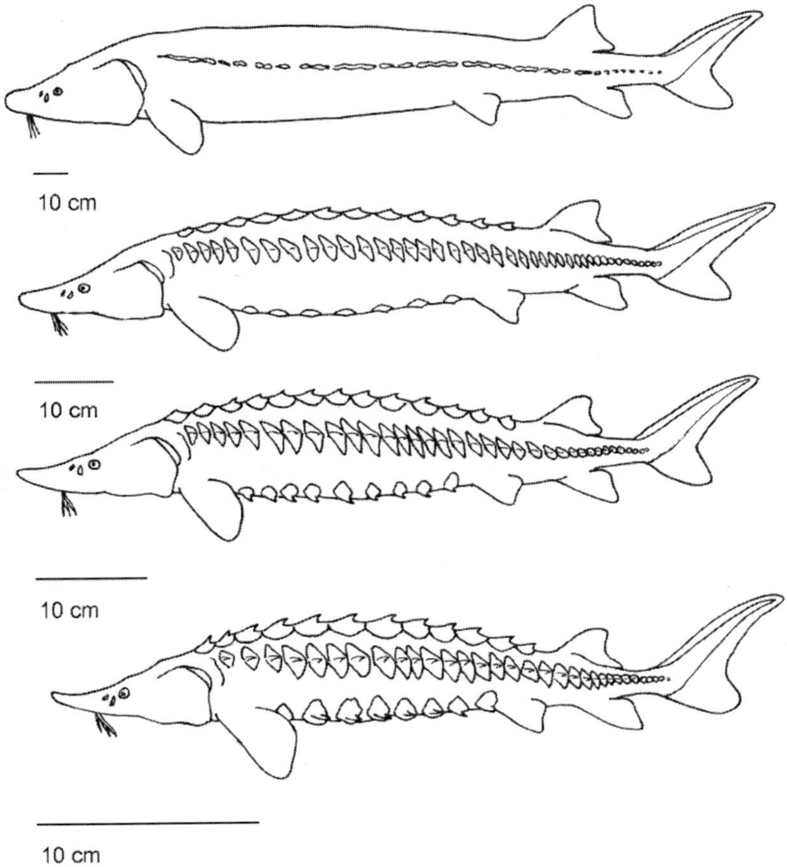

Figure. 3 Ontogenetic changes is body armouring of lake sturgeon

Body coloration of lake sturgeon is variable among stocks but is typically dark brown or dark gray dorsally with a similar but slightly lighter coloration on the lateral surfaces. The ventrum is typically white or cream-colored. Some individuals have gray or black pigmentation on the underside of the head, particularly on the lips and barbels (Harkness and Dymond 1961). Rarely, adults may exhibit white or milky blotches or spots on the lateral body surfaces. The dorsal and lateral scutes are typically the same color as the surrounding skin, although rare specimens may have slightly lighter lateral scutes or dark pigmentation on the lateral surfaces of the ventral scutes.

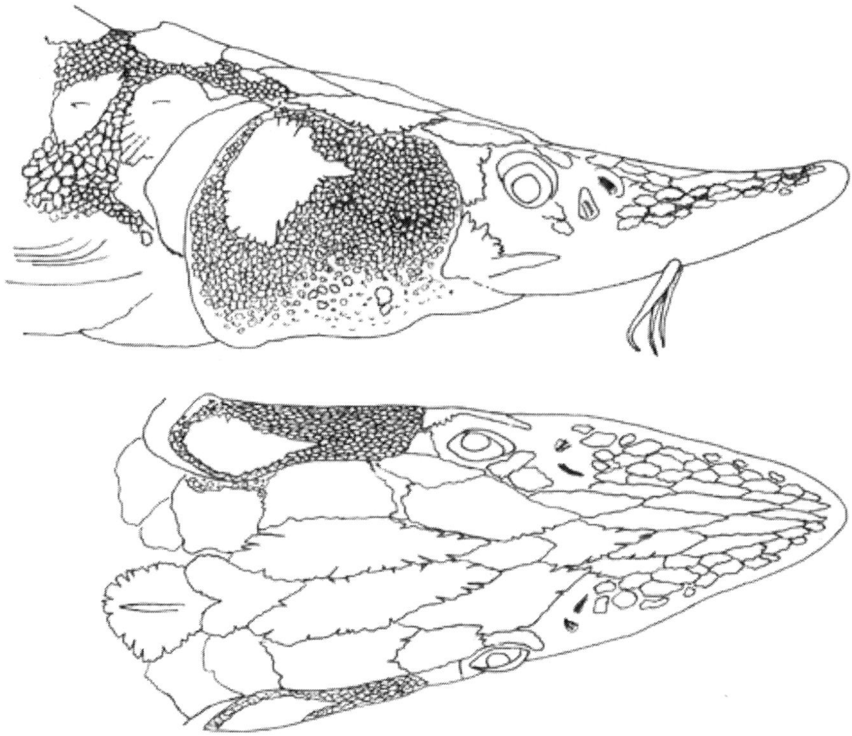

Fig. 4 Armouring of the lake sturgeon skull

Although lake sturgeon exhibit considerable morphological ontogeny, the changes in color pattern from early juvenile to adulthood are among the most pronounced (Vladykov and Greeley 1963; Priegel and Wirth 1971; Peterson et al. 2003). In juveniles < 30 cm, two large black saddles typically are present across the gray or brown dorsum and sides (Fig. 5). Black speckling on the upper surfaces of the body also is common, often producing a 'peppered' appearance on the juveniles. Scutes and other dermal ossifications of juveniles are usually of the same color as the surrounding skin, but lateral scutes may sometimes be lighter (as in the adults). In 2–4 year old juveniles (>60 cm) the large saddle marks are lacking but the black speckling may persist into early adulthood.

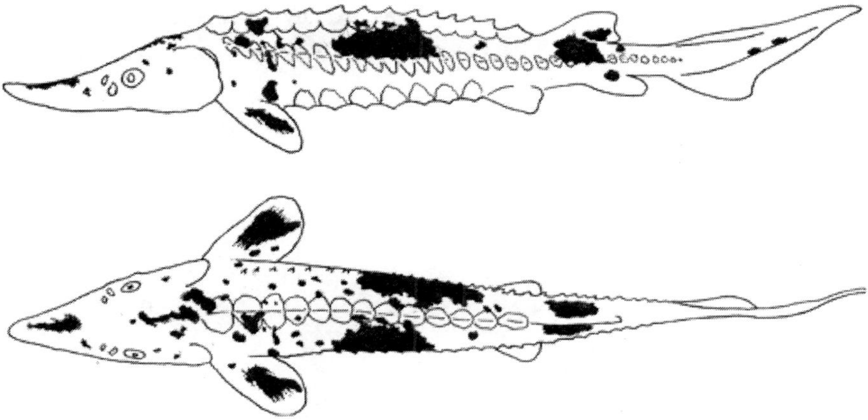

Figure. 5 Basic morphology and coloration of juvenile lake sturgeon (<30 cm)

DISTRIBUTION AND LEGAL STATUS

The current distribution of *A. fulvescens* includes three major North American drainages: the Mississippi, the Great Lakes, and the Hudson Bay (Priegel and Wirth 1971). The historic range of the species extended from the Canadian waters of the Hudson Bay in Saskatchewan and Manitoba, east to the St. Lawrence River estuary. To the south, US populations were found primarily in the Great Lakes and Mississippi River basins with smaller isolated populations occurring further south in the larger rivers of the Tennessee River and Ohio River drainages (Scott and Crossman 1973). Although found primarily in larger freshwater lakes and rivers, lake sturgeons are also native to the brackish waters of Hudson Bay and the St. Lawrence River estuary (Fig. 6) (Vladykov and Greeley 1963).

Currently, the lake sturgeon is not federally protected in either the US or Canada; however, the species was listed in 1975 under Appendix II of the Convention on International Trade of Endangered Species (CITES). Although this listing was temporarily suspended in 1983, it was reinstated in 1998. Within the US, the species receives various levels of protection at the state level. Populations in Iowa, Indiana, Illinois, Ohio, Missouri, Pennsylvania, Tennessee, and Vermont are considered endangered while those in Nebraska, New York, and Michigan are listed as threatened (Johnson 1987; Auer 2004; Leonard et al. 2004). In Canada, lake sturgeons are listed as threatened in Alberta, Manitoba, New Brunswick,

Newfoundland, Ontario, Quebec, and Saskatchewan (Freedman et al. 2001; TRAFFIC North America 2002).

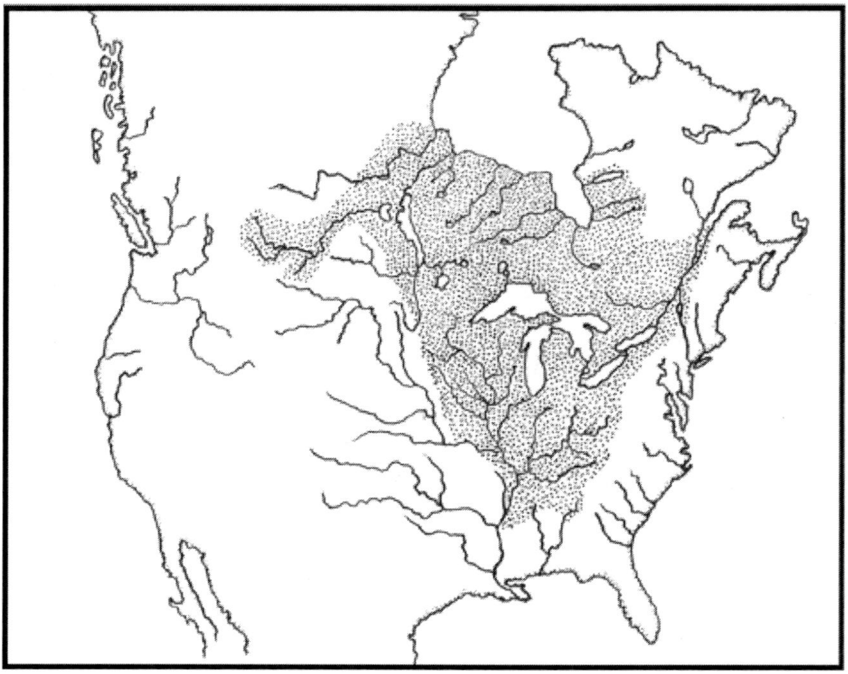

Figure. 6 Historic distribution of lake sturgeon in North America (adapted from Scott and Crossman (1973), and CITES (2000))

Despite the many anthropogenic factors that have decimated most populations, a few healthy populations still exist. The largest of these is probably that of the St. Lawrence River near Montreal where commercial fisheries in the 1980s and 1990s produced sustained annual harvests of 15,000–30,000 fish (Fortin et al. 1993). Recent studies by LaHaye et al. (1992) however, suggest that this population is actually comprised of at least three distinct stocks. In US waters, the largest remaining population is that of the Lake Winnebago System in central Wisconsin (Bruch 1999; Bruch and Binkowski 2002) where the species has been actively managed for the past 100 years. Recent studies by Thomas and Haas (2002) have shown that the largest remaining Great Lakes population is probably that of Lake St. Clair. All three of these populations support limited recreational fisheries, and some commercial fishing is permitted in both the St. Lawrence River and the Canadian waters of Lake St. Clair.

LIFE HISTORY AND ECOLOGY

A first comprehensive summary of lake sturgeon life history was provided by Harkness and Dymond (1961). More recently, studies of lake sturgeon biology and management have come from Wisconsin, where sizeable populations have been re-established through careful management, restoration of spawning habitat, and judicious stocking programs (Priegle 1973; Priegle and Wirth 1975; Thuemler 1985; Kempinger 1988; Larson 1988; Bruch 1999; Bruch and Binkowski 2002). Many of these management efforts have been aided by recent investigations of new propagation and habitat restoration techniques (Ceskleba et al. 1985; Conte et al. 1988).

Spawning periodicity and fecundity

Lake sturgeon life history is characterized by rapid growth during a protracted juvenile stage, with first spawning typically delayed until age 12–15 for males and 18–27 for females (Scott and Crossman 1973; Bruch 1999; Bruch et al. 2001). Studies of lake sturgeon physiology have shown that this delay in maturation results from an unusually disproportionate allocation of energy to somatic growth during the juvenile years (Beamish et al. 1996; LeBreton and Beamish 2004). This strategy has the evolutionary disadvantage of delaying reproduction, but it yields a compensatory advantage by providing both the time and energy needed to maximize size at first spawning. Because natural mortality is largely size-dependent, this reproductive strategy helps limit mortality to the earliest life stages, thereby increasing potential life span. As in other long-lived species, this strategy provides the adults with multiple reproductive opportunities spread over many years or even decades. In the absence of anthropogenic disturbance, this strategy conveys the selective advantage of minimizing any loss of fitness resulting from year-class failures in any one year when spawning conditions are poor (Crouse 1999).

Large size at first spawning also provides the advantage of increased fecundity in lake sturgeon, which is among the highest of all freshwater species in North America (Harkness and Dymond 1961; Scott and Crossman 1973). Although the number of eggs produced on a per-weight basis is variable, fecundity generally increases as a function of size with a typical adult female producing 49,000–667,000 eggs in each spawning year (Harkness and Dymond 1961; Priegel and Wirth 1971). Like many other demersal broadcast spawners, lake sturgeons benefit from their high fecundity by essentially overwhelming potential predators with the sheer numbers of offspring they produce. Unlike other fishes of this reproductive guild however, lake sturgeon also rely on their high fecundity to help maximize reproductive output during years of favorable spawning conditions (Beamesderfer and Farr 1997).

Although a large female lake sturgeon may spawn more than a million eggs in a single spawning season, the species' reproductive rate is inherently slow because of protracted spawning periodicity (Harkness and Dymond 1961). Typically, females spawn only once every 4–9 years, males every 1–3 years (Roussow 1957; Magnin 1966; Fortin et al. 1996). This low spawning frequency, coupled with a life span of up to 154 years (MacKay 1963), typically results in spawning runs consisting of 20 or more age classes in unexploited populations. The complex age-structure of such populations helps buffer them against short-term environmental disturbances—an important key to the species' evolutionary success. Unfortunately, this inherent population resilience is quickly eroded when spawning adults are exposed to excessive harvest.

Spawning behavior

Lake sturgeons spawn in spring, usually from mid-April to early June; however, several researchers have noted that males typically arrive at the spawning grounds before females. On the Wolf and Fox rivers for example, Bruch and Binkowski (2002) observed males at several spawning sites 1–2 days before the females. Once females are present, spawning may begin as soon as water temperatures reach 10–15°C (Harkness and Dymond 1961; Kempinger 1988; Auer 1996b; Smith and King 2005); however, specific temperatures that trigger spawning are variable and depend on the ovarian cycles of individual females (Webb et al. 2001).

Studies of courtship behavior in lake sturgeon are rare, but a few authors have noted that males often produce drumming sounds in the presence of gravid females (Priegel and Worth 1971, 1974; Bruch and Binkowski 2002). During spawning, 2–8 males typically crowd individual females on each flank, frequently pounding her with their tales to stimulate egg release. These spawning bouts can be intense, but typically last only 1–2 minutes (Bruch and Binkowski 2002). The process is repeated several times with different males maneuvering into the group during each subsequent bout until the female is spent. Although this behavior leaves most fish in poor post-spawning condition, Bruch and Binkowski (2002) suggest that this polygamous mating system may help guarantee the highest number of possible mates for both sexes while minimizing the energy expended in finding them.

During spawning, females scatter their adhesive eggs widely over gravel or cobble substrates in water depths of 0.1–2 m (Priegel and Wirth 1974;

Becker 1983; LaHaye et al. 1992; Auer and Baker 2002; Bruch and Binkowski 2002). Females usually complete spawning in 8–12 h, after which they leave the spawning area. Males however, may remain on the spawning site as long as a ripe female is present. Once all females have left a spawning site, males may move downstream until they find deeper water where they await the arrival of additional ripe females.

Successful spawning in lake sturgeon largely depends on the suitability of both flow and temperature regimes, but optimal spawning conditions may vary substantially among populations (Cooke et al. 2002; Jager et al. 2002). For most populations, optimal spawning habitat is found in the high-gradient reaches of large rivers with current velocities of 0.5–1.3 m/sec and substrates of coarse gravel or cobble (Auer 1996a; McKineley et al. 1998); although, a few populations are known to spawn on rocky, wave-washed lake shores (Harkness and Dymond 1961; Carlson 1995). Regardless of the specific spawning habitat chosen, no parental care is given, and the adults typically migrate back downstream as soon as spawning has concluded (Harness and Dymond 1961; Kempinger 1988).

Early life stages

'Although morphological changes are continuous, there are periods during which changes take place much more rapidly than at other times. Changes, especially marked during the period from one to two weeks after hatching, are almost in the nature of a metamorphosis.'

— Harkness and Dymond (1961) in: The lake sturgeon

Eggs

Immature lake sturgeon eggs are yellowish and attached to a fatty ovarian mass without a covering membrane (Detlaf et al. 1993; Bruch et al. 2001). The fully-developed eggs are olive-green, grey or black and measure 2.6–3.5 mm (Bajkov 1930; Harkness and Dymond 1961; Priegel and Wirth 1974; Becker 1983). Hatching of lake sturgeon eggs typically occurs after 8–14 days with rate of embryonic development depending on water temperature (Kempinger 1988). Developmental studies by Wang et al. (1985) show that at 15°C, gastrulation occurs at 37 h, neurulation at 77.7 h, and heart formation at 118 h. Few studies have documented predation on lake sturgeon eggs; however, observations of Bruch and Binkowski (2002) suggest that male lake sturgeon may consume fertilized eggs while on the spawning grounds. Although directed studies are rare, lake sturgeon eggs are probably eaten by many fishes and invertebrates. In the Muskegon River of western Michigan, for example, stomachs from several brown trout (*Salmo trutta*) harvested in the recreational fishery were found to contain lake sturgeon eggs (P. Vecsei, personal observation).

Larvae

At hatching, the 9–11 mm yolk-sac larvae (or prolarvae) are poorly developed and many basic anatomical structures are barely discernable. The mouth is apparent only as an inward fold, the barbels as tiny stubs. At 17–18 mm however, both structures are well developed and clearly discernable. Other structures, including gills, fins, and lateral lines also develop late in the yolk-sac stage (Kempinger 1988). On live specimens < 15 mm, the contrast between the transparent body and dark or olive-grey yolk sac is evident; however, dead or preserved specimens appear uniformly light grey (Vecsei and Peterson, personal observation). At 15 mm, the yolk-sac larvae develop dark pigmentation, which becomes most prominent along the lateral portion of the head and trunk. At this stage, a dark spiral valve and anal plug also become apparent (Wang et al. 1985).

Newly hatched larvae are pelagic, negatively phototactic, and move about actively in search of suitable hiding places within the interstitial spaces of the rocky substrates where they were spawned (Harkness and Dymond 1961; Wang et al. 1985; Kempinger 1988). Within 13–19 days after hatching, the 17–18 mm larvae emerge from the substrate at night and rapidly disperse downstream, often drifting with the current for several kilometres before again settling on the river bottom (Kempinger 1988; LaHaye et al. 1992). Although the exact timing of this downstream dispersal is somewhat variable, a minimum water temperature of 16°C seems to trigger this behavior (Smith and King 2005).

The onset of exogenous feeding signals the transition from the prolarval to the larval stage. Kempinger (1988) noted that the anal plug is shed just prior to this transition when the larvae are about 22 mm. The post-yolk sac larva is easily identified by a prominent lateral band that extends the entire length of its body, including the caudal fin. The trunk is pigmented in its entirety but is darkest along this mid-lateral band. The larval stage continues for several weeks until the start of the juvenile stage at about 400 mm.

Juveniles

In lake sturgeon, the transition from the larval to juvenile stage is marked by the formation of all definitive adult structures except for the gonads, which remain undifferentiated for several years. The juvenile period is protracted and immature individuals may vary greatly in size and age. Like the adults, juveniles are thought to feed primarily on benthic invertebrates such as small crustacea, insect larvae, leeches, molluscs, and isopods (Harkness 1923;Wallus 1990; Chiasson et al. 1997). Information on the movements and habitats of juvenile lake sturgeon is scant; however, available data suggest that yearlings may gather in large schools in shallow river mouths or adjacent bays during late summer and fall (Priegel and Wirth 1974; Becker 1983; Wallus 1990). After their first year, juveniles are found in the same habitats as the adults (Priegel and Wirth 1974).

ADULTS

Age and growth

Adult lake sturgeons are among the largest of North American freshwater fishes. Adult males typically measure 100–185 cm and weigh 11–30 kg; adult females 130–215 cm and 25–100 kg. Although larger specimens are now rare, numerous specimens > 100 kg were reported in historical records of Great Lakes fisheries. The largest documented specimen, taken from Lake Michigan in 1943, measured 241 cm and weighed 141 kg (Van Oosten 1956). Although lake sturgeon growth is quite variable even within a population, historical accounts suggest that average size tends to decrease in the southern portion of the range (Stearns and Atkinson 1953), presumably because water temperatures become too warm during summer months.

Age determination of most bony fishes is usually accomplished by counting inter-annual growth rings present in cross-sections of the sacular otoliths; however, this method is not preferred by most sturgeon researchers. Although the methodology is similar for sturgeons (Classen 1944; Currier 1951; Brennan and Cailliet 1991; Wilson 1987), the marginal pectoral fin ray is used instead because it is easier to collect and its removal is non-lethal (Roussow 1957; Rossiter et al. 1995). Typically, the annuli appear as widely separated bands in early years but these become increasingly crowded near the outer margins of the fin ray cross-sections. Hence, age determination in sturgeons becomes increasingly uncertain in older specimens (Keenlyne and Jenkins 1993). Rossiter et al. (1995) found that this method was easiest and most accurate for lake sturgeon up to age-15; and while age estimates of much older fish are frequently published, numerous studies have questioned their accuracy (Brennan and Cailliet 1991; Rien and Beamesderfer 1994).

Food habits and feeding

Lake sturgeons feed primarily on benthic invertebrates that they find using a combination of tactile, olfactory, chemosensory, and electrosensory receptors (Harkness and Dymond 1961; Binkowski and Doroshov 1985; Chiasson et al. 1997). Feeding is accomplished as the animal swims along the bottom with its barbels in contact with the substrate. As prey items are detected, they are sucked in by a rapid extension of the protractible mouth (Priegel and Wirth 1974; Vecsei and Peterson 2004). Inedible materials such as sand or silt, sucked are expelled through the mouth or gills, while food items are retained and crushed against the ridges of cartilaginous palate before being swallowed (Harkness and Dymond 1961; Priegel and Wirth 1974). Although benthic macroinvertebrates are the most important prey consumed, lake sturgeon diet varies considerably both spatially and temporally (Chiasson et al. 1997; Beamish et al. 1998). Prey items reported from lake sturgeon stomachs include leeches, snails, small clams, and small fishes, although when available, soft-bodied insect larvae may comprise up to 90% of the prey volume consumed (Bajkov 1930; Harkness 1923; Harkness and Dymond 1961; Priegel and Wirth 1974). Lake sturgeons forage actively throughout the year; however, feeding may slow during winter in northern portions of the range (Priegel and Wirth 1974).

Habits and movements

Habitat

Lake sturgeons often migrate over great distances in search of food, suitable spawning habitat or simply to avoid seasonally unfavorable conditions (Auer 1996b; Bemis and Kynard 1997). However, habitat selection depends on availability and the specific requirements of each life stage. Young juveniles for example, often use deep (>2 m) pools within their natal streams for feeding and over-wintering; whereas, adults typically inhabit deepwater habitats of large lakes. Although adult lake sturgeons are rarely observed in non-spawning habitats, several studies suggest that they prefer depths of < 9 m during cooler months but will readily move to much deeper water in summer (Harkness and Dymond 1961; Priegel and Wirth 1974). Other studies suggest that these deep-water habitats may also be used for over wintering (Bajkov 1930) or to avoid disturbance from intense boat traffic (Engel 1990).

Lake sturgeons can be found over a variety of substrate types, but prey abundance is undoubtedly an important factor in determining habitat selection (Harkness and Dymond 1961). In shallow lakes such as Lake Winnebago, Wisconsin, where water depths are < 7m, lake sturgeons occupy all depths (Priegel and Wirth 1971; Lyons and Kempinger 1992). In deeper lakes such as Black Lake, Michigan, however, adult fish are typically found at depths of 6–12 m (Hay-Chmielewski 1987). Although seasonal habitat selection in these inland systems is probably more influenced by water temperature, corroborative studies are lacking.

Migration

Throughout their life cycle, lake sturgeons exhibit both random and non-random movements. Several studies have shown that most individuals move about randomly within an established home range of 10–14 km; however, some individuals seem to make longer unidirectional movements indicative of emigration (Harkness and Dymond 1961; Priegel and Wirth 1974; Larson 1988; Engel 1990). Early tagging studies from Wisconsin suggested that individuals with established home ranges rarely leave these areas except to spawn (Harkness and Dymond 1961; Priegel and Wirth 1974). Although lake sturgeons are known to migrate up to 200 km when returning to their natal streams, spawning-site fidelity in lake sturgeons has not been well studied, and the environmental cues that trigger and guide the fish during these migrations are unknown. Studies of imprinting in lake sturgeon have not been attempted, but many biologists believe that juveniles are able to recognize their natal stream within only a few months after hatching.

Gerbilskiy (1957) and Bemis and Kynard (1997) characterized sturgeon spawning migrations as either a 'one-step' or 'two-step' pattern. Species exhibiting one-step migrations typically migrate in spring and spawn within a few days of reaching their natal spawning grounds. Those following the two-step migration pattern typically begin their migrations in fall but overwinter in deep pools before spawning in the subsequent spring. These two distinct migration patterns are not only exhibited by different sturgeon species, but also by different races or subpopulations (Bemis and Kynard 1997). Within the Great Lakes, one-step migrations are well documented for most lake sturgeon populations including the Manistee (Peterson et al. 2002), Muskegon (Peterson and Vecsei 2004) and the Sturgeon (Auer 1996b) rivers. In Wisconsin tributaries of Lake Michigan however, Bruch and Binkowski (2002) found that both one-step and two-step migrations are typical of lake sturgeon populations spawning in the Fox and Wolf rivers. Regardless of which upstream migration pattern is used, adult lake sturgeon typically move rapidly downstream after spawning has concluded, eventually returning to a larger river or lake to replenish energy stores over the next several years before the next spawning cycle.

MECHANISMS LEADING TO DECLINES

Commercial fisheries of the mid 19th Century dealt the first major blow to lake sturgeon populations throughout much of North America. Ironically, the species was widely regarded as 'nuisance bycatch' prior to 1860, and most individuals were either used as pig feed, fertilizer or were simply discarded (Tody 1974). After 1860, the popularity and value of lake sturgeon grew rapidly once European demand for North American caviar exceeded the production from Atlantic sturgeon fisheries operating on the Delaware and Hudson rivers (Bogue 2000; Saffron 2002). The first dedicated commercial facility for processing lake sturgeon was established in 1868 by German immigrants (Bogue 2000). Located on the shores of Lake Erie in Sandusky, Ohio, the new plant used the traditional European processing methods to establish and exploit lucrative markets for almost every part of the fish. The flesh was smoked or sold fresh and the roe processed into fine caviar. Even the swim bladder was used to produce isinglass, paint additives, and a number of other valuable commercial products (Priegel and Wirth 1971; Scott and Crossman 1973).

Once new markets for lake sturgeon products had been established, targeted commercial fisheries grew rapidly. By 1925, lake sturgeon had become the most valuable commercial species in the Great Lakes, but by this time many stocks had already collapsed (Tody 1974). By 1928, commercial harvest of lake sturgeons had been banned throughout the Great Lakes except for a limited fishery that has persisted in the Canadian waters of lakes St. Clair and Huron. The US ban on commercial fishing for lake sturgeon has remained in effect to the present, except for a limited re-opening of the fishery from 1950–1970 in the Michigan waters of Lake Michigan.

The rapid boom and bust cycle of commercial lake sturgeon fisheries in the Great Lakes was typical of the many smaller inland fisheries operating in both the US and Canada (Harkness and Dymond 1961). On the Mississippi River for example, commercial landings declined from 113,046 kg in 1894 to 55,842 kg in 1899—a decline of about 50% in only 5 years. By 1922, annual harvest had declined to only 3,178 kg; and by 1931, lake sturgeon had disappeared completely from the commercial catch (Carlander 1954). A similar scenario unfolded in Canada where most inland populations suffered similar declines during the first decade of the 20th Century, although careful management and conservation have allowed a few limited fisheries to continue.

MANAGEMENT APPROACHES

'Sturgeon: Unlawful to take from inland waters except with hook & line. Unlawful to take more than 50 in any one day or have more than 100 in possession at any one time'.

— From the Michigan Fish & Game regulations, 1913

Restoration of lake sturgeon populations to self-sustaining levels is a common goal shared by many contemporary state and provincial management agencies. Towards this end, the species receives varying levels of protection and, in some instances directed management, depending on the biological status of the various populations and local public support. Most traditional management approaches focus on regulations that prohibit or severely limit harvest (Johnson 1987). Such programs are based on the assumption that reductions in fishing mortality will result in higher numbers of spawners, higher recruitment of juveniles, and ultimately, increased abundance of adults. However, several authors suggest that this approach alone is inadequate to recover lake sturgeon stocks that have been severely depressed by overfishing because of the species' low reproductive rate and also because spawning habitat for many populations has been either lost or degraded (Harkness and Dymond 1961; Priegel and Wirth 1975; Thuemler 1988). Hence, some management agencies have developed comprehensive restoration plans based on a combination of management practices. Although regulations are typically included in these plans, habitat restoration, stocking, and public education are used increasingly to help expedite recovery.

Regardless of which management practices are chosen, successful restoration of lake sturgeon requires a long-term approach because the protracted reproductive cycle of the species requires that many juvenile and adult year classes be established before a population can become self-sustaining (Noakes et al. 1999). Where populations have been extirpated, managers must first ensure that suitable habitat is still available and then develop innovative approaches to re-introduce and protect the stock as it is rebuilt over many years or even decades. Once established, populations must be carefully monitored and managed to prevent overharvest. Management of recreational lake sturgeon fisheries in both Wisconsin and Michigan has included size limits, bag limits, harvest caps, and closed seasons (Priegel 1973; Priegel and Wirth 1971, 1975, 1978; Baker 1980; Larson 1988; Bruch 1999). Although these regulations may help stabilize extant populations, they have not yet proven to be effective at rebuilding depressed stocks (Becker 1983).

Lake sturgeon management is perhaps more complex that that of other North America sturgeons because the species' native range spans several different states, provinces, and international boundary waters between the US and Canada (Williamson 2003).

In Canada, lake sturgeons are protected and managed by the provinces under the Federal Fisheries Act (Houston 1987). Alberta closed all commercial fisheries between 1940 and 1968 as did Manitoba in 1995 (Ferguson and Duckworth 1997). In Saskatchewan, a moratorium has been enforced since 1996; however, small commercial fisheries still operate in the Ontario waters of Lake Huron, Lake Nipigon, Lake St. Clair, Namakan Lake, Rainy Lake, and the Seine River (Brousseau 1987). From 1998 to 2000, the combined catch quota from these fisheries was 11,553 kg. In Quebec, a commercial gill net season was restricted to the period of June 14–October 31 to avoid harvest of ripe females. The annual catch quota since the mid 1990s has been approximately 150,000–200,000 kg with a total allowable catch of approximately 20,000–30,000 fish (Williamson 2003). There are presently no commercial fisheries for lake sturgeon in U.S. waters, where individual states maintain jurisdiction over their respective stocks. The largest annual recreational harvest is on Lake Winnebago, Wisconsin, where approximately 1,000 fish are taken annually in a limited-harvest spear fishery (Folz and Meyers 1985; Bruch 1999).

RESTORATION

Because the success of the lake sturgeon's life history strategy depends on delayed maturation and infrequent spawning over a long life span, the species is particularly vulnerable to overfishing (Boreman 1997; Crouse 1999). Although most populations are now protected, loss of habitat and degraded water quality continue to threaten many remaining stocks. In the Great Lakes for example, dams have been constructed on every known spawning tributary. Effective sturgeon fishways have been constructed on some low-head impoundments, and artificial spawning habitats have been successfully introduced in some rivers (Bruch 1998). However, the lack of effective fish passage systems around hydropower facilities (and other high-relief dams) continues to fragment habitat and degrade water quality on many river systems (Baxter 1977; Jager et al. 2001). Consequently, ongoing studies of fish passage structures specifically designed for lake sturgeon may hold the greatest promise for restoring populations where dams limit access to suitable spawning habitat.

Because many remnant sturgeon populations have not recovered despite decades of legal protections, the use of stocking to either reestablish or supplement existing stocks has become widespread. In recent years, lake sturgeons have been stocked as both fry and fingerlings in several states including Wisconsin, Missouri, Michigan, New York, Georgia, and Tennessee. Although recent studies have demonstrated the importance of using only native specimens in sturgeon stocking programs (Secor et al. 2002; Paragamin and Beamesderfer 2004), lack of suitable brood stock has prompted new interest in alternative stocking strategies. Among the most promising of these is an experimental method known as 'head start', in which naturally-spawned lake sturgeon larvae are captured from the wild and then transferred to a protective hatchery environment where they can be reared for several months before release back into their natal streams. Because environmental conditions within the hatchery can be carefully controlled, first-year survival of these fry is typically much higher than in the wild. When used over many spawning seasons, this technique can dramatically amplify annual recruitment of *naturally* produced juveniles by *artificially* increasing critical-period survival. Unlike traditional stocking programs that typically rely on only a few wild adults to produce entire cohorts, the use of a head-start program can increase juvenile abundance while avoiding the potential problem of inbreeding depression. Although the new practice is not completely free of some artificial selection, the method may provide an important tool for restoration of depleted populations where at least some spawning still occurs. The first known use of the head-start method for lake sturgeon restoration occurred on the Black River, Michigan, in 2001 (D. Peterson, unpublished data). Since the inception of this program, annual releases of head-start lake sturgeon fingerlings have helped restore lake sturgeon populations in the Black, Sturgeon, and Pigeon rivers of northern Michigan (http://www.michigandnr.com/fishstock/default.asp 2006).

RESEARCH DIRECTION

The available literature on the biology and life history of lake sturgeon seems comprehensive. However, the lake sturgeon has been extirpated from the southern portion of its range and is uncommon or rare in much of its remaining range. Numerous studies are currently underway to assess remaining populations as a first step toward restoration. While these studies will undoubtedly provide the scientific basis for future management, identification of limiting factors is currently the most critical research need. Accordingly, investigations focusing on recruitment mechanisms and ecological links between habitat requirements and successful reproduction of lake sturgeons are high priority. To ensure the success of current restoration efforts, additional studies are needed to address existing knowledge gaps regarding critical habitat requirements of the various life stages. This will be of particularly importance in regions where lake sturgeon populations are being reintroduced through stocking.

OUTLOOK

The outlook for lake sturgeon recovery range-wide is guardedly optimistic, thanks in part to renewed interest in the species, novel approaches to management, new opportunities to eliminate long-standing data gaps, and continued progress in habitat restoration. Recent emphasis on maintaining biodiversity has prompted several new management initiatives to 'bring back the natives'. This has been especially true for large, charismatic species and has contributed to the renewed interest in enhancing or restoring lake sturgeon populations range-wide. As a result, many state and federal agencies have initiated or are planning to initiate lake sturgeon restoration programs.

Novel approaches such as 'sturgeon head start' and similar initiatives have increased recruitment success of naturally spawned lake sturgeon. In instances where natural reproduction is insufficient to support head start programs, stream-side rearing facilities are now being used to help ensure that juveniles produced for stocking are properly imprinted on their receiving waters prior to their release. Improved husbandry techniques such as converting hatchery-reared lake sturgeon to 100% commercial feed (Kornberg and Peterson 2005) also promises to help restoration efforts by increasing pre- and post-stocking growth and survival of such individuals.

The emerging application of conservation genetics techniques to fishery science also may improve artificial propagation programs. For example, mating protocols that manipulate parental crosses to maximize available genetic diversity have been developed and used for other species. The adoption of these approaches for artificial propagation of lake sturgeons will increase the effective population size in instances where brood stock is limited; and in turn, produce larger numbers of individuals for stocking than would have been available otherwise.

The increase in stocking programs to augment or re-establish lake sturgeon populations has provided new opportunities to study lake sturgeon in the wild. These new opportunities will provide data that were unavailable previously because lake sturgeon populations were limited in size or distribution. As a result, many known gaps in our knowledge about lake sturgeon biology, ecology, and life history could not be previously evaluated because of a scarcity of fish and the inability to study them in the wild. This limitation is being removed as an increasing number of populations are established throughout the species' range.

Finally, as state and federal regulations intended to protect the environment are enacted and enforced, there has been a gradual improvement in lake sturgeon habitat quality and quantity. Though not yet available at historic levels, suitable habitat has increased, and with it the probability of success of the aforementioned restoration activities. Further, as increasing numbers of lake sturgeon populations are enhanced or restored and then studied, our understanding of how the species responds to habitat changes (positive and negative) should increase and allow for refinement of habitat mitigation and protection.

Clearly, the recent interest in lake sturgeon has helped the species and its long-term prospects for survival. Our guarded optimism, however, is not intended as an "all clear" regarding threats facing the species. Obviously, many threats still remain, and they will be diminished only by continued awareness of how anthropogenic activities affect lake sturgeon populations and continued efforts toward habitat preservation and restoration.

REFERENCES

1. Antoniu-Murgoci A (1936a) Particularit_s anatomiques caract_risant les esp_ces des genres *Huso* et *Acipenser* des eaux roumaines. Compt Rend S_anc Acad Sci Roum 1:1–3
2. Antoniu-Murgoci A (1936b) Particularit_s anatomiques qui diff_rencient le genre Huso du genre *Acipenser* des eaux roumaines. Ann Sci Univ Jassy 23:94–103
3. Antoniu-Murgoci A (1942) Contributions a l'_tude des Acipenserid_s de Roumanie. Ann Sci Univ Jassy 2 Sci Natur 28:289–385
4. Auer NA (1996a) Response of spawning lake sturgeon to change in hydroelectric facility operation. Trans Am Fish Soc 125(1):66–77
5. Auer NA (1996b) Importance of habitat and migration to sturgeons with emphasis on lake sturgeon. Can J Fish Aquat Sci 53:152–160
6. Auer NA (1999) Lake sturgeon: A unique and imperilled species in the Great Lakes. In: Taylor WW, Ferreri CP (eds), Great lakes fisheries policy and management. A binational perspective. Michigan State University Press, pp 515–536
7. Auer NA (2004) Conservation. In: LeBreton GTO, Beamish FWH, McKinley RS (eds) Sturgeon and Paddlefish of North America. Kluwer Academic Publishers, pp 252–276
8. Auer NA, Baker EA (2002) Duration and drift of larval lake sturgeon in the Sturgeon River. Michigan J Appl Ichthyol 18(4–6):557–564
9. Bajkov A (1930) Fishing industry and fisheries investigations in the prairie provinces. Trans Am Fish Soc 60:215–237
10. Baker JP (1980) The distribution, ecology, and management of the lake sturgeon (*Acipenser fulvescens* Rafinesque) in Michigan. Michigan Department of Natural Resources, Fisheries Division. Fisheries Research Report. No. 1883, 95 pp
11. Baxter RM (1977) Environmental effects of dams and impoundments. Annu Rev Ecol Syst 8:255–283
12. Beamish FWH, Jebbink JA, Rossiter A, Noakes DLG (1996) Growth strategy of juvenile lake sturgeon (*Acipenser fulvescens*) in a northern river. Can J Fish Aquat Sci 53:481–489
13. Beamish FWH, Noakes DLG, Rossiter A (1998) Feeding ecology of juvenile Lake Sturgeon, *Acipenser fulvescens*, in Northern Ontario. Can Field Nat 112(3):459–468
14. Beamesderfer RCP, Farr RA (1997) Alternatives for the protection and restoration of sturgeons and their habitat. Environ Biol Fish 48(1–4):407–417

15. Becker GC (1983) Fishes of Wisconsin. The University of Wisconsin Press, Madison, WI, 1052 pp
16. Bemis WE, Kynard B (1997) Sturgeon rivers: an introduction to acipenseriform biogeography and life history. Environ Biol Fish 48:167–183
17. Bemis WE, Findeis EK, Grande L (1997) An overview of Acipenseriformes. Environ Biol Fish 48:25–71
18. Binkowski FP, Doroshov SI (eds) (1985) North American sturgeons: biology and aquaculture potential. Dr. W. Junk Publishers, Dordrecht, 163 pp
19. Birstein VJ, Doukakis P, DeSalle R (2002) Molecular phylogeny of Acipenseridae: nonmonophylyof Scaphirhynchinae. Copeia 2:287–301
20. Bogue MB (2000) Fishing the Great Lakes: An environmental history, 1783–1933. The University of Wisconsin Press, Madison WI, 443 pp
21. Boreman J (1997) Sensitivity of North American sturgeon and paddlefish to fishing mortality. Environ Biol Fish 48:399–405
22. Brennan JS, Cailliet GM (1991) Age determination and validation studies of white sturgeon, *Acipenser transmontanus*, in California. In: Williot P (ed) The Proceedings of the 1st International Symposium on the Sturgeon. CEMAGREF, Bordeaux, France
23. Brousseau CS (1987) The lake sturgeon (*Acipenser fulvescens*) in Ontario. In: Oliver CH (ed) Proceedings of a Workshop on the Lake Sturgeon (*Acipenser fulvescens*). Ontario Fisheries Technical Report Series No. 23. Ontario Ministry of Natural Resources, Toronto, ON, pp 2–9
24. Bruch RM (1998). Management and trade of lake sturgeon in North America. In: Williamson DFG, Benz G, Hoover CM (eds) Caviar and conservation: Status, management and trade of North American sturgeon and paddlefish. TRAFFIC North America Washington D.C. World Wildlife Fund
25. Bruch RM (1999) Management of lake sturgeon on the Winnebago System-long term impacts of harvest and regulations on population structure. J Appl Ichthyol 15:142–152
26. Bruch RM, Dick TA, Choudhury A (2001) A field guide for the identification of stages of gonad development in lake sturgeon, *Acipenser fulvescens Rafinesque*, with notes on lake sturgeon reproductive biology and management implications. Publication of Wisconsin Department of Natural Resources. Oshkosh and Sturgeon for Tomorrow. Appleton, WI, 38 pp
27. Bruch RM, Binkowski FP (2002) Spawning behaviour of lake sturgeon (*Acipenser fulvescens*). J Appl Ichthyol 18:570–579
28. Buckland F (1883) Natural history of British Fishes. London
29. Carlander HB (1954) A history of fish and fishing in the upper Mississippi River. Upper Mississippi River Conservation Committee, Rock Island, IL
30. Carlson DM (1995) Lake sturgeon waters and fisheries in New York State. J Great Lakes Res 21:35–41
31. Ceskleba DG, AveLallemant S, Thuemler TF (1985) Artificial spawning and rearing of lake sturgeon, *Acipenser fulvescens*, in Wild Rose State Fish Hatchery, Wisconsin, 1982–1983. Environ Biol Fish 14(1):79–85
32. Chiasson WB, Noakes DLG, Beamish FWH (1997) Habitat, benthic prey, and distribution of juvenile lake sturgeon (*Acipenser fulvescens*) in northern Ontario rivers. Can J Fish Aquat Sci 54: 2866–2871
33. Classen TEA (1944) Estudio Bio-Estadistico del Esturion o-Sollo del Quadalquivir (*Acipenser sturio* L.) Instituto Espanol de Oceanografia, Madrid, Trabajo numero 19
34. Conte FS, Doroshov SI, Lutes PB, Strange EM (1988) Hatchery manual for the white sturgeon. Publication #3322. University of California Press, Oakland, 104 pp
35. Cooke DW, Leach SD, Isely JJ (2002) Behaviour and lack of upstream passage of shortnose sturgeon at a hydroelectric facility and navigation lock complex. In: Van Winkle W, Anders PJ, Secor DH, Dixon DA (eds) Biology, management, and protection of North American Sturgeon. American Fisheries Society, Symposium 28, Bethesda, MD, pp 101–110

36. Crouse DT (1999) The consequences of delayed maturity in a human dominated world. In: Musick JA (ed), Life in the slow lane: ecology and coservation of long-lived marine animals. Am Fish Soc Symp 23, Bethesda, MD, pp 195–202
37. Currier JP (1951) The use of pectoral fin rays for determining age of sturgeon and other species of fish. Can Fish Culturist 11:10–18
38. Detlaf TA, Ginsburg AS, Schmalhausen OI (1993) Sturgeon fishes: developmental biology and aquaculture. Springer-Verlag, Berlin, 300 pp
39. Engel MP (1990) Population parameters of lake sturgeon in the St. Croix River with special reference to movements. Draft Summary Report, Wisconsin Department of Natural Resources, Madison, WI, 44 pp
40. Ferguson MM, Duckworth GA (1997) The status and distribution of lake sturgeon, *Acipenser fulvescens*, in the Canadian provinces of Manitoba, Ontario and Quebec: a genetic perspective. Environ Biol Fish 48:299–309
41. Findeis EK (1997) Osteology and phylogenetic interrelationships of sturgeons (Acipenseridae). Environ Biol Fish 48:73–126
42. Folz DJ, Meyers LS (1985) Management of the lake sturgeon, *Acipenser fulvescens*, population in the Lake Winnebago system, Wisconsin. In: Binkowski FP, Doroshov SI (eds) North American sturgeons: biology and aquaculture potential. Dr. W. Junk Publishers, Dordrecht, The Netherlands, pp 135–146
43. Fortin R, Mongeau JR, Desjardins G, Dumont P (1993) Movements and biooigcal statistics of lake sturgeon (*Acipenser fulvescens*) populations from the St. Lawrence and Ottawa River system, Quebec. Can J Zool 71:638–650
44. Fortin R, Dumont P, Guénette S (1996) Determinants of growth and body condition of lake sturgeon (*Acipenser fulvescens*). Can J Fish Aquat Sci 53:1150–1156
45. Freedman B, Roger L, Ewins P, Green DM (2001) Species at risk in Canada. In: Beazley K, Boardman R (eds) Politics of the wild: Canada and endangered species. Oxford University Press, ON, pp 26–48
46. Gerbilskiy NL (1957) Ways of development of the intraspecies biological differentiation, types of anadromous migrants, and a question of migration impulse in sturgeons. Uchenye Zapiski Leningradskogo Gosudarstvennogo Universiteta 228:11–31 (in Russian)
47. Grande L, Bemis WE 1996. Interrelationships of Acipenseriformes, with Comments on Chondrostei. In: Stiassny MLJ et al. (eds) Interrelationships of Fishes. Academic Press, pp 85–116
48. Harkness WJK (1923) The rate of growth and food of the lake sturgeon (*Acipenser fulvescens* LeSueur). University of Toronto Biological Studies Series 24, Publication of the Ontario Fisheries Research Lab 18
49. Harkness WJK, Dymond JR (1961) The lake sturgeon. Ontario Department of Lands and Forests, Fish and Wildlife Branch, Toronto, 97 pp
50. Hay-Chmielewski EM (1987) Habitat preferences and movement patterns of the lake sturgeon (*Acipenser fulvescens*) in Black Lake, Michigan. Michigan Department of Natural Resources, Fisheries Research Report 1949, Ann Arbor, MI
51. Hilton EJ, Bemis WE (1999) Skeletal variation in shortnose sturgeon (Acipenser brevirostrum) from the Connecticut River: Implications for comparative osteological studies of fossil and living fishes. In: Arratia G, Schultze HP (eds) Mesozoic FishesII-Systematics and Fossil Record. Munchen, Germany, pp 69–94
52. Hochleitner M, Vecsei P (2004) Identification key to sturgeons and paddlefish of North America. In: LeBreton GTO, Beamish FWH, McKinley RS (eds) Sturgeons and Paddlefish of North America. Fish and Fisheries Series, Kluwer Acadmeic Publishers, pp 315–318
53. Holey ME, Trudeau TN (eds) (2005) The state of Lake Michigan in 2000. Great Lakes Fisheries Commission Spec. Pub. 05-01
54. Houston JJ (1987) Status of lake sturgeon, *Acipenser fulvescens*, in Canada. Can Field-Nat 101:171–185
55. Jager HI, Van Winkle W, Chandler JA, Lepla KB, Bates P, Counihan TD (2002) A simulation study of factors Controlling White sturgeon recruitment in the Snake

River. In: Van Winkle W, Anders PJ, Secor DH, Dixon DA (eds) Biology, management, and protection of North American sturgeon. American Fisheries Society, Symposium 28, Bethesda, MD, pp 127–150

56. Jager HI, Lepla K, Chandler J, Van Winkle W (2001) A theoretical study of river fragmentation by dams and its effects on other riverine fishes. Environ Biol Fish 60:347–361
57. Johnson JE (1987) Protected fishes of the United States and Canada. American Fisheries Society, Bethesda MD, 42 pp
58. Jollie M (1980) Development of head and pectoral girdle skeleton and scales in *Acipenser*. Copeia 1980:226–249
59. Keenlyne KD, Jenkins LG (1993) Age at sexual maturity of the pallid sturgeon. Trans Am Fish Soc 122(3):393–396
60. Kempinger JJ (1988) Spawning and early life history of lake sturgeon in the Lake Winnebago system, Wisconsin. Am Fish Soc Symp 5:110–122
61. Kornberg J, Peterson D (2005) New methods for intensive culture of lake sturgeon in Georgia. Technical report to the Georgia Department of Natural Resources, Project F-75, 13 pp
62. LaHaye M, Branchaud A, Gendron M, Verdon R, Fortin R (1992) Reproduction, early life history, and characteristics of the spawning grounds of the lake sturgeon (*Acipenser fulvescens*) in Des Prairies and L'Assomption rivers, near Montreal, Quebec. Can J Zool 70:1681–1689
63. Larson T (1988) The lake sturgeon fishery of Lake Wisconsin, 1978–1985. Wisconsin Department of Natural Resources, Fish Management Report No. 36, Madison, 34 pp
64. LeBreton GTO, Beamish FWH (2004) Growth, Bioenergetics and age In: LeBreton GTO, Beamish FWH, McKinely RS (eds) Sturgeon and Paddlefish of North America. Kluwer Academic Publishers, pp 217–230
65. Leonard NJ, Taylor WW, Goddard C (2004) Multijurisdictional management of Lake Sturgeon in the Great Lakes and St. Lawrence River. In: LeBreton GTO, Beamish FWH, McKinley RS (eds) Sturgeons and Paddlefish of North America, Kluwer Academic Publishers, pp 231–251
66. Lyons J, Kempinger JJ (1992) Movements of adult lake sturgeon in the Lake Winnebago system. Wisconsin Department of Natural Resources. Publication No. RS-156-92
67. MacKay HH (1963) Fishes of Ontario. Ontario Department of Lands and Forests. Bryant Press Ltd., Toronto, 300 pp
68. Magnin E (1966) Recherches sur les cycles de reproduction des sturgeons *Acipenser fulvescens* Rafinesque. Annales de la Station Centrale d'Hydrobiologie appliquee 9:8–242
69. McKinley RS, Van Der Kraak G, Power G (1998) Seasonal migration and reproductive patterns in the lake sturgeon, *Acipenser fulvescens*, in the vicinity of hydroelectric stations in northern Ontario. Environ Biol Fish 51:245–256
70. Noakes DLG, Beamish FWH, Rossiter A (1999) Conservation implications of behaviour and growth of the lake sturgeon, *Acipenser fulvescens*, in northern Ontario. Environ Biol Fish 55:135–144
71. Paragamian VL, Beamesderfer RC (2004) Dilemma on the Kootenai River – the risk of extinction or When does the hatchery become the best option? Am Fish Soc Symp 44:377–385
72. Peterson DL, Gunderman B, Vecsei P (2002) Lake Sturgeon of the Manistee River: a current assessment of spawning stock size, age, and growth. In: Biology, management, and protection of North American sturgeon. Am Fish Soc Symp 28:175–183
73. Peterson DL, Vecsei P, Noakes DLG (2003) Threatened fishes of the world: *Acipenser fulvescens* Rafinesque, 1817 (Acipenseridae). Environ Biol Fish 68:174
74. Peterson DL, Vecsei P (2004) Lake Sturgeon of the Muskegon River: population dynamics and life history. Final Report to the Great Lakes Fishery Trust, 33 pp

75. Priegel GR (1973) Lake sturgeon management on the Menominee River. Wisconsin Department of Natural Resources, Madison. Technical Bulletin No. 67. 20 pp
76. Priegel GR, Wirth TL (1971) The lake sturgeon: it's life history, ecology and management. WI Dept. Nat. Res. Publication 240-7, Madison, WI, 17 pp
77. Priegel GR, Wirth TL (1975) Lake sturgeon harvest, growth, and recruitment in Lake Winnebago, Wisconsin. Wisconsin Department of Natural Resources, Madison. Technical Bulletin No. 83, 25 pp
78. Priegel GR, Wirth TL (1978) Lake sturgeon populations, growth, and exploitation in lakes Poygan, Winneconne, and Lake Butte Des Morts, Wisconsin. Wisconsin Department of Natural Resources, Madison. Technical Bulletin No. 107. 23 pp
79. Priegel GR, Wirth TL (1974) The lake sturgeon: it's life history, ecology and management. WI Department of Natural Resources Publication 4-3600 (74), Madison, WI
80. Rien TA, Beamesderfer RC (1994) Accuracy and precision of white sturgeon age estimates from pectoral fin rays. Trans Am Fish Soc 123:255–265
81. Rossiter A, Noakes DLG, Beamish FWH (1995) Validation of age estimation for the lake sturgeon. Trans Am Fish Soc 124:777–781
82. Roussow G (1957) Some considerations concerning sturgeon spawning periodicity. J Fish Res Board Canada 14:553–572
83. Saffron I (2002) Caviar: the strange history and uncertain future of the World's most coveted delicacy. Broadway Books, 288 pp
84. Scott WB, Crossman EJ (1973) Freshwater fishes of Canada. Bulletin 184, Fisheries Research Board of Canada, Ottawa, 966 pp
85. Secor DH, Anders PJ, Van Winkle W, Dixon DA (2002) Can we study sturgeons to extinction? What we do and don't know about the conservation of North American sturgeons. Am Fish Soc Symp 28:3–12
86. Smith KM, King DK (2005) Dynamics and extent of larval lake sturgeon *Acipenser fulvescens* drift in the Upper Black River, Michigan. J Appl Ichthyol 21:161–168
87. Stearns JL, Atkinson B (eds) (1953) Etowah Sturgeon, Hitchhicking Lampreys. Georgia Game and Fish Commission 1:9
88. Thomas MV, Haas RC (2002) Abundance, age structure, and spatial distribution of lake sturgeon *Acipenser fulvescens* in the St. Clair System Michigan. Department of Natural Resources Fisheries Research Report 2076
89. Thuemler TF (1985) The lake sturgeon, *Acipenser fulvescens*, in the Menominee River, Wisconsin-Michigan. Environ Biol Fish 14(1):73–78
90. Thuemler TF (1988) Movements of young lake sturgeons stocked in the Menominee River, Wisconsin. Am Fish Soc Symp 5:104–109
91. Tody WH (1974) Whitefish, sturgeon, and the early Michigan commercial fishery. Michigan Department of Natural Resources, Lansing MI, pp 45–60
92. TRAFFIC North America (2002) Fishy business. TRAFFIC North America bulletin, Washington, DC, Volume 5(1)
93. Van Oosten J (1956) The lake sturgeon in our endangered wildlife. Nat Wildl Fed Washington, DC, pp 9–10
94. Vecsei P, Peterson DL (2004) Sturgeon ecomorphology: a descriptive approach. In: LeBreton GTO, Beamish FWH, McKinley RS (eds) Sturgeons and Paddlefish of North America. Kluwer Academic Publishers, pp 103–133
95. Vladykov VD, Greeley JR (1963) Order Acipenseroidei. In: Bigelow HB, Breeder CM, Cohen DM, Mead GW, Merriman D, Olsen YH, Schroeder WC, Schultz LP, Teevan J (eds) Fishes of the Western North Atlantic, Part Three, New Haven, pp 24–60
96. Wang YL, Binkowski FP, Doroshov SI (1985) Effect of temperature on early development of white and lake sturgeon, *Acipenser transmontanus* and *A. fulvescens*. Environ Biol Fish 14(1):43–50
97. Wallus R (1990) Family Acipenseridae. In: Wallus R, Simon TP, Yeager BL (eds) Reproductive biology and early life history of fishes in the Ohio River drainage, vol. 1. Acipenseridae through Esocidae. Tennessee Valley Authority, Chattanooga, TN, pp 29–46

98. Webb MAH, Van Eenennaam JP, Doroshov SI, Moberg GP (2001) Preliminary observations on the effects of holding temperature on reproductive performance of female white sturgeon, *Acipenser transmontanus* Richardson. Aquaculture 176:315–329

99. Williamson DF (2003) Caviar and conservation: status, management and trade of North American sturgeon and paddlefish. TRAFFIC North America, Washington DC, World Wildlife Fund

100. Wilson JA, McKinley RS (2004) Distribution, habitat and movements In: LeBreton, GTO, Beamish FWH, McKinley RS (eds) Sturgeon and paddlefish of North America. Kluwer Academic Publishers, pp 40–69

101. Wilson NC (1987) Age determination of lake sturgeon (*Acipenser fulvescens*) by use of the marginal pectoral fin ray. In: Proceeding of a workshop on the lake sturgeon (*Acipenser fulvescens*). Ontario Ministry of Natural Resources Fisheries Technical Report Series No.23, Toronto, ON, pp 77–83

CHAPTER 5

Trophic Ecology of Introduced Populations of Alaska Blackfish (*Dallia Pectoralis*) in The Cook Inlet Basin, Alaska

Dona M. Eidam, Frank A. von Hippel, Matthew L. Carlson, Dennis R. Lassuy, J. Andrés López

Department of Biological Sciences, University of Alaska Anchorage, 3211 Providence Dr, Anchorage, AK 99508, USA

Department of Biological Sciences, Northern Arizona University, 617 S. Beaver St., PO Box 5640, Flagstaff, AZ 86011, USA

Alaska Center for Conservation Science, University of Alaska Anchorage, 3211 Providence Dr, Anchorage, AK 99508, USA

North Slope Science Initiative, 222 West 7th Ave., #13, Anchorage, AK 99513, USA

University of Alaska Museum, 907 Yukon Drive, Fairbanks, AK 99775, USA

School of Fisheries and Ocean Sciences, University of Alaska, 905 N. Koyukuk Drive, Fairbanks, AK 99775, USA

ABSTRACT

Introduced non-native fishes have the potential to substantially alter aquatic ecology in the introduced range through competition and predation. The Alaska blackfish (*Dallia pectoralis*) is a freshwater fish endemic to Chukotka and Alaska north of the Alaska Range (Beringia); the species was introduced outside of its native range to the Cook Inlet Basin of Alaska in the 1950s, where it has since become widespread. Here we characterize the diet of Alaska blackfish at three Cook Inlet Basin sites, including a lake, a stream, and a wetland. We analyze stomach plus esophageal contents to assess potential impacts on native species via competition or predation. Alaska blackfish in the Cook Inlet Basin consume a wide range of prey, with major prey consisting of epiphytic/benthic dipteran larvae, gastropods, and ostracods. Diets of the introduced populations of Alaska blackfish are similar in composition to those of native juvenile salmonids and stickleback. Thus, Alaska blackfish may affect native fish populations via competition. Fish ranked third in prey importance for both lake and stream blackfish diets but were of minor importance for wetland blackfish.

Keywords: Biological invasion Competition Index of relative importance Predation Stomach content analysis

INTRODUCTION

Invasive species are second only to habitat loss as a cause of extinction of native species in the United States, and pose a similar threat worldwide (Lassuy 1995; Wilcove et al. 1998; Strong and Pemberton 2000; Mooney and Cleland 2001). Biological invasions may result in alterations of natural ecosystems, including nutrient cycling, disturbance regimes, habitat structure, and community composition (Drake et al. 1989). The establishment of non-native fish populations is of particular concern because fish introductions often correlate with reduction or extinction of native fishes due to predation (Brown 1989). Additionally, invasive fish establishment can result in dramatic changes in invertebrate communities, including reductions in potential prey resources for native fishes (Macan 1966, 1977; Brown 1989; Gerking 1994; Knapp et al. 2001; Byström et al. 2007). The success of invasive fishes is often attributed to low dietary specialization and the ability to capitalize on diverse prey resources (Moyle and Marchetti 2006). In the United States, many native fish species are listed as threatened, endangered, or negatively impacted by introduced fishes (Wilcove and Bean 1994) and resulting human economic losses are conservatively estimated at more than one billion U.S. dollars annually (Pimentel 2007).

Alaska lists 14 introduced fishes within its boundaries (McClory and Gotthardt 2008), including two species that are native to some regions of the state but introduced to others: northern pike (*Esox lucius*) and Alaska blackfish (*Dallia pectoralis*; hereafter blackfish; Fig. 1). The blackfish is endemic to fresh waters of Beringia, the subcontinent of eastern Siberia to western Yukon that remained ice-free during the Pleistocene (Fig. 2). The native range of blackfish extends from 55° to 72° N latitude on the Chukchi

Peninsula of eastern Siberia, across western Alaska from the Arctic Coastal Plain to the Alaska Peninsula, and inland through the Yukon-Tanana drainage (Mecklenburg and Mecklenburg 2002). Introduced blackfish populations are found on St. Paul Island in the Bering Sea and in numerous lakes, ponds, and streams of the Kenai Peninsula and Matanuska-Susitna Valley within the Cook Inlet Basin of south Alaska (Fig. 2; Morrow 1980; Stratton and Cyr 1997; Mecklenburg and Mecklenburg 2002; Eidam 2015).

Figure. 1 Adult blackfish in laboratory aquarium at the University of Alaska Anchorage. Fish is feeding on previously frozen *Chironomid* larvae. Photograph by Dr. Thomas C. Kline, Jr., salmonography.com, copyrighted and used with permission

Blackfish are well known for their adaptations to life in arctic and subarctic waters, adaptations that have facilitated their spread in the Cook Inlet Basin. They can tolerate living in high densities in small tundra pools (Morrow 1980). They exhibit extreme cold tolerance (Scholander et al. 1953; Scholander et al. 1957). A vascularized esophagus functions as an air-breathing organ and is known in only two other teleosts, the shanny (*Lipophrys pholis*) of northern European marine waters (Laming et al. 1982) and the Asian swamp eel (*Monopterus albus*), which is invasive in freshwater habitats of the southeastern United States (Liem 1967; Liem 1987; Fuller et al. 1999). Facultative air breathing enables blackfish to survive under hypoxic conditions by gulping air at the surface, such as in

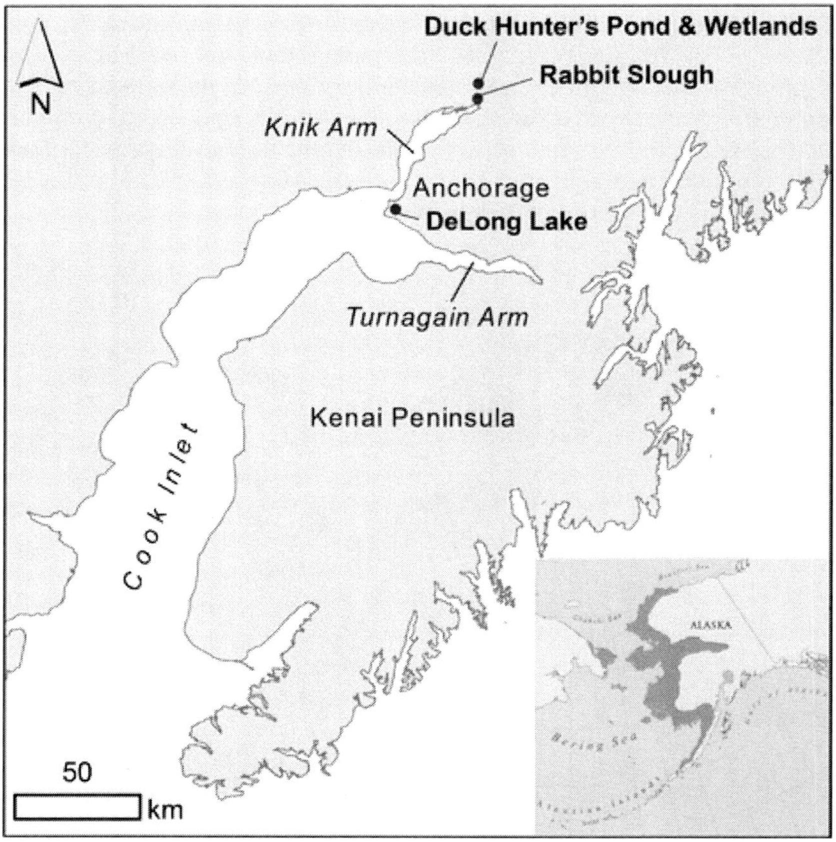

Figure. 2 Map of study sites in the Cook Inlet Basin, Alaska. The map of Alaska (*lower right*) shows the native range of the Alaska blackfish (*green*) and the introduced range within the Cook Inlet Basin (*orange*)

warm shallow wetlands in summer when oxygen levels drop below 2.3 mg/L, and also via surface holes in iced-over waters during winter (Ostdiek and Nardone 1959; Crawford 1974; Morrow 1980).

Fisheries managers in Alaska have expressed concern over possible competition and predation by blackfish impacting native and stocked salmonids (Trent and Kubik 1974; Hepler and Bowden 1986). A previous study in their introduced range reported substantial numbers of salmonids in blackfish gut contents (Chlupach 1975).

Fish diet analysis is an effective tool for understanding the potential impacts of introduced fishes on aquatic ecosystems and native fishes (Garvey et al. 1998; Vander Zanden et al. 2000; Chipps and Garvey 2007).

Diet composition reveals the trophic position of a species within the food web as well as the prevalence of predation on other fish species (Pauly et al. 1998; Stergiou and Karpouzi 2002). Dietary overlap can indicate potential resource competition between introduced and native species. Prey type diversity helps to define specialist versus generalist feeders, while spatial and temporal diet shifts highlight opportunistic, flexible feeding strategies. Trophic interactions including feeding habits can help to reveal the extent of ecosystem alteration by introduced fishes, thereby providing useful information for ecosystem-based management (Pauly et al. 1998; Bachok et al. 2004; Stobberup et al. 2009).

Management actions have been taken to reduce or eliminate non-native blackfish populations in south Alaska based on the assumption that these introduced populations represent a significant risk to the native and stocked fishes and natural resources. In this study we explore the implicit hypothesis that introduced blackfish populations pose a measureable risk to the ecology of the invaded waterbodies and native and stocked fishes of south Alaska. Specifically, we examine the predictions that fish constitute a major prey component in blackfish diets, and that blackfish diets overlap greatly with those of native fishes and stocked sportfish. Additionally, because most successful fish invaders show low dietary specialization and a high capacity to use available resources (Moyle and Light 1996a, 1996b; Marchetti et al. 2004; Gido and Franssen 2007), we predict that blackfish stomach contents vary significantly by waterbody, season, sex, and body size.

MATERIALS AND METHODS

Study sites

Three freshwater habitats—a wetland, stream, and lake—were selected within the Cook Inlet Basin of south Alaska, based on year-round presence of blackfish (Fig. 2). Duck Hunter's Pond and surrounding wetland (61.53920° N, 149.25460° W) within the Matanuska-Susitna Valley lowlands consist of a marsh, constructed rectangular pool, and narrow drainage ditch containing a large blackfish population (Eidam 2015). Blackfish and the occasional threespine stickleback (*Gasterosteus aculeatus*) and ninespine stickleback (*Pungitius pungitius*) are the only fish species we documented from the site. The water in this site is shallow and heavily vegetated.

Rabbit Slough (61.53750° N, 149.25460° W) is a stream within the Palmer Hayflats State Game Refuge located 0.15 km south of Duck Hunter's Pond, separated from it by a road. The stream drains into the Knik Arm of upper Cook Inlet and is characterized by slow-flowing, tannic-colored water with abundant overhanging forb, graminoid, and willow and alder shrub vegetation. Soft silty benthos support rooted macrophytes with some floating macrophytes as well. Rabbit Slough serves as a coho salmon (*Oncorhynchus kisutch*) nursery and a popular coho fishery, and is home to other native fishes including Dolly Varden char (*Salvelinus malma*), threespine stickleback, ninespine stickleback, slimy sculpin (*Cottus cognatus*), sockeye salmon (*Oncorhynchus nerka*), and chinook salmon (*Oncorhynchus tshawytscha*).

DeLong Lake (61.16390°N, 149.95550°W) in Anchorage is an 8 ha lake with a mean depth of 4 m and maximum depth of 7 m. Invasive waterweed (*Elodea canadensis* and likely hybrids with *E. nuttallii*) forms dense stands of long-stemmed, rooted macrophytes. The urban lake is a popular sport fishery stocked annually with hatchery rainbow trout (*Oncorhynchus mykiss*) and chinook salmon. Catchable fish usually average 18 cm total length or longer; in 2010, 15,000 fingerling rainbow trout (mean total length 6.6 cm) were stocked (ADF&G 2011).

Sampling methodology

Blackfish were captured once a month over a 12-month period at each location using 0.32 cm and 0.64 cm wire mesh unbaited minnow traps. Optimal trap-soaking times were three hours or less to avoid digestion of stomach contents while fish were in the traps; however, soaking times were increased when trapping yielded low numbers of fish. Blackfish were euthanized with an overdose of pH-neutral MS-222 anesthetic and then blotted and wet-weighed to the nearest 0.1 g. Fish were measured with digital calipers to the nearest 1.0 mm for total length and standard length (SL) and then injected through the mouth with buffered 37 % formaldehyde solution to halt digestion. Whole specimens were placed into labeled teabags and fixed in buffered 10 % formalin for three weeks prior to rinsing with water; they were then transferred to 70 % ethanol.

Gastrointestinal tracts were dissected and esophageal and stomach contents of each fish washed with 70 % ethanol into a Petri dish. Prey items protruding into the mouth were included in the analyses. Prey organisms were viewed under a dissecting microscope, sorted, identified to major taxonomic category (McCafferty 1998; Thorp and Covich 2001; Merritt et al. 2008), and counted. Mean dry mass for each prey type was obtained by drying and weighing a representative number of organisms. Prey were loaded into 3.5 × 5 mm or 4 × 6 mm pre-weighed pressed tin capsules, dried at 60 °C in a drying oven, and weighed to the nearest 0.001 mg on a Sartorius microscale. Larger prey were placed in Petri dishes, dried, and weighed to the nearest 0.001 g on an analytical balance. Weights were computed from a subsample (80 % to 100 % for rare prey and 10 % to 50 % for other prey) of the total prey items for each prey category. Diptera, Coleoptera, Trichoptera, and Gastropoda were divided into subgroups for weighing in order to account for variable sizes and morphologies. Digested prey without identifiable parts were excluded.

Stomach content analysis

To obtain a measure of overall importance of each prey category, we used the index of relative importance (*IRI*): $IRI = (\%N + \%M) * (\%F)$, where *N* equals the actual count of individual prey items and highlights the importance of small prey such as zooplankton; *M* is the dry mass of prey items and emphasizes large, bulky prey; and *F* is the number of stomachs containing a specific food organism (Pinkas et al. 1971; Cailliet et al. 1986). *IRI* values were computed for the following eight sample groupings: combined sites and seasons, combined sites by season, combined seasons by site, each site by season, males and females by combined sites and seasons, and size classes by combined sites and seasons.

Statistical analysis

All statistical analyses were performed using SPSS v.22. A General Linear Model MANCOVA was used to test for differences in diet among waterbodies, among seasons, and by sex. For grouping of seasons, winter (November–March) corresponded to the period when waterbodies were covered in ice and snow; spring (April–May) was during ice and snow thaw; summer (June–August) was the warmest period free of ice and snow; and fall (September–October) was when snow and ice began forming again. Diet response variables consisted of prey masses of nine prey categories equal to or greater than 1 % *IRI* for combined sites and seasons (thus seven prey categories were omitted). Covariates were SL and trapping hours. Additional MANCOVAs were performed to test for seasonal differences in diet within each site. A binary logistic regression was also performed to analyze presence of fish in diet, using sex and size class as predictor variables. Size differences among sites were analyzed with ANOVA and between sexes were analyzed with a two-sample Student's *t*-test assuming equal variances using a pooled estimate of the variance.

We used non-metric multidimensional scaling (NMS) ordination to visualize patterns of diet among the 277 captured blackfish in the three waterbodies and seasons. We used raw prey mass from ten taxonomic diet categories (Ostracoda, Copepoda, Cladocera, Diptera, Trichoptera, Odonata, Gastropoda, Bivalvia, Teleostei, and Angiospermae) and included a categorical variable for fish with an empty gut, because it is not possible to generate IRI scores for individual fish. NMS is a standard free ordination technique in ecology to explore patterns in multivariate data (Peck 2010). NMS has the advantages over linear ordination techniques (e.g., PCA) of not requiring linear relationships among variables or normal distributions of response variables to ordination axes; its use of ranked distances aids in linearizing the relationship between distances measured in species space and distances in environmental space; and it can be used with any distance measure (Fasham 1977; Peck 2010). The original data were used to generate a Bray-Curtis dissimilarity matrix. NMS was first run using the "Autopilot" mode, with six starting dimensions twice to determine dimensionality in PC-Ord (McCune and Mefford 2006). Once the dimensionality of the solution was determined, NMS was run in manual settings with two axes, 250 runs, and instability criterion of 10^{-6}. This analysis was conducted three times and results compared to ensure consistency (Peck 2010).

RESULTS

Trapping times and body size comparisons

Trapping times ranged from 2.5–13.5 h with a mean of 5.7 h; processing times ranged from 0.5–5.5 h with a mean of 1.9 h. Total trapping plus processing times (time from trap placed in water to specimen placed into formalin) ranged from 3.5–14.5 h with a mean of 7.6 h. Trapping times in excess of 10 h occurred overnight during summer in the lake and stream. Overall prey consumption and trapping hours were not significantly related (F (9, 244) = 1.779, p = 0.073; Table 1), indicating a representative sampling.

From a total of 470 blackfish collected and fixed in formalin, 302 fish were dissected for stomach content analysis: 84 from the lake, 104 from the stream, and 114 from the wetland (Suppl. Table 1). In spring, only wetland fish were captured and analyzed due to unsafe ice conditions at the other sites. Seventy-eight percent of dissected blackfish were trapped during daytime compared to 22 % trapped overnight, in summer, from the lake and stream. Among specimens examined, males (SL mean = 98.90 mm, SD = 15.51, n = 145) were larger (t (297) = 7.23, p < 0.001) than females

Table 1 Multivariate effects on diet of blackfish. The MANCOVA is based on three factors (waterbody, season, and sex), two covariates (SL, trapping hours), and nine response variables (nine prey categories whose % *IRI* ≥ 1). SL = standard length, *ns* = not significant

Variable	Pillai's Trace	*F*	Hypothesis *df*	Error *df*	*p*
SL	0.063	1.82	9	244	*ns*
Trapping hrs.	0.062	1.78	9	244	*ns*
Site	0.287	4.56	18	490	< 0.001
Season	0.231	3.56	18	490	< 0.001
Sex	0.061	1.76	9	244	*ns*
Site × season	0.413	3.16	36	988	< 0.001

(SL mean = 91.67 mm, *SD* = 15.44, *n* = 154), though sizes broadly overlapped (Suppl. Fig. 1). Seventy-five percent of all fish were in 75–115 mm SL size range. The three sites varied significantly (ANOVA, F = 7.14, $p < 0.001$) in SL (lake mean SL = 90.96 mm, stream mean SL = 97.61 mm, wetland mean SL = 94.61 mm). Blackfish did not vary significantly in SL by season.

Prey categories and dietary diversity

Blackfish gut contents from the three study sites contained prey from 20 taxonomic groups distributed among five animal phyla (mollusks, annelids, arthropods, bryozoans, and chordates) and one plant division (Suppl. Table 2). In the stomach content analysis, insects were classified as Insecta adults (Order unknown), Diptera larvae, Trichoptera larvae, Odonata larvae, and Ephemeroptera larvae. Order Coleoptera included both larvae and adults. Bryozoa/Plumatellida from DeLong Lake consisted of over-wintering cysts (statoblasts). Angiosperms in blackfish gut contents were represented by seeds of unknown plants. Occasional plant stem and leaf tissue found in the guts of blackfish were excluded from analysis because of their rarity.

Consistent with the prediction of high diet diversity, blackfish in these populations typically had many prey types in their guts. The number of different prey types, based on 20 broad prey categories (Suppl. Table 2) found in each nonempty gut (esophagus + stomach + intestines) ranged from 1 to 10 (Fig. 3). The mode for number of prey categories per examined gut was three in the lake and stream samples, and four in the wetland samples. Waterbody and season were significant predictors of the nine response variables (nine prey categories whose % *IRI* ≥ 1), as was the interaction between them (Table 1). Blackfish size and sex were not significant predictors of prey types across all nine prey categories, although the frequency of fish in diet was related to blackfish size and sex (Chi-square = 9.487, $df = 2$, $p = 0.009$).

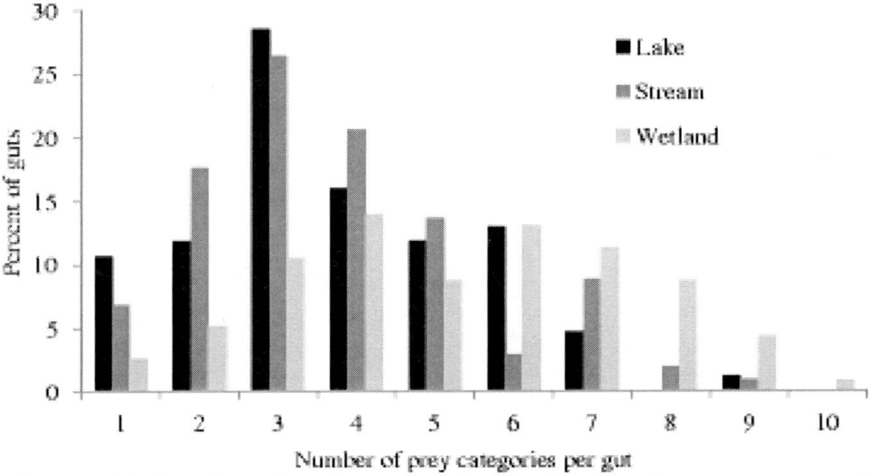

Figure. 3 Diet diversity histogram. For each site, the percentage of esophagus + stomach + intestines containing 1–10 prey categories is given. This is the only analysis that included intestine contents

Stomach contents by combined sites and seasons

Figure 4 provides a graphic representation of the *IRI* from all samples combined for prey categories with ≥3 % *IRI*. For combined waterbodies and seasons, gastropods were by far the dominant prey group (51 % *IRI*). Ostracods contributed 28 % of the percent *IRI*, while all other prey taxa recorded 7 % *IRI* or less. Twelve other categories made minor contributions to blackfish diets at these sites (%*IRI* < 0.5 %; Table 2). Ostracods were the primary prey in terms of total number, while gastropods and teleosts were the two most important prey in terms of biomass. Fish as prey ranked fourth in importance by percent *IRI*, while frequency of fish consumption was less than 10 %. Dipterans were found most frequently in guts (62 % frequency), followed by gastropods (46 % frequency). Overall, six prey types were consumed by at least 25 % of all blackfish: dipteran larvae, gastropods, copepods, trichopteran larvae, ostracods, and bivalves. Plant seeds were found in 14 % of all guts.

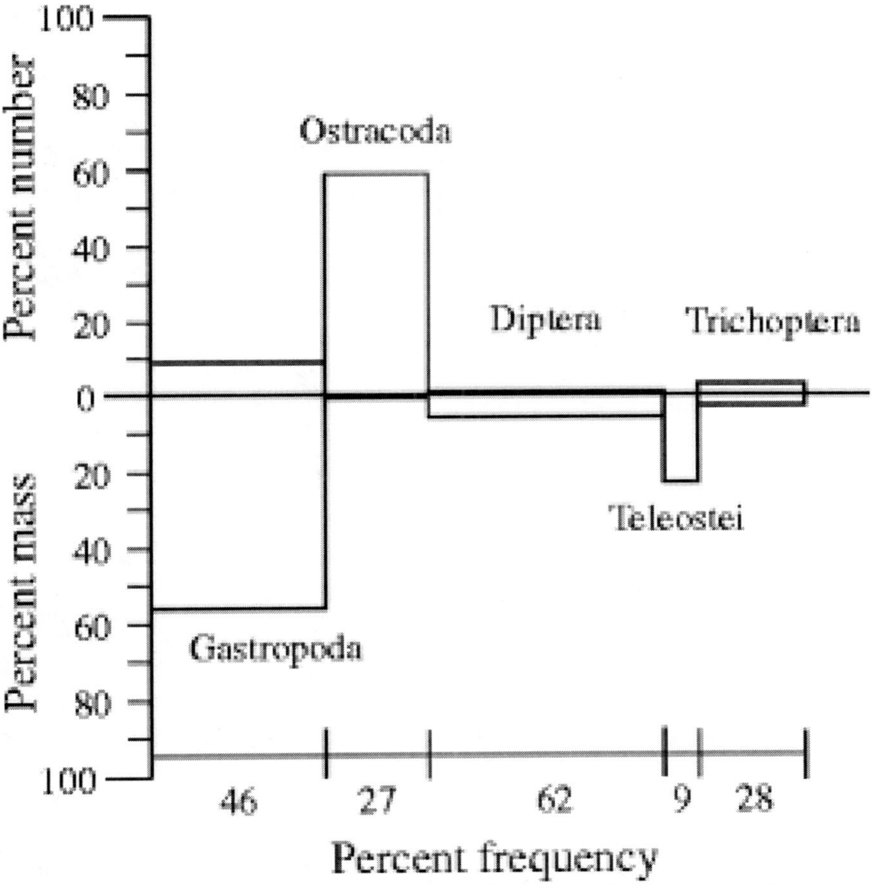

Figure. 4 Prey *IRI* diagram for combined sites and seasons. Only major prey categories ≥3 % *IRI* are shown. The *IRI* for each prey category is represented by a rectangle with area proportional to the values used in calculating the *IRI* = (%*N* + %*M*) * %*F*, where where *N* equals the number of individual prey items; *M* is the dry mass of prey items; and *F* is the number of stomachs containing a specific food organism. Frequency of occurrence axis begins at zero for each prey category

Stomach contents by site

Prey consumption differed by waterbody for seasons pooled (F (18, 490) = 4.56, $p < 0.001$; Table 1). A different prey taxon dominated gut contents for each site: dipterans (66 % *IRI*) in lake fish, gastropods (78 % *IRI*) in stream fish, and ostracods (59 % *IRI*) in wetland fish (Suppl.

Table 2 Prey values for combined sites and seasons ($n = 302$). Major and minor prey categories are given in descending order by percent *IRI*. Grey-colored rows are prey categories shown in *IRI* diagram (Fig. 4). Diptera, Odonata, Trichoptera, and Ephemeroptera represent larvae

Prey	Total no.	% no.	Total mass (mg)	% mass	% freq	IRI	% IRI
Gastropoda	1307	8.7	6697.817	56.2	46.4	3009	50.5
Ostracoda	8921	59.1	142.214	1.2	27.5	1658	27.8
Diptera	187	1.2	657.617	5.5	61.9	419	7.0
Teleostei	35	0.2	2729.331	22.9	9.3	216	3.6
Trichoptera	512	3.4	354.070	3.0	28.1	179	3.0
Bivalvia	262	1.7	495.639	4.2	25.2	148	2.5
Copepoda	643	4.3	10.288	0.1	30.5	132	2.2
Cladocera	1123	7.4	23.583	0.2	13.2	101	1.7
Angiospermae	172	1.1	172.860	1.5	14.2	37	0.6
Odonata	95	0.6	207.100	1.7	10.6	25	0.4
Coleoptera	37	0.2	237.826	2.0	7.6	17	0.0
Plumatellida	429	2.8	38.610	0.3	3.6	12	0.0
Corixidae	30	0.2	108.330	0.9	4.0	4	0.0
Gammaridae	28	0.2	12.852	0.1	5.3	2	0.0
Insecta adult	6	0.0	11.005	0.1	1.7	0	0.0
Araneae	6	0.0	7.533	0.1	2.0	0	0.0
Hydracarina	10	0.1	0.560	0.0	2.3	0	0.0
Ephemeroptera	6	0.0	1.304	0.0	1.7	0	0.0
Hirudinea	2	0.0	0.998	0.0	0.7	0	0.0
Unknowns	2	0.0	0.337	0.0	0.7	0	0.0
Collembola	1	0.0	0.052	0.0	0.3	0	0.0

Fig. 2, Suppl. Tables 3, 4, 5). All other prey values at each site were less than 20 % *IRI*. Teleosts contributed at least 60 % of the percent biomass in gut contents from lake fish, while gastropods contributed 70 % of overall biomass in gut contents for stream fish. No single taxon dominated by percent biomass for wetland fish. Dipteran larvae were the most frequently consumed prey of lake fish, compared to gastropods and dipteran larvae for stream fish. Wetland blackfish had the greatest variety of prey in their guts, consuming three prey categories at 50 % or greater frequency and three additional taxa at frequencies of 24 %–35 %. Fish ranked third in prey importance for both lake and stream blackfish but were of minor importance for wetland blackfish.

Stomach contents by season

Prey consumption differed by season (spring excluded) for waterbodies pooled (F (18, 490) = 3.56, $p < 0.001$; Table 1). During summer, gastropods were the single major prey (69 % *IRI*), while all other groups contributed less than 10 % *IRI* (Suppl. Tables 6). Gastropods also dominated in summer by biomass (79 % mass), followed by teleosts (10 % mass). More than half of all guts in summer contained gastropods and dipterans. Small ostracods and cladocerans outnumbered larger prey.

Dominant prey groups in autumn consisted of gastropods and ostracods, which contributed 48 % and 24 % of the percent *IRI* values, respectively (Suppl. Table 7). Gastropods were the most important prey in terms of biomass (61 % mass). At least 45 % of all guts in autumn contained dipterans, gastropods, trichopterans, and ostracods.

Prey in winter switched to Ostracoda (39 % *IRI*) followed by Diptera (24 % *IRI*) and Teleostei (19 % *IRI*; Suppl. Table 8). Fish were the most dominant winter prey by biomass (72 % mass), although their total count (13) was small compared to ostracods (4452). The most frequently consumed winter prey were dipteran larvae and copepods.

We chose to interpret a significant (randomization test $p = 0.004$) two-dimensional NMS solution with mean stress of 12 after verifying consistency of interpretation among the three NMS solutions. Blackfish with a greater mass of gastropods in their gut are associated with greater positive Axis 1 values; blackfish with a greater mass of teleosts in their gut are associated with greater positive Axis 2 values; blackfish with a greater mass of dipterans in their gut are associated with lower Axis 2 values (Fig. 5). Blackfish with more mixed diets and low mass of the six arthropod categories (including ostracods, copepods, dipterans, and trichopterans) are located in the center of the biplots, while blackfish with an empty gut are represented by the lowest Axis 1 value.

The biplots of waterbodies across the three seasons (summer, autumn, winter) show overlap in diet for many blackfish (Fig. 5). However, lake blackfish are concentrated on the upper-left side (greater mass of teleosts, and mixed arthropods); stream blackfish extend to the right side (greater mass of gastropods and teleosts, lower mixed arthropods), and wetland blackfish occupy the center portion of the biplots (greater mass of mixed arthropods, low mass of gastropods and teleosts). In general, diets varied considerably within each waterbody, though wetland blackfish had a less variable diet, especially in summer. Little difference was evident among the summer and fall biplots. In winter, however, diets for the stream and

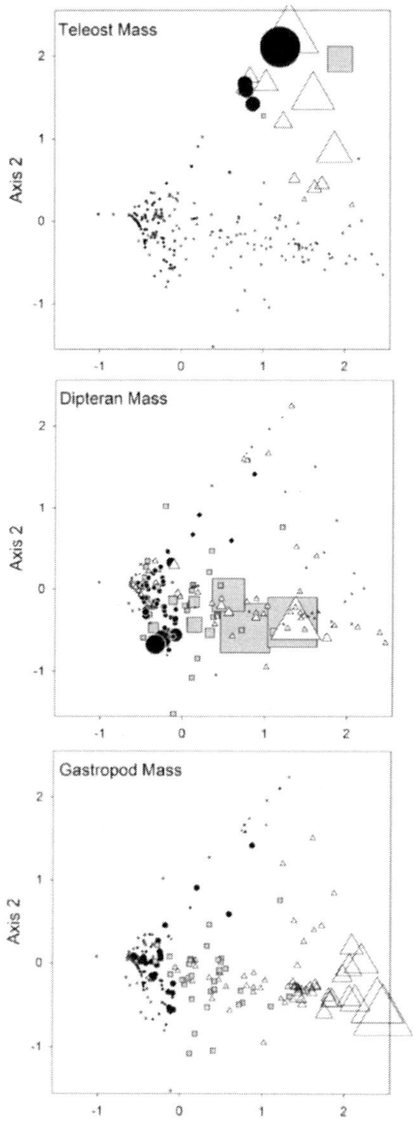

Figure. 5 Non-metric multidimensional scaling ordination biplots of blackfish diets by waterbody. Lake blackfish are shown as *black circles*, stream blackfish are shown as *triangles*, wetland blackfish are shown as *gray squares*. The three panels on the left show diets for all three seasons combined with symbol size scaled to mass of three prey categories (top = teleosts, middle = dipterans, bottom = gastropods). The three panels on the right show diets in the three waterbodies for summer (*top*), fall (*middle*), and winter (*bottom*). Minimum convex polygons are drawn around each waterbody

wetland blackfish constricted along Axis 1 and expanded along Axis 2. This is associated with a reduction in gastropods and increase in teleosts and copepods in their diets. Diets among fish in the three waterbodies converged more in winter. Lake blackfish showed little change in diet composition across seasons in the ordination.

Fish in diet

Gut contents from pooled waterbodies and seasons showed that 9 % of all blackfish had recently consumed a total of 35 fish (Table 2, Suppl. Table 9). Seven percent of the lake blackfish had fish in their guts—four threespine stickleback, one juvenile blackfish, and two unidentifiable fish (Suppl. Table 3). Stream blackfish were the most piscivorous (17.3 % frequency; Suppl. Table 4); fish prey included 10 threespine stickleback, four ninespine stickleback, one coho, and three unidentified fish. Piscivory among blackfish in the wetland was lowest at 3.5 % frequency across the year (2.2 % when spring samples are excluded; Suppl. Table 5), with threespine stickleback, juvenile blackfish, and one unidentifiable fish. Overall, the frequency of cannibalism for all waterbodies and seasons was 1.3 %, excluding unidentifiable prey fish. Less than 1 % of blackfish had salmonids in their guts.

The likelihood of fish in blackfish diet was related to both sex and blackfish size (Chi-square = 9.487, $df = 2$, $p = 0.009$). The Wald statistic showed that both sex ($p = 0.043$) and size ($p = 0.045$) made significant contributions to the model. Fish consumption first appeared in blackfish with a SL of at least 78 mm, although few blackfish this small had fish in their guts (Suppl. Table 9). Fish were more important in the diets of blackfish greater than 105 mm SL. Males tended to consume more fish than did females. Approximately 13 % of males consumed fish, while 6 % of females consumed fish.

DISCUSSION

Dietary diversity

Introduced populations of blackfish in the Cook Inlet Basin displayed an opportunistic, generalist diet consisting primarily of benthic epiphytic invertebrates—gastropods, ostracods, and dipteran larvae (Table 2). While their diets were diverse, we did not detect surface feeding in these populations, and consumption of adult insects was rare.

Blackfish diets varied by site and by season (Fig. 5). Lake blackfish consumed mainly dipterans, stream blackfish consumed mainly gastropods, while ostracods were the major prey of wetland blackfish. Important prey based on biomass included teleosts among lake fish and gastropods among stream and wetland fish. Such spatial variation in the food habits of blackfish supports the trophic model of a generalist feeder, whose diet consists mostly of a diversity of benthic invertebrates selected in part based on availability. The occasional consumption of fish by this species is also indicative of an opportunistic strategy (Gerking 1994), and the seasonal importance of fish prey in terms of biomass indicates that piscivory may be important energetically for some larger blackfish.

Diets varied significantly across seasons for combined sites, as expected (Suppl. Tables 6–8). Temporal prey shifts support a trophic model of adaptability based on seasonal prey availability. The high trophic flexibility of these blackfish may be partially responsible for their successful colonization of multiple freshwater habitats.

Gut contents of 77 blackfish collected during summer in its native range on the Arctic Coastal Plain of Alaska contained 17 prey categories including nematodes and algae (Ostdiek and Nardone 1959). In the current study, Cook Inlet Basin blackfish guts from summer contained 16 prey categories, with no evidence of nematodes or algae. Most frequently consumed prey of the Arctic Coastal Plain blackfish were cladocerans (91 %), dipteran larvae (90 %), and ostracods (88 %); fish consumption occurred at less than 3 % frequency. By comparison, Cook Inlet Basin blackfish during summer most frequently ate gastropods (57 %) and dipteran larvae (54 %), while fish consumption occurred at a frequency of 11 %.

A study of 320 Meadow Lakes (Anchorage) blackfish harvested during September reported gut contents as follows: major prey by relative frequency, Cladocera (59 %) and Copepoda (32 %), and six minor prey valued at less than 5 % frequency (Hemiptera, Diptera, Odonata, Teleostei, Mollusca, and Ephemeroptera; Chlupach 1975). By comparison, 67 Cook Inlet blackfish collected in autumn in the current study ate Diptera (79 %), Gastropoda (69 %), and Trichoptera (63 %) as well as six other prey types valued between 10 and 40 %. These results support the model of the blackfish as a generalist opportunist that feeds on a wide range of prey species, with variable sizes and morphologies, from more than one trophic level. Such low dietary specialization is often characteristic of successful fish invaders (Moyle and Marchetti 2006).

Management implications

Fish consumption

Fish consumption was infrequent among these three introduced blackfish populations (Table 2), indicating that blackfish are unlikely to impact native or stocked fish populations in the studied sites. In contrast, other studies reported dominance of fishes in blackfish diet. Stomachs of adult blackfish from native populations in Western Alaska contained mostly small blackfish and northern pike (Baxter 1973; unpublished, cited in Chlupach 1975). Stomach contents of 320 introduced blackfish from Meadow Lake (Anchorage) contained 132 fish identified as Salmoniformes (Chlupach 1975). Such spatial heterogeneity in blackfish diet suggests that although blackfish may not pose a major threat to native fishes or sportfishes in the three populations investigated in this study, they may impact fishes through predation in other Cook Inlet Basin sites and may have seasonal impacts that vary by site.

This study found a significant relationship between blackfish size and fish consumption. As expected, ontogenetic diet shifts were observed in that prey size increased with predator size; smaller juvenile blackfish consumed small invertebrates and no fish (Eidam 2015), while adult blackfish fed on a greater variety of both small and large prey including fish (Suppl. Table 9). Therefore, in habitats containing abundant small fishes and blackfish that grow to large sizes, the impact on fishes is predicted to be greater.

Fish consumption was also based on the sex of the blackfish. In these populations, males were more than twice as likely as females to consume fish (Suppl. Table 9). Blackfish males are slightly larger than females (Suppl. Fig. 1). Such sexual size dimorphism is frequently documented in vertebrates and can result in niche divergence (Shine 1989) in which diets of males and females differ (Holtby and Healey 1990; Houston and Shine 1993; Laufer et al. 2009; Keppeler et al. 2013).

Dietary overlap

Diets of introduced blackfish in Cook Inlet Basin fresh waters overlap with those of native fishes and stocked sportfish. Threespine stickleback feed on small benthic invertebrates including dipteran larvae, ostracods, mollusks, copepods, cladocerans, and amphipods (Hynes 1950; Greenbank and Nelson 1959), while slimy sculpin typically feed on slightly larger organisms on or just below the sediment—amphipods and larvae of dipterans, trichopterans, and odonates (Morrow 1980; Flecker 1984; Hershey 1985).

Juvenile Dolly Varden char forage on small crustaceans, insect larvae, snails, clams, spiders, and fish (Morrow 1980). Coho salmon fry consume microzooplankton, mites, Collembola, and spiders, while larger juveniles also eat adult beetles (Morrow 1980). Similarly, blackfish consume diverse epiphytic benthic prey across a wide size range. In contrast to blackfish, coho salmon fry feed heavily on surface insects including winged dipterans and trichopterans, and large adults can also become primarily piscivorous (Morrow 1980). Blackfish swim to the surface to breathe atmospheric air but are not known to eat surface insects.

Rainbow trout feeding habits also overlap to some extent with those of blackfish. Rainbow trout diet shifts ontogenetically from cladocerans for small juveniles to dipteran larvae and winged adults, leeches, amphipods, gastropods, water beetles, and fishes for large adults (Scott and Crossman 1973; Morrow 1980; Beauchamp 1990). Rainbow trout feed at the surface, in mid current, and sometimes at the bottom. Blackfish feed demersally by picking organisms off of benthic macrophytes or by probing sediment in search of buried clams and large dipteran larvae, using their protruding lower jaw like a scoop (Eidam 2015). In contrast, rainbow trout do not burrow for prey (Frost and Brown 1967; Knapp et al. 2001).

These comparisons indicate that the potential for resource competition between blackfish and some native fish species warrants further research, including quantitative analysis of dietary overlap on a species by species basis. The dietary breadth of blackfish detected in the current study could also be quantitatively compared to dietary breadth of both native and non-native species to test the hypothesis that a generalist feeding strategy facilitates invasion.

Blackfish as an invasive species

Alterations of aquatic invertebrate communities associated with the introduction of non-native fishes are a common occurrence (e.g., Goldschmidt et al. 1993; Gerking 1994; Knapp et al. 2001; Byström et al. 2007). Introduced brown trout (*Salmo trutta*) significantly decreased overall benthic biomass including larvae of dipterans and trichopterans (Macan 1966, 1977). Brook trout (*Salvelinus fontinalis*) stocked in a fishless lake in New York immediately impacted the benthic fauna, including eliminating *Chaoborus* dipteran larvae (Gloss et al. 1989). Selective feeding by introduced yellow perch (*Perca flavescens*) in a lake in Quebec changed the overall community structure, resulting in reduced populations of larger invertebrates and more abundant populations of smaller invertebrates (Berglund 1968; Crowder and Cooper 1982; Post and Cucin 1984). Removal of a fish predator in some cases caused measurable increases in benthic organisms (Gerking 1994). The magnitude of the alterations to the aquatic community by invasive fishes, however, is likely to be reflected in the abundance of the invaders.

Introduced blackfish are documented in large numbers in many Cook Inlet Basin lakes (K. Dunker, ADF&G, personal communication; Eidam 2015). However, to our knowledge no systematic studies of blackfish population sizes exist from either native or introduced locations. Although direct predation of salmonids was infrequent in this study, the substantial dietary overlap with native fishes and sportfishes indicates that blackfish may impact the ecology of invaded freshwater systems. We suggest that introduced populations of blackfish likely reduce prey availability to other fishes, as well as affect broader ecological interactions, notably competitive interactions, among aquatic invertebrate guilds, and potentially alter community composition (e.g., Chase and Knight 2003). To establish the scale of that impact, however, it will be necessary to produce robust estimates of the size of introduced blackfish populations, increase our understanding of competitive and facilitative relationships among aquatic invertebrates in these systems, and conduct manipulative experiments to estimate the effect of blackfish on growth rates of native fishes.

ACKNOWLEDGMENTS

Research reported in this publication was supported by an Institutional Development Award (IDeA) from the National Institute of General Medical Sciences of the National Institutes of Health (NIH) under grant number P20GM103395. The content is solely the responsibility of the authors and does not necessarily reflect the official views of the NIH. This work was also supported by the Union of Students of the University of Alaska Anchorage. All collections were undertaken in accordance with the University of Alaska Anchorage Animal Care and Use Committee (protocols 2007vonHil and 159870-1) and were authorized by permits from the Alaska Department of Fish and Game (SF2009-080, SF2009-256, SF2010-034, SF2011-026, SF2012-006). We appreciate the assistance of biologists and staff of the Alaska Department of Fish and Game. We greatly thank Stormy Haught, Christoff Furin, and James Willacker for valuable discussions; Justin Bullock for generous field assistance; and Emily Eidam for help with formatting maps and IRI diagrams.

REFERENCES

1. Alaska Department of Fish and Game (ADF&G) (2011) Fish stocking update. ADF&G Division of Sport Fish. http://www.adfg.alaska.gov /index.cfm?adfg=SportStockingHatcheriesSearch.main.
2. Bachok Z, Mansor MI, Noordin RM (2004) Diet composition and feeding habits of demersal and pelagic marine fishes from Terengganu waters, east coast of peninsular Malaysia. NAGA Worldfish Center Quarterly 27:41–47
3. Beauchamp DA (1990) Seasonal and diel food habits of rainbow trout stocked as juveniles in Lake Washington. Trans Am Fish Soc 119(3):475–482
4. Berglund T (1968) The influence of predation by brown trout on *Asellus* in a pond. Institute of Freshwater Research Drottningholm Report 48:77–101
5. Brown JH (1989) Patterns, modes and extents of invasions by vertebrates. In: Drake JA, Mooney HA, di Castri F, Groves RH, Kruger FJ, Rejmánek M, Williamson M (eds) Biological invasions: a global perspective. Wiley & Sons, Chichester, UK, pp. 85–109
6. Byström P, Karlsson J, Nilsson P, van Kooten T, Ask J, Olofsson F (2007) Substitution of top predators: effects of pike invasion in a subarctic lake. Freshw Biol 52:1271–1280
7. Cailliet G, Love M, Ebeling A (1986) Fishes—a field and laboratory manual on their structure, identification and natural history. Waveland Press, Prospect Heights, Illinois
8. Chase JM, Knight TM (2003) Drought-induced mosquito outbreaks in wetlands. Ecol Lett 6:1017–1024
9. Chipps SR, Garvey JE (2007) Assessment of food habits and feeding patterns. In: Brown ML, Guy CS (eds) Analysis and interpretation of freshwater fisheries data. American Fisheries Society, Bethesda, Maryland, pp. 473–514
10. Chlupach RS (1975) Studies of introduced blackfish in waters of south Alaska. Alaska Department of Fish and Game (ADF&G) Federal Aid Fish Restoration Annual Performance Report. Project F-9-7. Study G-II 16:62–78
11. Crawford RH (1974) Structure of an air-breathing organ and the swim bladder in the Alaska blackfish, *Dallia pectoralis* Bean. Can J Zool 52:1221–1225
12. Crowder LB, Cooper WE (1982) Habitat structural complexity and the interaction between bluegills and their prey. Ecology 63:1802–1813
13. Drake JA, Mooney HA, di Castri F, Groves RH, Kruger FJ, Rejmánek M, Williamson M (eds) (1989) Biological invasions. A global perspective. Wiley & Sons, Chichester, UK
14. Eidam DM (2015) Trophic ecology of introduced populations of Alaska blackfish (*Dallia pectoralis*) in the Cook Inlet Basin, Alaska. University of Alaska Anchorage, Thesis
15. Fasham MJR (1977) A comparison of nonmetric multidimensional scaling, principal components and reciprocal averaging for the ordination of simulated coenoclines, and coenoplanes. Ecology 58:551–561
16. Flecker AS (1984) The effects of predation and detritus on the structure of a stream insect community: a field test. Oecologia 64(3):300–305
17. Frost WE, Brown ME (1967) The trout. Collins, London
18. Fuller PL, Nico LG, Williams JD (1999) Nonindigenous fishes introduced into inland waters of the United States. Special Publication 27, American Fisheries Society, Bethesda, Maryland
19. Garvey JE, Wright RA, Stein RA (1998) Overwinter growth and survival of age-0 largemouth bass: revisiting the role of body size. Can J Fish Aquat Sci 55:2414–2424
20. Gerking SD (1994) Feeding ecology of fish. Academic Press, San Diego, California

21. Gido KB, Franssen NR (2007) Invasion of stream fishes into low trophic positions. Ecol Freshw Fish 16(3):457–464
22. Gloss SP, Schofield CL, Spateholts RL, Plonski BA (1989) Survival, growth, reproduction, and diet of brook trout (*Salvelinus fontinalis*) stocked into lakes after liming to mitigate acidity. Can J Fish Aquat Sci 46:277–286
23. Goldschmidt T, Witte F, Wanink J (1993) Cascading effects of the introduced Nile perch on the detritivorous/phytoplanktivorous species in the sublittoral areas of Lake Victoria. Conserv Biol 7(3):686–700
24. Greenbank J, Nelson PR (1959) Life history of the threespine stickleback *Gasterosteus aculeatus* Linnaeus in Karluk Lake and Bare Lake, Kodiak Island, Alaska. Alaska Fisheries Bulletin US Fish and Wildlife Service 59(153):537–559
25. Hepler KR, Bowden CF (1986) Annual performance report for Anchorage urban trout studies: Cook Inlet trout. Study T-3. Alaska Department of Fish and Game (ADF&G) Sport Fish Division 27:20–38
26. Hershey AE (1985) Effects of predatory sculpin on the chironomid communities in an arctic lake. Ecology 66(4):1131–1138
27. Holtby LB, Healey MC (1990) Sex-specific life history tactics and risk-taking in coho salmon. Ecology 7:678–690
28. Houston D, Shine R (1993) Sexual dimorphism and niche divergence: feeding habits of the Arafura file snake. J Anim Ecol 62:737–748
29. Hynes HBN (1950) The food of fresh-water sticklebacks (*Gasterosteus aculeatus* and *Pygosteus pungitius*), with a review of methods used in studies of the food of fishes. J Anim Ecol 19:36–58
30. Keppeler FW, Lanés LEK, Rolon AS, Stenert C, Maltchik L (2013) The diet of *Cynopoecilus fulgens* Costa, 2002 (Cyprinodontiformes: Rivulidae) in southern Brazil wetlands. Italian Journal of Zoology 80(2):291–302
31. Knapp RA, Matthews KR, Sarnelle O (2001) Resistance and resilience of alpine lake fauna to fish introductions. Ecol Monogr 71(3):401–421
32. Laming PR, Funston CW, Armstrong MJ (1982) Behavioural, physiological, and morphological adaptations of the shanny (*Blennius pholis*) to the intertidal habitat. J Mar Biol Assoc U K 62:329–338
33. Lassuy DR (1995) Introduced species as a factor in extinction and endangerment of native fish species. Am Fish Soc Symp 15:391–396
34. Laufer G, Arim M, Loureiro M, Piñeiro-Guerra JM, Clavijo-Baquet S, Fagúndez C (2009) Diet of four annual killifishes: an intra and interspecific comparison. Neotropical Ichthyology 7(1):77–86
35. Liem KF (1967) Functional morphology of the integumentary, respiratory and digestive systems of the synbranchoid fish, *Monopterus albus*. Copeia 2:375–388
36. Liem KF (1987) Functional design of the air ventilation apparatus and overland excursions by teleosts. Fieldiana: Zoology 37:1–29
37. Macan TT (1966) The influence of predation on the fauna of a moorland fishpond. Arch Hydrobiol 61:432–452
38. Macan TT (1977) The influence of predation on the composition of freshwater animal communities. Biol Rev 52:45–70
39. Marchetti MP, Moyle PB, Levine R (2004) Alien fishes in California watersheds: characteristics of successful and failed invaders. Ecol Appl 14:587–596
40. McCafferty WP (1998) Aquatic entomology: the fisherman's and ecologist's illustrated guide to insects and their relatives. Jones and Bartlett Publishers, Sudbury, Massachusetts
41. McClory J, Gotthardt T (2008) Non-native and invasive animals of Alaska: a comprehensive list and select species status reports. Alaska Natural Heritage Program Environment and Natural Resources Institute, University of Alaska Anchorage, Final report
42. McCune B, Mefford MJ (2006) PC-ORD. Multivariate analysis of ecological data, version 5. MjM Software Design, Gleneden Beach

43. Mecklenburg CW, Mecklenburg TA (2002) Fishes of Alaska. American Fisheries Society, Bethesda, Maryland
44. Merritt RW, Cummins KW, Berg MB (2008) An introduction to the aquatic insects of North America, 4th edn. Kendall/Hunt Publishing Company, Dubuque, Iowa
45. Mooney HA, Cleland EE (2001) The evolutionary impact of invasive species. Proc Natl Acad Sci 98(10):5446–5451
46. Morrow JE (1980) The freshwater fishes of Alaska. Alaska Northwest Publishing Company, Anchorage, Alaska
47. Moyle PB, Light T (1996a) Fish invasions in California: do abiotic factors determine success? Ecology 77:1666–1670
48. Moyle PB, Light T (1996b) Biological invasions of freshwater: empirical rules and assembly theory. Biol Conserv 78:149–161
49. Moyle PB, Marchetti MP (2006) Predicting invasion success: freshwater fishes in California as a model. Bioscience 56(6):515–524
50. Ostdiek JL, Nardone RM (1959) Studies on the Alaskan blackfish *Dallia pectoralis*. 1. Habitat, size and stomach analyses. Am Midl Nat 61(1):218–229
51. Pauly D, Christensen V, Dalsgaard J, Froese R, Torres F Jr (1998) Fishing down marine food webs. Science 279:860–863
52. Peck J (2010) Mulivariate analysis for community ecologists: step-by-step using *PC-ORD*. MjM Software Design, Gleneden Beach, Oregon
53. Pimentel D (2007) Environmental and economic costs of vertebrate species invasions into the United States. Managing vertebrate invasive species. Paper 38. USDA National Wildlife Research Center Symposia, University of Nebraska
54. Pinkas L, Oliphant MS, Iverson ILK (1971) Food habits of albacore, bluefin tuna and bonito in California waters. California Department of Fish and Game Fish Bulletin 152:1–105
55. Post JR, Cucin D (1984) Changes in the benthic community of a small Precambrian lake following the introduction of yellow perch, *Perca flavescens*. Can J Fish Aquat Sci 41(10):1496–1501
56. Scholander PF, Flagg W, Hock RJ, Irving L (1953) Studies of the physiology of frozen plants and animals in the arctic. J Cell Comp Physiol 42:1–56Suppl. 1
57. Scholander PF, Van Dam L, Kanwisher JW, Hammel HT, Gordon MS (1957) Supercooling and osmoregulation in arctic fish. Journal of Cellular Comparative Physiology 49:5–24
58. Scott WB, Crossman EJ (1973) Freshwater fishes of Canada. Bulletin 184. Fisheries Research Board of Canada, Ottawa
59. Shine R (1989) Ecological causes for the evolution of sexual dimorphism: a review of the evidence. Q Rev Biol 64(4):419–461
60. Stergiou KI, Karpouzi VS (2002) Feeding habits and trophic levels of Mediterranean fish. Rev Fish Biol Fish 11:217–254
61. Stobberup KA, Morato T, Amorim P, Erzini K (2009) Predicting weight composition of fish diets: converting frequency of occurrence of prey to relative weight composition. The Open Fish Science Journal 2:42–49
62. Stratton B, Cyr P (1997) Annual management report for the Anchorage area, 1995. Alaska Department of Fish and Game. Fishery Management Report No. 97–1. Anchorage
63. Strong DR, Pemberton RW (2000) Biological control of invading species—risk and reform. Science 288(5473):1969–1970
64. Thorp JH, Covich AP (2001) Ecology and classification of North American freshwater invertebrates, 2nd edn. Academic Press Incorporated, San Diego, California
65. Trent TT, Kubik SW (1974) Studies of introduced blackfish in waters of south Alaska. Alaska Department of Fish and Game. Federal Aid Fish Restoration, annual report of performance. Study G-II 15:81–85

66. Vander Zanden MJ, Shuter JB, Lester PN (2000) Within-and among-population variation in the trophic position of a pelagic predator, lake trout (*Salvelinus namaycush*). Can J Fish Aquat Sci 57:725–731
67. Wilcove DS, Bean MJ (1994) The Big Kill: declining Biodiversity in America's Lakes and Rivers. Washington, DC: Environmental Defense Fund
68. Wilcove DS, Rothstein D, Dubow J, Phillips A, Losos E (1998) Quantifying threats to imperiled species in the United States. Bioscience 48:607–615

CHAPTER 6

Ecology and Fish Biodiversity of Man-Made Lakes of Southern Benin (West Africa): Implications for Species Conservation and Fisheries Management

Houehanou M. A. G. Gbaguidi[1], Alphonse Adite[1*], Edmond Sossoukpe[2]

[1]Laboratoire d'Ecologie et de Management des Ecosystèmes Aquatiques (LEMEA), Département de Zoologie, Faculté des Sciences et Techniques, Université d'Abomey-Calavi, Cotonou, Bénin

[2]Unité de Recherches sur les Zones Humides, Département de Zoologie, Faculté des Sciences et Techniques, Université d'Abomey-Calavi, Cotonou, Bénin

ABSTRACT

Man-made lakes are alternative and potential habitats for biodiversity conservation, fisheries and extensive aquaculture. We investigated the ecology and the fish community structure of two (2) sand-dragged man-made lakes, Lake Ahozon and Lake Bewacodji of Southern Benin, with implications for species conservation, fisheries management and aquaculture valorization. From August 2014 to July 2015, habitats conditions were evaluated and fishes were sampled monthly with seine, cast net, experimental gill net and hooks in the open water and aquatic vegetation habitats of both lakes. Overall, the water quality of Lake Ahozon was globally favorable for the growth and the survival of the fish resources whereas Lake Bewacodji exhibited a poor water quality indicated mainly by an acid pH (mean: 6.32 ± 0.58) and low dissolved oxygen concentrations (mean: 3.52 ± 1.25 mg/l) caused by dense floating plants, Nymphea sp mainly and huge daily dumping of domestic wastes. The study revealed low species richness, d = 5.89 and d = 3.87, and low species diversity, H' = 0.76 and H' = 0.48 for Lakes Ahozon and Bewacodji, respectively, with Lake Ahozon more diverse than Lake Bewacodji. The fish community of Lake Ahozon comprised six (6) species, 3 cichlids Sarotherodon galilaeus, Oreochromis niloticus and Tilapia guineensis, the silver catfish, Chrysichthys nigrodigitatus (Claroteidae), the African bonytongue, Heterotis niloticus (Osteoglossidae), and the African catfish, Clarias gariepinus (Clariidae). Numerically, S. galilaeus dominated Lake Ahozon and made 85.21% of the sample. In Lake Bewacodji, the fish composition comprised four (4) species, Sarotherodon galilaeus multifasciatus, the dominant species making numerically 91.58% of the total sample, T. guineensis, C. gariepinus and C. nigrodigitatus. With regard to trophic structure, the fish assemblages of both lakes were numerically dominated by planktinovores/ detritivores, mainly S. galilaeus, O. niloticus, T. guineensis and C. nigrodigitatus making together 99.46% of Lake Ahozon fish community, and S. galilaeus multifasciatus, T. guineensis and C. nigrodigitatus accounting together for about 98.59% of Lake Bewacodji. In Lake Ahozon, standard length (SL) frequencies histograms showed an unimodal size distribution for H. niloticus, the nile tilapia O. niloticus and C. gariepinus whereas the two cichlids, S. galilaeus and T. guineensis exhibited a bimodal size distribution. In Lake Bewacodji, S. galilaeus multifasciatus, C. nigrodigitatus and C. gariepinus exhibited an unimodal size distribution. A sustainable exploitation of both man-made lakes requires the implementation of an integrated management scheme which should include habitat restoration and protection plan, fisheries/aquacultural valorization, ecological sound agriculture/ecotourism and environmental monitoring.

Keywords: Conservation, Diversity, Domestic Wastes, Fisheries, Habitat Degradation, Management, Man-Made Lakes, Sand-Dragging, Sarotherodon galilaeus

1. INTRODUCTION

In most region of Africa, fisheries production and yields have drastically decreased and catches per unit of effort (CPUE) have become low in natural water bodies such as rivers, stream, lakes, lagoons etc. [1] - [3] . Major threats were overfishing caused by the increasing fishermen population, grassroots poverty, the use of sophisticated and prohibited fishing gears and fishing techniques, pollution, alteration of water quality, mangrove destructions, habitat losses and degradation, introduction of non-native species, multiple utilizations (agriculture, aquaculture, sand dragging etc.) of water bodies, invasion of floating plants (Eichhornia crassipes), climate changes and modification of flooding regimes [4] - [6] . Therefore, an alternative response is probably the use of various non conventional medium-environments such as the numerous man-made lakes to develop the sector of fisheries and aquaculture in order to reduce the fishing pressures in natural inlands waters and to contribute to ecosystem recovery and species conservation [7] - [9] .

In Benin, man-made lakes include 1) few lakes generated from electrical and agricultural projects such as the hydro electrical dams of Nangbeto belonging to Benin and Togo and the hydro agricultural dams of Savè Sugar Company (SOCUBE), 2) constructed small ponds that hold running waters used to supply rearing animals, mainly cows, with drinking water, and 3) man-made lakes originated from sand-dragging activities that supply building enterprises with sand [10] - [12] . This last category appears to be widespread and dominated the man-made lakes in Benin. Indeed, following the excavation of grasslands and dry lands for sand provisions, running waters progressively accumulated in the resulting holes to generate some man-made lakes that were generally neglected, abandoned and considered as useless or lost lands by the owners. Thus, a benefic management of these lakes was scant and most of them underwent daily dumping of domestic wastes that caused lake pollution along with profound changes in water quality and fish community structures [13] - [15] . However, these artificial lakes could be managed to increase inlands fisheries, to develop aquaculture industries, to conserve fish species and to implement other various economical activities that could alleviate poverty and improve grassroots revenues in Benin [16] - [18] . In particular, unlike constructed classic small fishponds of reduced sizes, man- made lakes exhibited relatively large area and could be used as alternative breeding grounds for some species in order to produce juveniles (seeds) to supply fish culture centers with fingerlings. Indeed, some delicate species, such as the African bonytongues, Heterotis niloticus (Osteoglossidae), the silver catfish, Chrysichthys nigrodigitatus (Claroteidae) and the African catfish, Clarias gariepinus (Clariidae) etc., having a high potential in aquaculture, do not reproduce or barely reproduce in captivity in small fishponds [19] - [21] . In particular, to date, no scientific data is available on the captive breeding and on the artificial reproduction of Heterotis niloticus. Therefore, the man-made lakes could be an alternative appropriate medium-environment to experiment the breeding of such species, in order to get fingerlings for aquaculture centers. In general, these species required a relatively large water space under a special hydrologic regime, that is, the succession of low-water and high- water seasons that stimulate gonad maturations and egg hatchings.

Furthermore, these lakes can serve as fish conservation biota or as a biological reserve, especially for threatened and endangered species [22] - [24] . Such species could be reproduced and the juveniles reintroduced in their natural habitats [25] [26] . Though artificial, these lakes harbor an important naturally-colonized or introduced biodiversity such as aquatic plants, fishes, amphibians, micro invertebrates, insects, mollusks etc. that could serve for scientific researches and environmental education tools [27] . From a quick exploration during the wet season of 2014, Djihouessi [28] reported some fisheries data dominated in both lakes by S. galilaeus. When well managed, these lakes could serve for sustainable and ecological agriculture and ecotourism.

However, despite these various possible utilizations and valorizations of the man-made lakes, little is known about the ecology and the fish community structure of the sand-dragged man-made lakes of Benin. This information is badly needed to implement an integrated management (including fisheries and aquaculture valorization) of the multitude abandoned and neglected man-made lakes of Benin.

The current research was undertaken to study the ecology and the fish community structure of some man- made lakes of Southern Benin (West Africa) with implications for species conservation, fisheries management and aquaculture valorization. Specific objectives were 1) to evaluate the environmental quality of these lakes; 2) to investigate the fish biodiversity and community structures; 3) to evaluate relationships between the environment traits and the fish community indices; and 4) to design an integrated management scheme of the man-made lakes investigated.

2. MATERIAL AND METHODS

2.1. Study Region

The study area comprised two (2) man-made water bodies, Lake Ahozon (06°22'52"N; 002°10'34"E) and Lake Bewacodji (06°23'10"N; 002°12'11"E), both located in Southern Benin, Ouidah City, Pahou County (Figure 1).

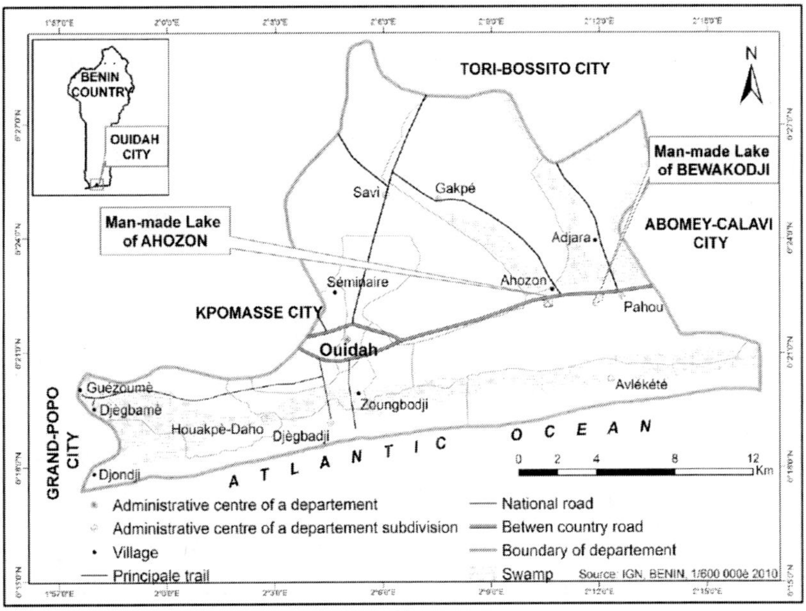
TORI-BOSSITO CITY
N
BENIN COUNTRY
OUIDAH CITY
Man-made Lake of BEWAKODJI
Savi
Gakpé
Man-made Lake of AHOZON
ABOMEY-CALAVI CITY
Adjara
Séminaire
Ahozon
KPOMASSE CITY
Pahou
Ouidah
Guézoumè
Djègbamè
Avlékété
Zoungbodji
GRAND-POPO CITY
Houakpè-Daho
Djègbadji
ATLANTIC OCEAN
0 2 4 8 12 Km
Djondji
Administrative centre of a departement
Administrative centre of a departement subdivision
Village
Principale trail
National road
Betwen country road
Boundary of departement
Swamp
Source: IGN, BENIN, 1/600 000è 2010

(a)

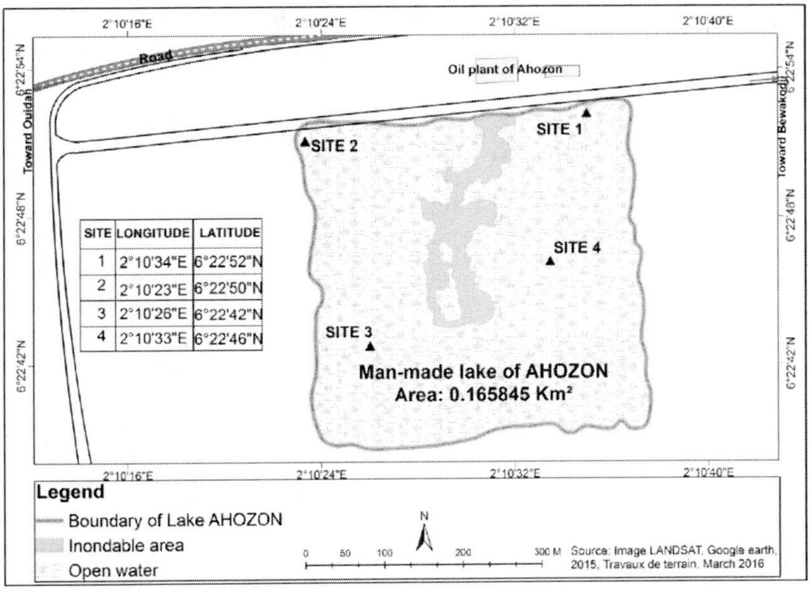
2°10'16"E
2°10'24"E
2°10'32"E
2°10'40"E
Road
Oil plant of Ahozon
6°22'54"N
Toward Ouidah
Toward Bewakodji
SITE 1
SITE 2
6°22'48"N
SITE LONGITUDE LATITUDE
1 2°10'34"E 6°22'52"N
2 2°10'23"E 6°22'50"N
3 2°10'26"E 6°22'42"N
4 2°10'33"E 6°22'46"N
SITE 4
SITE 3
6°22'42"N
Man-made lake of AHOZON
Area: 0.165845 Km²
Legend
Boundary of Lake AHOZON
Inondable area
Open water
N
0 50 100 200 300 M
Source: Image LANDSAT, Google earth, 2015, Travaux de terrain, March 2016

(b)

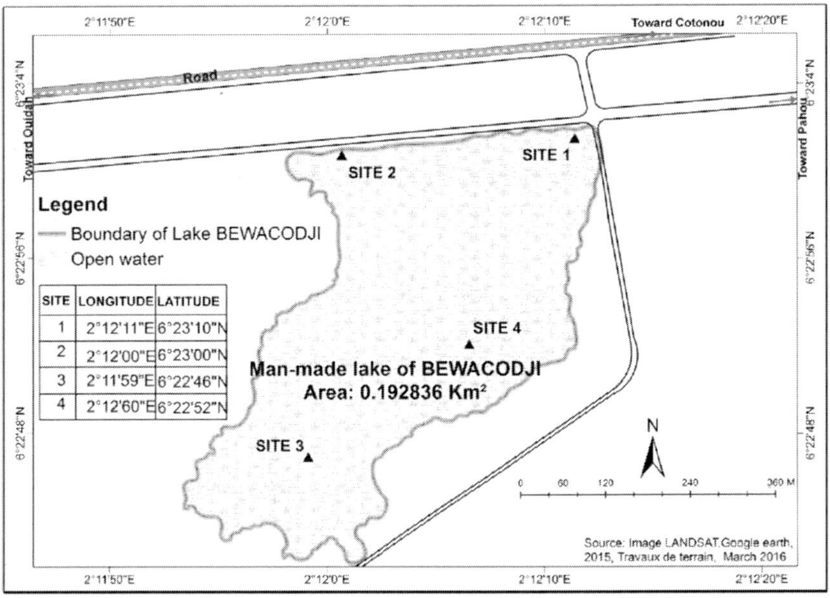

(c)

Figure 1. Map showing: (a) Ouidah City in South-Benin and the two study locations; (b) the man-made lake of Ahozon and (c) the man-made lake of Bewacodji. Site 1 and Site 2 = aquatic vegetation habitats; Site 3 and Site 4 = Open water habitats.

Ouidah City covers about 364 km^2 and has on its South, the Atlantic Ocean and the mangrove-lined coastal lagoon [25] . The climate is sub equatorial with two (2) wet seasons (April to July; mid-September to October) and two (2) dry seasons (December to March; mid-August to mid-September) with a peak usually recorded in June. Yearly mean rainfall reached 1307.3 mm and ambient temperatures ranged between 25°C and 33.6°C. Monthly evaporation varied between 59.2 - 145 mm [26] [27] . The study region comprised sandy soils, red ferric soils and a vast swampy land located at the coastal zone and extended from Cotonou City (South-Benin) to Ouidah. Currently, this swamp undergoes permanent sand-dragging activities, considered as an ecological disaster. Plantations such as Elaeis guineensis (palm tree) and Coco nucifera (coconut tree), producing oil for industries, were common in the study region [26] [27] . Also, intense agriculture dominated by corn, tomatoes, beans, groundnuts etc. occurred in the study area. Furthermore, multispecies fisheries dominated Ouidah City and took place in the coastal lagoon and in some freshwater lakes such as Toho-Todougba, Ahouangan and Dati. Though not widespread, fish culture was encountered in the study region with couple of fish ponds constructed in the swamps and wetlands.

2.2. Description of the Man-Made Lakes

The two man-made lakes, Lake Ahozon and Lake Bewacodji, about 5 km apart, are located between two water bodies, Toho-Todougba (15 km^2), a freshwater lake, and the coastal lagoon (60 km^2), a brackish water. These man-made lakes originated from sand-dragging activities to supply building enterprises with sand. Indeed, because the government has prohibited the exploitation of marine sand, source of severe ecological disasters at the Atlantic seashore, the alternative was the excavation of grasslands, dry lands or even wetlands to get sand. Thus, running waters from rainfall progressively accumulated in the resulting holes to generate artificial lakes, also called man-made lakes. These kinds of lakes are currently widespread, but generally neglected, abandoned, and considered as useless or lost lands by most owners. GPS tracking reveal that Lake Bewacodji (Figure 1) covered about 0.1928 km^2 and showed dense floating plants on the surface. In particular, this lake undergoes a daily dumping of domestic wastes that negatively affect the water quality and its productivity. The artificial lake of Ahozon extended on about 0.1658 km^2 (Figure 1) and in contrast with Lake Bewacodji, was exempt of floating plants and no domestic waste dumping was recorded during the study period. Dominant phytoplankton genera in Lake Ahozon were Navicula, Peridinium, Scennedesmus, Pinnularia, Spirogyra, Cosmarium, Melosira, Synechocystis, Microcystis, Oscillatoria, Euglena, Phacus, Surirella and Lychmophora. Zooplanktons in Lake Ahozon were dominated by Copepods, Trichocerca, Keratella, Brachiomus and. In Lake Ahozon, benthic macro invertebrates recorded were chironomid larvae and a Gasteropod mollusk, Melanoïde tubercularis. Aquatic vegetation in Lake Ahozon was dominated by Cyperus crassipes, Cyperus rotundus, Fuirena umbellata, Andropogon gayanus, Ludwigia perennis, Emilia praetermissa, Eleocharis complanata, Enydra fluctuans et Mariscus ligularis. In Lake Bewacodji, dominant phytoplanktons were Lychmophora, Spirogyra, Cosmarium, Euglena, Microcystis and Oscillatoria and zooplanktons comprised mainly Daphnia, Tropocyclops, Keratella, Brachiomus and Copepods. Also, aquatic vegetation in lake Bewacodji comprised mainly Ludwigia stolonifera, Utricularia reflexa, Ludwigia octovalois, Ipomea aquatica, Paspalum vaginatum, Nymphaea lotus, Fuirena umbellata, Cyperus articulatus and Andropogon gayanus.

In both lakes, subsistence fisheries occurred and were practiced by a couple of migrant fishermen. Around Lake Ahozon, subsistence agriculture involving maize, salads, beans, carrots lettuce, tomatoes and cucumbers sporadically take place.

2.3. Sampling Sites

In both Lakes, Lake Ahozon and Lake Bewacodji, four (4) sampling sites were selected for this study, two sites in the "open water" habitat and two sites in the "aquatic vegetation" habitat (Figure 1). The "open water" habitat exhibited a relatively high depth and high water velocity, but exempt of vegetation. In contrast, the "aquatic vegetation" habitat, the edge of the lake, was shallow and characterized by a low water velocity and relatively dense vegetation.

2.4. Evaluation of Habitat Characteristics

The environment conditions of each man-made lake were evaluated. Evaluation were made on soil type, aquatic vegetation, water types, dominant terrestrial habitats, utilization of the lakes and adjacent lands, substrates, pollutions, and water physicochemical conditions [28] . Water samples were collected with a hydrobios sampler. Water depth was measured to the nearest millimeter using a graduated rope and turbidity was measured to the nearest millimeter with a secchi disk. Temperature and dissolved oxygen were measured to the nearest 0.1°C and 0.1 $mg \cdot l^{-1}$respectively, with a digital oxythermometer (Model HANNA), and the pH was evaluated to the nearest 0.1 with a waterproof pH meter (Model HANNA). Conductivity was measured to the nearest 1.0 µs/cm with a conductimeter and salinity was measured to the nearest 1‰ with a model VISTA refractometer. Aquatic vegetation was sampled and identifications were made in the Department of Botany of the Faculty of Sciences and Technics at the "Université d'Abomey-Calavi". Seasonally (flood, dry, wet), phytoplankton and zooplankton were sampled with a planktonic net and the macro invertebrates were collected with a dip net. Phytoplankton, zooplankton and macro invertebrates were identified to the lowest possible taxonomic level using the identification keys of Needham [29].

2.5. Fish Sample Collection

Fishes were sampled in the "aquatic vegetation" habitat and in the "open water" habitat using appropriate set of fishing gears and methods so that common fish species would be well represented and samples reflect the relative abundance of the fish community. Likewise, the use of various collecting gears helped not only to increase the likelihood of uncommon species to occur in the fish sample, but also to assess the whole fish biodiversity of these man-made lakes [30] . At the "open water" and "aquatic vegetation" habitats, fishes were sampled twice a month from August 2014 to July 2015 with cast nets (9.80 m-diameter, 4.90 m-height,

40 mm-mesh), seines (4.15 m-length × 1.77 m-width, 3 mm-mesh), hooks (90 m-length) and experimental gill net (40 m × 1.05 m, 40 mm mesh). Cast nets were used in the "open water" with the help of fishermen. The net, once casted covers a defined area, then the cast net was pulled out delicately in the boat and all trapped fishes were then removed by hands. Seine hauls in marginal aquatic vegetation were made by setting the seine stationary, and kicking the vegetation to drive the fish in to the net before lifting it [5] . At each sampling site, ten rounds of seining were done. Hooks and gill nets were set in the "open water", yet, contribute very little to the total sample. Samplings with gill nets and hooks were made by attaching the net to the sticks and left it for 12 hours. Aggregated samples from cast nets, seining, gill nets and hooks were gathered to assess the whole fish assemblages of these man- made lakes.

Once caught, the fish samples were identified, measured, weighted and preserved in situ in 10% formalin. Species identification was based on references such as Leveque et al. [31] , Van Thielen et al. [32] , and Lowe McConnell [33] . The fish samples were then transported to the "Laboratoire d' Ecologie et de Management des Ecosystèmes Aquatiques" and preserved in 70% ethanol for further observation on the feeding ecology and some aspects of the reproductive ecology of keys species.

2.6. Data Analysis

Both physicochemical and fish community data were recorded in SPSS spreadsheet [34] . Means, ranges and standard deviations of physicochemical features were computed and recapitulated on a table. Relative to fish community data, size ranges and means of fish standard length (SL) and weight (W) were computed using SPSS computer software [34] . Species abundance and relative abundance were computed to indicate the numerical importance of each species in the sample. Species richness (d) was determined following Margalef index [35] :

$$d = S - 1/\ln N$$

where S is the number of species, and N the number of individuals in the sample. Species diversity (H') was determined following Shannon & Weaver index of diversity [36] :

$$H' = -\sum(p_i) * \log_2(p_i)$$

where H' is the index of species diversity, $p_i = n_i/N$, the proportion of total sample belonging to ith species, n_i the number of individuals of each species in the sample, N the total number of individuals of all species in the sample. The evenness measure (J' of Shannon & Weaver [36] was computed following the formula:

$$J' = H'/\log_2 S$$

where ***H'*** is the Shannon & Weaver index [36] of diversity, S is the number of species in the sample. For this study, these three indexes (richness, diversity, evenness) were computed for each habitat. I used SPSS computer program [34] to generate species diversity, species richness and evenness indexes.

The trophic structure of the fish community was determined by categorizing fishes in five trophic categories, 1) detritivores, 2) planktinovores/microcarnivores, 3) herbivores, 4) intermediate carnivores and 5) top-carnivores [3] [12] . Previous studies from Halliday &Young [37] , Adite & Winemiller [38] , on diet were used for this classification. The importance of each trophic category was appreciated using the relative abundance of fish species. Relationships between the fish community traits and the physicochemical features were explored through linear regression models and F-test. Frequency histograms of fish standard length were constructed by habitat type for dominant species.

3. RESULTS

3.1. Hydrology and Water Characteristics

The two man-made lakes, Lake Ahozon and Lake Bewakodji, originated from the excavation of grasslands and dry lands, consequently, water accumulation and hydrological regime depended solely on rainfall and running waters. Both water bodies were freshwater lakes and lacked connection with coastal brackish waters and any other freshwater lakes. However, these man-made lakes are located about 5 km from a natural freshwater lake, Toho-Todougba Lagoon (16 km^2). Both lakes showed sandy-muddy bottoms with dominance of sand in Lake Ahozon. In general, significant variations ($p < 0.05$) were recorded between the water quality of both lakes.

In the man-made lake of Ahozon, depths varied from 16.2 to 240 cm (mean: 80.19 cm), transparencies from 16.2 to 60.5 cm (mean: 36.63 cm). Water temperatures ranged between 28.2°C and 38.7°C (mean: 33.25°C), pH between 6.7 and 9.7 (mean: 7.51), dissolved oxygen concentrations between 0.73 and 11.8 mg/l (mean: 5.43 mg/l), and conductivitiy between 50 and 560 μ/cm (mean: 240 μ/cm) (Table 1). In Lake of Bewacodji, depths varied from 31.1 to 95.5 cm (mean: 67.99 cm), transparencies from 31.1 to 95.5 cm (mean: 53.83 cm). Water temperatures ranged between 28.8°C and 37.9°C (mean: 33.41°C), pH between 5.2 and 7.1 (mean: 6.32), dissolved oxygen between 1.32 and 5.53 mg/l (mean: 3.52 mg/l), and conductivity between 30 and 410 μ/cm (mean: 146 μ/cm) (Table 1).

In both man-made lakes, significant spatial variations ($p \leq 0.001$) of physicochemical parameters were recorded. Indeed, except conductivity, the remaining water features, namely depth, transparency, temperature, dissolved oxygen, percent saturation, and pH values were higher in the open water habitat than in the aquatic vegetation habitat (Table 1). Likewise, seasonal variations of the water features were recorded in both Lakes with depths and transparencies higher in wet season in Lake Ahozon where they reached 240 cm and 60.5 cm, respectively. In Lake Bewacodji, depths and transparencies were higher in the flood season and reached 95.5 cm and 65.5 cm, respectively. Inversely, temperatures were higher during the dry periods in both lakes and reached 38.7°C and 37.8°C in Lake Ahozon and Lake Bewacodji, respectively.

3.2. Fish Community Composition and Abundance

A total of 8845 individual fishes were sampled from both artificial lakes during the study period, August 2014 - July 2015.

Table 1. Ranges and mean values (±SD) of the water parameters measured from August 2014 to July 2015 in the man-made lakes of Ahozon and Bewacodji in Ouidah city, Southern Benin.

Physicochemical parameters	Artificial lake of Ahozon			Artificial lake of Bewacodji		
	Mean	Range	±SD	Mean	Range	±SD
Depth (cm)	80.19	16.2 - 240	56.82	67.99	31.1 - 95.5	15.33
Transparency (cm)	36.63	16.2 - 60.5	11.18	55.83	31.1 - 95.5	14.69
Temperature (°C)	33.25	28.2 - 38.7	2.50	33.41	28.8 - 37.9	2.87
pH	7.51	6.7 - 9.7	0.78	6.32	5.2 - 7.1	0.58
Dissolved oxygen (mg/l)	5.43	0.73 - 11.8	2.52	3.52	1.32 - 5.53	1.25
O_2 percent of saturation (%)	82.69	10.5 - 208.8	41.63	51.75	19.2 - 84.9	19.62
Conductivity (μ/cm)	240	50 - 560	118.40	146	30 - 410	99.76

Man-made lake of Ahozon: In this lake, 6506 individual fishes comprising 6 species belonging to 6 genera and 4 families were collected (Table 2). Among them, the family Cichlidae was the most speciose and comprised Sarotherodon galilaeus, Oreochromis niloticus, Tilapia guineensis. In addition, the fish community of Lake Ahozon comprised the silver catfish, Chrysichthys nigrodigitatus (Claroteidae), the African bonytongue, Heterotis niloticus (Osteoglossidae), and the African catfish, Clarias gariepinus (Clariidae). Spatially, the "open water" habitat, with six (6) species, was more diverse than the aquatic vegetation harboring four (4) species (Table 3 and Table 4).

Likewise, numerically, cichlids dominated the sample and accounted together for 88.26% with S. galilaeus, a naturally-colonized species of Lake Ahozon, making alone 85.21% of the total sample, followed by the newly introduced cichlid O. niloticus (2.18%), and T. guineensis (0.75%), an occasional tilapine species. The second species numerically most important in Lake Ahozon was the introduced claroteid, C. nigrodigitatus accounting for about 11.22% of the total sample. The newly introduced osteoglossid, H. niloticus (0.52%) and the African catfish C. gariepinus (0.02%) were sporadically recorded in the sample.

Table 2. Fish species composition, abundance, size range and mean, weight range and mean, of the fishes in the man-made lake of Ahozon, Ouidah city, Southern Benin.

Family & Species	Abundance	Relative abundance (%)	SL mean (mm)	SL range (mm)	Weight mean (g)	Weight range (g)
Cichlidae						
Oreochromis niloticus	142	2.18	64.25	28 - 182	15.64	0.2 - 110
Sarotherodon galilaeus	5550	85.31	56.57	8 - 192	14.38	0.01 - 190
Tilapia guineensis	49	0.75	47.49	20 - 96	9.71	0.3 - 448
Clariidae						
Clarias gariepinus	1	0.02	506	506 - 506	1328	1328 - 1328
Claroteidae						
Chrysichthys nigrodigitatus	730	11.22	127.10	55 - 262	57.95	4 - 469.7
Osteoglossidae						
Heterotis niloticus	34	0.52	332.24	54 - 660	523.69	2.5 - 3100
Total number of individuals: 6506						
Total number of families: 4						
Total number of species: 6						

In term of biomass, the 6506 individual fishes sampled in lake Ahozon weighted 143,913.11 grams with 132,498.58 (92.07%) in the "open water" habitat and 11,414.53 grams (7.93%) in the "aquatic vegetation" habitat. S. galilaeus dominated the biomass and alone contributed for about 55.46% of the total catches, followed by C. nigrodigitatus (29.40%), H. niloticus (12.37%). The remaining biomass (2.78%) was shared by three (3) species O. niloticus (1.54%), C. gariepinus (0.92%), and T. guineensis (0.33%).

Table 3. Fish species composition, abundance, size range and mean, weight range and mean, of the fishes in the "open water" habitat of the man-made lake of Ahozon, Ouidah city, Southern Benin.

Family & Species	Abundance	Relative abundance (%)	SL mean (mm)	SL range (mm)	Weight mean (g)	Weight range (g)
Cichlidae						
Oreochromis niloticus	109	2.40	67.20	29 - 182	17.28	0.2 - 110
Sarotherodon galilaeus	3674	88.89	75.93	17 - 192	21.50	0.4 - 190
Tilapia guineensis	18	0.40	78.56	66 - 96	23.83	12.9 - 44.8
Clariidae						
Clarias gariepinus	1	0.02	506	506	1328	1328
Claroteidae						
Chrysichthys nigrodigitatus	730	16.07	127.10	55 - 262	57.95	4 - 469.7
Osteoglossidae						
Heterotis niloticus	10	0.22	396.1	302 - 660	757.56	230 - 3100

Total number of individuals: 4542
Total number of families: 4
Total number of species: 6

Table 4. Fish species composition, abundance, size range and mean, weight range and mean, of the fishes in the "aquatic vegetation" habitat of the man-made lake of Ahozon, Ouidah city, Southern Benin.

Family & Species	Abundance	Relative abundance (%)	SL mean (mm))	SL range (mm	Weight mean (g)	Weight range (g)
Cichlidae						
Oreochromis niloticus	33	1.68	54.51	28 - 100	10.25	0.6 - 36.9
Sarotherodon galilaeus	1876	95.52	18.65	8 - 76	0.43	0.01 - 17.9
Tilapia guineensis	31	1.58	29.45	20 - 49	1.51	0.3 - 6.2
Osteoglossidae						
Heterotis niloticus	24	1.22	305.75	54 - 605	426.25	2.5 - 2700

Total number of individuals: 1964
Total number of families: 2
Total number of species: 4

Table 5. Fish species composition, abundance, size range and mean, weight range and mean, of the fishes in the man-made lake of Bewacodji, Ouidah city, Southern Benin.

Family & Species	Abundance	Relative abundance (%)	SL mean (mm))	SL range (mm	Weight mean (g)	Weight range (g)
Cichlidae						
Sarotherodon galilaeus m.[*]	2142	91.58	44.90	20 - 129	4.89	0.2 - 59.8
Tilapia guineensis	1	0.04	54	54 - 54	7.40	7.4 - 7.4
Clariidae						
Clarias gariepinus	33	1.41	153.48	110 - 264	51.79	13.1 - 217.3
Claroteidae						
Chrysichthys nigrodigitatus	163	6.97	100.90	41 - 148	28.39	2.3 - 90.7

Total number of individuals: 2339
Total number of families: 3
Total number of species: 4

[*]*m. = multifasciatus*

Man-made lake of Bewacodji: In the man-made lake of Bewacodji, 2339 individual fishes comprising 4 species belonging to 4 genera and 3 families were collected (Table 5). The fish composition included two (2) cichlids, Sarotherodon galilaeus multifasciatus and Tilapia guineensis, the African catfish, C. gariepinus (Clariidae) and the silver catfish, C. nigrodigitatus (Claroteidae). Like Lake Ahozon, cichlids numerically dominated the fish community of Lake Bewacodji, and accounted together for 91.62%, with S. galilaeus multifasciatus, the dominant species, making alone 91.58% of the fish assemblages, followed by the introduced claroteid, C. nigrodigitatus (6.97%), C. gariepinus (1.41%) and T. guineensis (0.04%) sporadically recorded in the samples.

In term of biomass, the 2339 individual fishes collected in Lake Bewacodji weighted 16,823.7 grams with 8563 grams (50.90%) in the "open water" and 8260.7 grams (49.10 %) in the "aquatic vegetation" habitat (Tables 5-7). S. galilaeus multifasciatus dominated the biomass and alone contributed for about 62.26% of the total biomass followed by C. nigrodigitatus (27.51%), C. gariepinus (10.16%), and T. guineensis (0.04%).

3.3. Diversity, Richness and Evenness Indices

In general, species richness and diversity indices were low in both man-made lakes. In Lake Ahozon, the Margalef index (d) [35] of species richness computed from aggregated habitats and samples was d = 5.89. The species richness was higher in the open water (d = 5.88) than in the aquatic vegetation (d = 3.87). Also, the Shannon-Weaver index (H') [36] of species

Table 6. Fish species composition, abundance, size range and mean, weight range and mean, of the fishes in the "open water" of the man-made lake of Bewacodji, Ouidah city, Southern Benin.

Family & Species	Abundance	Relative abundance (%)	SL mean (mm))	SL range (mm)	Weight mean (g)	Weight range (g)
Cichlidae						
*Sarotherodon galilaeus m.**	1193	95.21	43.00	20 - 129	4.72	0.2 - 59.8
Clariidae						
Clarias gariepinus	2	0.16	203.5	143 - 264	126.85	36.4 - 217.3
Claroteidae						
Chrysichthys nigrodigitatus	58	4.63	117.86	94 - 148	40.94	23.2 - 90.7

Total number of individuals: 1253
Total number of families: 3
Total number of species: 3

*m. = *multifasciatus*

Table 7. Fish species composition, abundance, size range and mean, weight range and mean, of the fishes in the "aquatic vegetation" of the man-made lake of Bewacodji, Ouidah city, Southern Benin.

Family & Species	Abundance	Relative abundance (%)	SL mean (mm))	SL range (mm)	Weight mean (g)	Weight range (g)
Cichlidae						
*Sarotherodon galilaeus m.**	949	87.38	47.28	29 - 104	5.11	13 - 52.2
Tilapia guineensis	1	0.09	54	54 - 54	7.4	7.4 - 7.4
Clariidae						
Clarias gariepinus	31	2.85	150.26	110 - 196	46.95	13.1 - 109
Claroteidae						
Chrysichthys nigrodigitatus	105	9.67	91.52	41 - 123	21.45	2.3 - 48.4
Total number of individuals: 1086						
Total number of families: 3						
Total number of species: 4						

**m.* = *multifasciatus*

diversity of the whole community was low (H' = 0.76) along with a low Evenness (J' = 0.43) (Table 8). However, the H' = 0.85 record in the open water was higher than that of the aquatic vegetation (H' = 0.33). With regard to season, the computed Shannon-Weaver index (H' = 0.55) for the dry period was higher than those of wet and flood period with reduced evenness indexes varying between 0.11 - 0.24 (Table 9). Also, the Shannon-Weaver indexes [36] computed for the fishing gears showed H' ranging between 0.03 for hooks and 1.68 for gillnets with Evenness (J') ranging between 0.11 and 0.98 (Table 10).

In Lake Bewacodji, the Margalef index (d) [35] of species richness computed from aggregated habitats was d = 3.87. Unlike Lake Ahozon, the species richness in Lake Bewacodji was higher in the "aquatic vegetation" (d = 3.86) compared to that of the "open water" (d = 2.86). Likewise, the Shannon-Weaver index [36] of species diversity of the whole community was low (H' = 0.48) with a low Evenness (J' = 0.35) (Table 8). However, the H' = 0.65 record in the aquatic vegetation was higher than that of the open water (H' = 0.28) (Table 8). With regard to season, the Shannon-Weaver index H' = 0.18 [36] for the dry period was lower than those of wet and flood period with reduced Evenness indexes varying between 0.18 - 0.37 (Table 11). Also, the computed Shannon- Weaver index [36] for the fishing gears showed H' ranging between 0.45 for trap and 1.00 for hooks with Evenness (J') between 0.23 and 0.99 (Table 12).

3.4. Trophic Structure of Fishes

The fish communities of both artificial lakes were numerically dominated by planktinovores/detritivores. In the man-made lake of Ahozon,

Table 8. Fish abundances and diversity indices by habitat type in the man-made lakes of Ahozon and Bewacodji, Ouidah city, Southern Benin.

Community features	Artificial lake of Ahozon			Artificial lake of Bewacodji		
	Aquatic vegetation	Open water	Total	Aquatic vegetation	Open water	Total
Number of families	2	4	4	3	3	4
Number of species	4	6	6	4	3	4
Abundance (Sample size)	1964	4542	6506	1086	1253	2339
Relative abundance (%)	30.19	69.81	100	46.43	53.57	100
Total weight (g)	11.414.53	132.498.58	143.913.11	8563	8260.7	16.823.7
Percentage weight (%)	7.93	92.07	100	50.90	49.10	100
Margalef index (d)	3.87	5.88	5.89	3.86	2.86	3.87
Shannon & weaver index (H')	0.33	0.85	0.76	0.65	0.28	0.48
Evenness (J')	0.24	0.47	0.43	0.47	0.25	0.35

Table 9. Fish abundances and diversity indices by season in the man-made lake of Ahozon, Ouidah city, Southern Benin.

Season	Number of species	Abundance (Number of individuals)	Shannon-Weaver diversity indices (H')	Evenness (J')
Flood	5	1575	0.28	0.12
Dry	5	1637	0.55	0.24
Wet	6	3294	0.29	0.11
Total	**6**	**6506**	**0.76**	**0.24**

Table 10. Fish abundances and diversity indices by fishing gears in the man-made lake of Ahozon, Ouidah city, Southern Benin.

Fishing gears	Number of species	Abundance (Number of individuals)	Shannon-Weaver diversity indices (H')	Evenness (J')
Cast net	4	3731	0.22	0.11
Gill net	4	100	1.68	0.84
Seine	2	1907	0.12	0.12
Longline	3	711	0.03	0.02
Trap	2	57	0.98	0.98
Total	**6**	**6506**	**0.76**	**0.29**

Table 11. Fish abundances and diversity indices by season in the man-made lake of Bewacodji, Ouidah city, Southern Benin.

Season	Number of species	Abundance (Number of individuals)	Shannon-Weaver diversity indices (H')	Evenness (J')
Flood	4	929	0.70	0.35
Dry	2	505	0.18	0.18
Wet	2	905	0.37	0.37
Total	**4**	**2339**	**0.48**	**0.24**

Table 12. Fish abundances and diversity indices by fishing gear in the man-made lake of Bewacodji, Ouidah city, Southern Benin.

Fishing gears	Number of species	Abundance (Number of individuals)	Shannon-Weaver diversity indices (H')	Evenness (J')
Longline	2	19	1.00	0.99
Trap	4	2320	0.45	0.23
Total	**4**	**2339**	**0.48**	**0.24**

planktinovores/detritivores species, mainly S. galilaeus, O. niloticus, Tilapia guineensis and C. nigrodigitatus made together 99.46% of the total biomass (Table 13). The omnivore and intermediate carnivore were trivial in the fish community and made numerically together only 0.54%. Also, in terms of biomass, planktinovores/detritivores dominated the fish community with 86.71% of the total biomass followed by the omnivore, H. niloticus (12.37%) and the intermediate carnivore, C. gariepinus, accounted only for 0.92% of the total biomass.

Like Lake Ahozon, the fish community of the man-made lake of Bewacodji was numerically dominated by planktinovores/detritivores species, mainly S. galilaeus multifasciatus, Tilapia guineensis and C. nigrodigitatus making together 98.59% of the total sample (Table 14). The intermediate carnivore, C. gariepinus, was trivial in the fish community and made numerically only 1.41% of the sample. Also, in terms of biomass, planktinovores/detritivores dominated the fish community with 89.83% of the total biomass. The intermediate carnivore, C. gariepinus, though not numerically abundant, accounted for 10.17% of the total biomass.

3.5. Size Distributions

In both lakes, individual fish size varied according to species and habitats. In the artificial lake of Ahozon the fish standard lengths (SL) ranged from 8 mm for S. galilaeus to 660 mm for H. niloticus and their corresponding weights were 0.3 g and 3100 g, respectively (Table 2). Larger sizes were mostly found in the "open water" habitat whereas smaller sizes were mostly recorded in the "aquatic vegetation". Three (3) larger species, the African bonytongue H. niloticus, the African catfish C. gariepinus and the silver catfish, C. nigrodigitatus were recorded in the fish community. Standard length (SL) frequencies histograms established for dominant species showed an unimodal size distribution for H. niloticus, the nile tilapia O. niloticus and C. gariepinus whereas the two cichlids, S. galilaeus and T. guineensis exhibited a bimodal size distribution (Figures 2-6).

Table 13. Trophic categories, relative abundance and commercial importance of the fish caught in the man-made lake of Ahozon, Ouidah city, Southern Benin. J: juvenile; A: adult; HC: highly commercial.

Feeding category & Species	Relative abundance (%)	Percent weight (%)	Life history stage	Fisheries importance
Planktinovores/Detritivores	**99.46**	**86.71**		
Sarotherodon galilaeus	85.31	55.45	J & A	HC
Oreochromis niloticus	2.18	1.54	J & A	HC
Tilapia guineensis	0.75	0.33	J & A	HC
Chrysichthys nigrodigitatus	11.22	29.39	J & A	HC
Omnivore	**0.52**	**12.37**		
Heterotis niloticus	0.52	12.37	J & A	HC
Intermediate carnivores	**0.02**	0.92		
Clarias gariepinus	0.02	**0.92**	A	HC
Total number of individuals:	**6506**	**100**		

Table 14. Trophic categories, relative abundance and commercial importance of the fish caught in the man-made lakes of Bewacodji, Ouidah city, Southern Benin. J: juvenile; A: adult; HC: highly commercial.

Feeding category & Species	Relative abundance (%)	Percent weight (%)	Life history stage	Fisheries importance
Planktinovores/Detritivores	98.59	89.83		
*Sarotherodon galilaeus m.**	91.58	62.28	J & A	HC
Tilapia guineensis	0.04	0.04	J & A	HC
Chrysichthys nigrodigitatus	6.97	27.51	J & A	HC
Intermediate carnivores	**1.41**	10.17		
Clarias gariepinus	1.41	**10.17**	A	HC
Total number of individuals:	**2339**	**100**		

* *m. = multifasciatus*

In the artificial lake of Bewacodji, the SL of fishes ranged from 20 mm for S. galilaeus to 264 mm for C. gariepinus and their corresponding weights were 0.20 g and 217.3 g, respectively. Larger sizes were mostly found in the "open water" habitat whereas smaller sizes were mostly recorded in the "aquatic vegetation". Two (2) larger species, C. gariepinus and C. nigrodigitatus were recorded in the fish community of Lake Bewacodji. Standard length (SL) frequencies histograms established for dominant species exhibited an unimodal size distribution for S. galilaeus, C. nigrodigitatus and C. gariepinus (Figures 7-9).

3.6. Environmental Correlates

In both lakes, the total abundance for each fish species collected were plotted against water features to diagnose the relationships between fish attributes and abiotic factors. In lake Ahozon, though not significant, the total abundance of fish collected was positively correlated with depth (r = 0.59; regression slope b = 1.59; N = 20), temperature (r = 0.37; regression slope b = 21.60; N = 20) and conductivity (r = 0.47; regression slope b = 0.40; N = 20). Though not significant, the abundance of the dominant species, S. galilaeus, was positively correlated with depth (r = 0.44; b = 0.94; N = 20), temperature (r = 0.48; b = 23.42; N = 20) and conductivity (r = 0.33; b = 0.295; N = 20). No specific trends were observed in the man-made lake of Bewacodji.

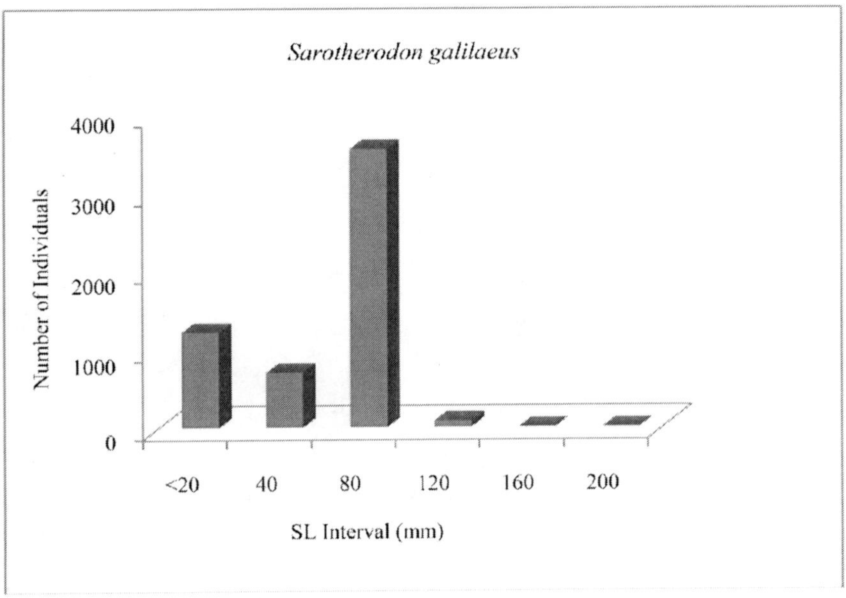

Figure 2. Size structure of Sarotherodon galilaeus (n = 5550) in the man-made lake of Ahozon (South-Benin).

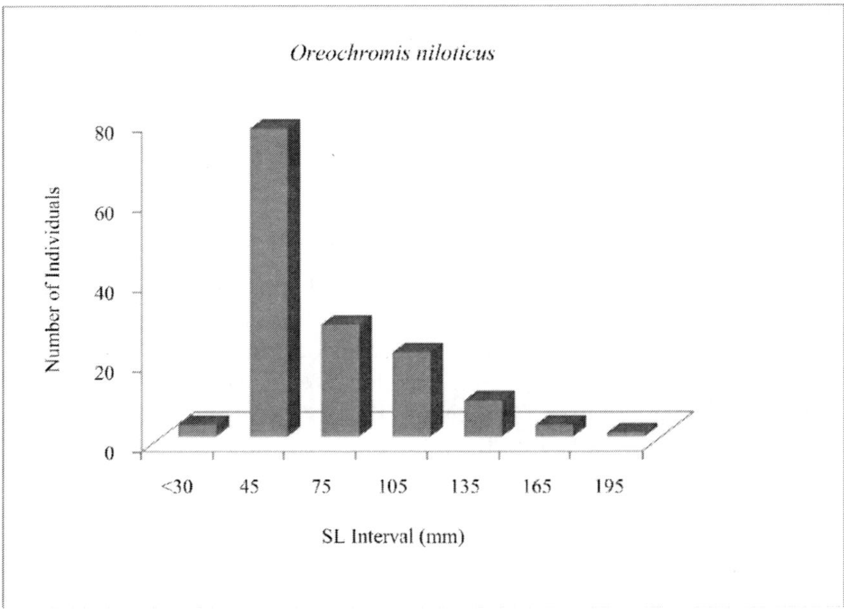

Figure 3. Size structure of Oreochromis niloticus (n = 142) in the man-made lake of Ahozon (South-Benin).

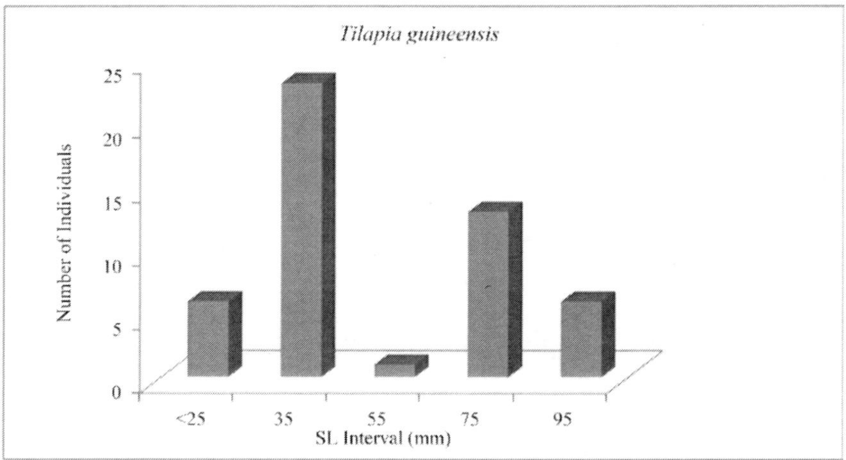

Figure 4. Size structure of Tilapia guineensis (n = 49) in the man-made lake of Ahozon (South-Benin).

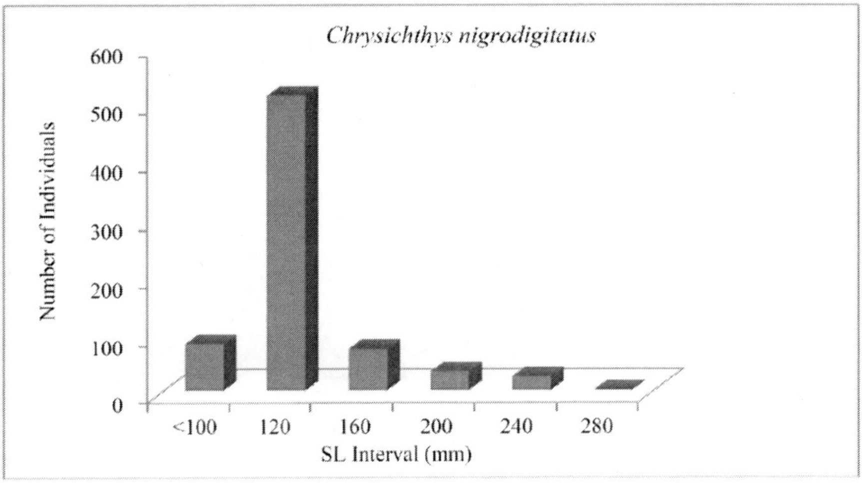

Figure 5. Size structure of Chrysichthys nigrodigitatus (n = 730) in the man-made lake of Ahozon (South-Benin).

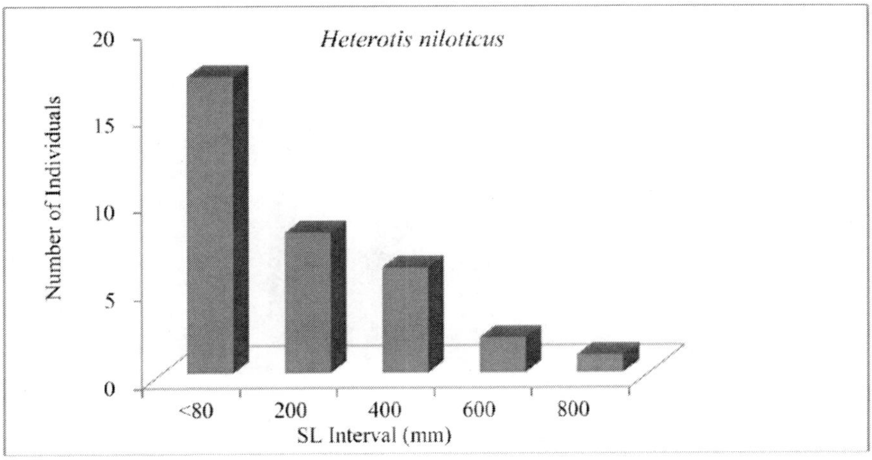

Figure 6. Size structure of Heterotis niloticus (n = 34) in the man-made lake of Ahozon (South-Benin).

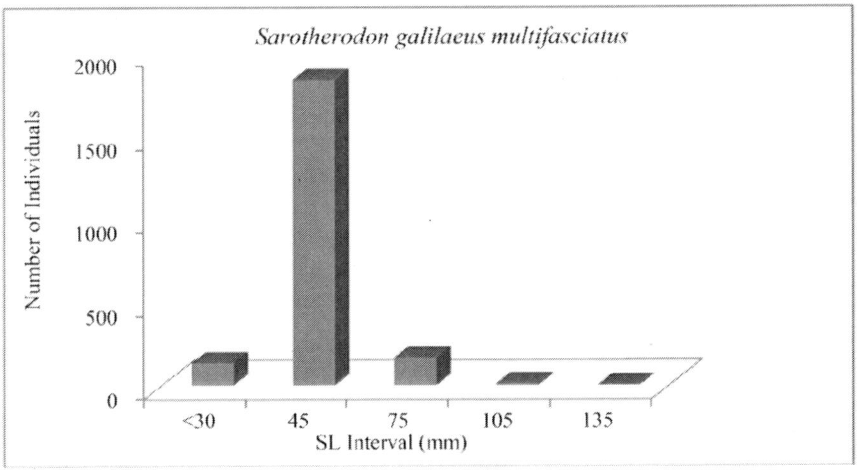

Figure 7. Size structure of Sarotherodon galilaeus multifasciatus (n = 2142) in the man-made lake of Bewacodji (South-Benin).

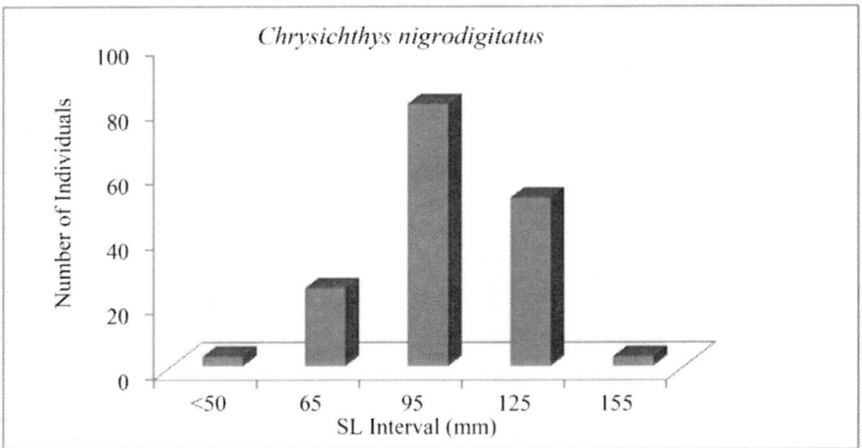

Figure 8. Size structure of Chrysichthys nigrodigitatus (n = 163) in the man-made lake of Bewacodji (South-Benin).

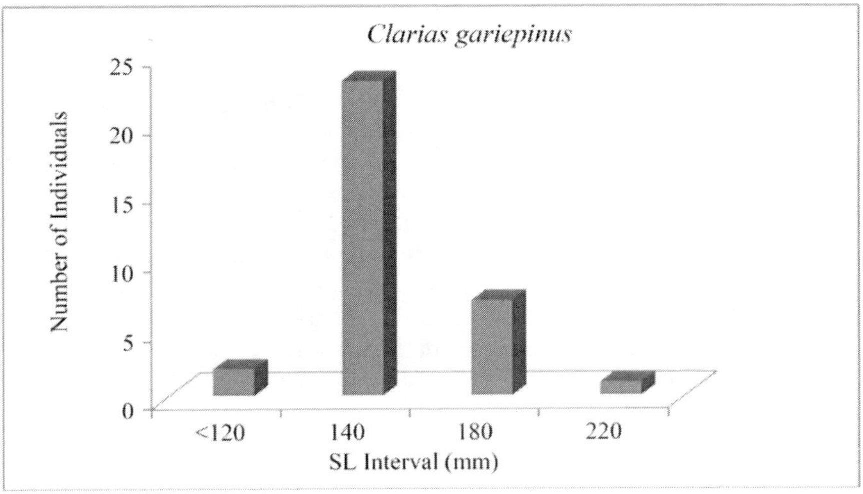

Figure 9. Size structure of Clarias gariepinus (n = 33) in the man-made lake of Bewacodji (South-Benin).

3.7. Lake's Resource Exploitation and Degradation Sources

After the sand-dragging activities, the resulting two man-made lakes, Lake Ahozon and Lake Bewacodji, were abandoned, neglected, and considered as useless or lost lands by the owners. Thus, fisheries and aquacultural management of these man-made lakes by the owners were scant. Nevertheless, sporadic fisheries exploitation implemented by couple of local and migrant fishermen occurred on both lakes. Fishing techniques and gears used were cast nets, traps, gillnets and hooks. On Lake Bewacodji, cast nets were not used because of the dense floating plant covering the lake's surface. Relative to the fish resources, all the species inventoried (6 in Lake Ahozon, 4 in Lake Bewacodji) and dominated by S. galilaeus and C. nigrodigitatus, were of high commercial and economic values.

In addition to fisheries, adjacent wetlands were utilized to grow legumes, salads, carrots, red papers, tomatoes, etc. causing pesticides and fertilizers intrusion and habitat pollution. In particular, Lake Bewacodji undergoes serious environmental degradation such as the daily dumping of huge domestic wastes, source of organic pollution, and the proliferation of floating vegetation mainly, Nymphea, covering the entire lake.

4. DISCUSSIONS

The urgent needs to protect degrading natural aquatic ecosystems, to conserve dwelling fish biodiversity and to feed the growing human population, require a search of alternative environment mediums for a sustainable fish production [21] [26] [39] . In Benin, because of the permanent exploitation of inland sand bed, the resulting man-made has become widespread, but generally unmanaged, neglected, abandoned and considered as lost wetlands. The current ecological study of the man-made lakes assessed water quality and habitat condition, fish assemblages and community structure, utilization of the lakes, degradation impacts and management schemes.

4.1. Habitat Conditions

Overall, the water quality of the two man-made lakes studied was significantly different. Indeed, one-way analysis of variance on the water physicochemical traits (transparency, dissolved oxygen concentrations, conductivities, pH) across the two lakes showed significant variations ($P \leq 0.01$) of the physicochemical parameters between Lake Ahozon and Lake Bewacodji. The calculted F-values, along with degrees of freedom and p-values were $F_{1,53} = 29.765$, $p = 0.0001$ for transparency, $F_{1,53} = 9.968$, $p = 0.003$ for dissolved oxygen, $F_{1,53} = 8.954$, $p = 0.004$ for conductivity, and $F_{1,53}= 35.138$, $p = 0.0001$ for pH. Nevertheless, one-way ANOVA fail to show any significant difference ($P > 0.05$) between the two lakes for depth ($F_{1,53} = 0.880$, $p = 0.353$) and temperature ($F_{1,53} = 0.044$, $p = 0.834$).

In general, the water quality of Lake Ahozon was globally favorable not only for a high primary production, but also for the growth and the survival of the fish resources. This favorable environmental quality was indicated by the high water transparencies, the alkaline pH, the high dissolved oxygen and the high conductivity (Table 1). Inversely, the man-made lake of Bewacodji exhibited a weak water quality indicated mainly by an acid pH (mean: 6.32 ± 0.58) and low dissolved oxygen concentrations (mean: 3.52 ± 1.25 mg/l). Indeed, during the study, Lake Bewacodji was almost totally covered by a floating plant, Nymphea sp mainly, coupled with the huge daily dumping of domestic waste that caused pollution and degradation of the water quality, critical for an optimal growth and survival of the fish resources. Similar trends of poor water quality were reported for some water bodies such as Lake Hlan, Sô River floodplain, Oueme River floodplain etc. invaded by water hyacinth (Ecchornia crassispes) and showing acid ph (mean: 5.3 ± 0.20) and low dissolved oxygen ranging between 0.1 mg/l and 4.8 mg/l [38] .

Also, when considering a single lake, significant spatial variations in water quality were recorded. In the man-made lake of Ahozon, one-way analysis of variance on the water physicochemical parameters indicated that depth, transparency and dissolved oxygen across the two habitats (open water versus aquatic vegetation) were significantly different ($p \leq 0.05$). The computed F-values, along with degrees of freedom and p-values were $F_{1,32} = 7.399$, $p = 0.010$ for depth, $F_{1,32} = 6.749$, $p = 0.014$ for transparency, $F_{1,32} = 11.398$, $p = 0.002$ for dissolved oxygen. Nevertheless, one-way ANOVA fail to show any significant difference ($p > 0.05$) between the open water habitat and aquatic vegetation habitat of the artificial lake of Ahozon for temperature ($F_{1,32} = 0.016$, $p = 0.899$) for conductivity ($F_{1,32} = 0.595$, $p = 0.446$) and for pH ($F_{1,32} = 0.495$, $p = 0.487$)

Inversely, in the man-made lake of Bewacodji, one-way ANOVA on the water physicochemical features fail to show any significant difference ($p > 0.05$) between the open water habitat and aquatic vegetation habitat for dissolved oxygen ($F_{1,18} = 1.826$, $p = 0.193$), temperature ($F_{1,18} = 0.18$, $p = 0.893$), conductivity ($F_{1,18} = 0.002$, $p = 0.966$) and pH ($F_{1,18} = 0.144$, $p = 0.709$). This result suggested that in Lake Bewacodji the water quality of the open water and the aquatic vegetation was similar. Indeed, Lake Bewacodji was almost totally covered by a floating plant, mainly Nymphea sp, and hence, the open water habitat and the aquatic vegetation habitat were less distinctive. Nevertheless, one-way analysis of the depth and transparency across the two habitats showed significant differences indicated by $F_{1,18} = 9.984$, $p = 0.005$ and $F_{1,18} = 6.537$, $p = 0.020$ respectively.

4.2. Community Structures of the Fish Assemblages

Both man-made water bodies are freshwater lakes with salinities nearly 0‰. Consequently, only typically freshwater species such as Sarotherodon galilaeus, Sarotherodon galilaeus multifasciatus Heterotis niloticus, and brackish water-tolerant species such as Chrysichthys nigrodigitatus, Clarias gariepinus, Oreochromis niloticus and Tilapia guineensis were encountered in these freshwater habitats [21] [23] [31] [40] . The naturally colonized S. galilaeus, a dominant species in both lakes, probably originated from lagoon Toho-Todougba, a freshwater lake located about 5 km, where this species was also dominant [26] . The natural colonization of S. galilaeus, Tilapia guineensis and Clarias gariepinus may originate from the intrusion of fish eggs probably introduced by running waters in these man-made lakes. In contrast, the silver catfish C. nigrodigitatus has been introduced by fishermen and though less prominent compared to S. galilaeus, appeared to be well-established. Also, the nile tilapia O. niloticus and the African bonytongue, H. niloticus has been introduced during the current research to appreciate their establishment, propagation and contribution to species conservation, fisheries and aquaculture valorization.

Both man-made lakes exhibited low species richness, d = 5.89 in Lake Ahozon, and d = 3.87 in Lake Bewacodji, with six (6) and four (4) fish species, respectively. Likewise, the Shannon-Weaver index (H') [36] of species diversity were low, H' = 0.76 for Lake Ahozon and H' = 0.48 for Lake Bewacodji. These trends of low species richness and diversity were due to the fact that both lakes were isolated and totally separated from speciose habitats such as rivers, streams, coastal lagoons and marine habitats from which some species could migrate and increase the ichthyofaunal composition of these man-made lakes [41] [42] .

In Lake Ahozon, the relatively higher species richness (d = 5.88) and species diversity (H' = 0.85) recorded in the "open water" compared to those of the "aquatic vegetation" habitat (d = 3.87; H' = 0.33), were the result of absence of floating plants in this habitat. The opposite trends were observed in Lake Bewacodji almost totally covered with densely patches floating vegetation (Nymphea sp) with reduced "open water" area that showed low species richness (d = 2.86) and low species diversity (H' = 0.28) compared to the "aquatic vegetation" habitat that exhibited relatively higher species richness (d = 3.86) and species diversity (H' = 0.65). The relatively higher species diversity (H' = 0.55) observed during the dry period was probably due to the concentration of the fish resources in a relatively low volume. The reduced water volume probably increased the likelihood to capture the fish compared to the wet and flood period exhibiting higher water volume. Among fishing gears used to exploit the fish resources, the gill net, a less selective fishing device, showed a higher Shannon-Weaver [36] diversity index. The low evenness, J' = 0.47 recorded for Lake Ahozon and J' = 0.35 recorded for Lake Bewacodji indicated unequal representation of the species in the fish assemblages [43] - [45] . Nevertheless, in Lake Ahozon, some higher evenness, J' = 0.84 and J' = 0.98 were respectively observed for non selective fishing gears such as gillnet and Trap. In the man-made lake of Bewacodji, a higher evenness (J' = 0.99) were recorded for hooks indicating an equal representation of each species in the fish assemblages.

With regard to species abundance, the freshwater species, S. galilaeus, numerically dominated both habitats and accounted for about 85.31% and 91.58% for Lake Ahozon and Lake Bewacodji, respectively. In addition, S. galilaeus dominated the biomass and alone constituted about 55.46% and 62.26% of the total biomass of lakes Ahozon and Bewacodji, respectively. This trend is typical for most isolated African freshwater bodies mostly dominated by well-established cichlids species, mainly tilapia, capable of rapid colonization because of their high breeding potential coupled with a degree of trophic plasticity that enable them to adapt and to invade various freshwater ecosystems [46] - [49] . In Benin, Adite and Van Thielen [26] reported similar trends for Lagoon Toho-Todougba dominated by S. galilaeus making 73% of the total catches. Also, Lake Toho, a flooding lake of the Mono River (South-Benin), was dominated by Sarotherodon melanotheron accounting for about 68.46% of the total sample [50] . The relatively high biomass of the introduced claroteid, C. nigrodigitatus (29.40% in Lake Ahozon, 27.51% in Lake Bewacodji) indicated a substantial contribution of this species to the fish production in both lakes, and hence, indicated a relatively good establishment of this catfish. In Lake Ahozon, the African bonytongue H. niloticus, although newly introduced, contributed substantially to the fish production with a percent biomass of 12.37%. Currently, the three remaining species C. gariepinus, T. guineensis and O. niloticus were trivial and require continued follow-up to evaluate their establishment.

Though not significant, the positive correlations ($0.33 \leq r \leq 0.59$) between the physicochemical factors and the abundance of the fish assemblages and that of the dominant species, S. galilaeus indicated that the increase in the fish abundance of Lake Ahozon was controlled by depth, temperature and conductivity. No specific trends were observed in the polluted man-made lake of Bewacodji showing a water surface almost totally covered by densely patched floating plants.

In term of trophic structure, the fish community of both lakes was dominated by planktonovores/detritivores species, mainly S. galilaeus, O. niloticus (introduced only in Lake Ahozon), T. guineensis and C. nigrodigitatus making numerically 99.46% in Lake Ahozon and 98.59% in Lake Bewacodji. During this study, the introduction of H. niloticus, an omnivore and O. niloticus, a planktinovore/detritivore is an advantage because their establishment will allow to appropriately and efficiently exploit the foods resources in order to increase the fish yield and production. The introduction of intermediate carnivores and top carnivores should be under control to avoid high predation on larvae and juveniles.

Though perfectly established in Lake Ahozon, the dominant species, S. galilaeus, showed moderate mean size (56.57 ± 28.41 mm-SL). The early maturation and the continuous reproduction strategy (adults spawn all seasons) shown by most cichlids, especially, the group of tilapia, coupled with the low abundance of predator (C. gariepinus: 0.02% in Lake Ahozon; 1.41% in Lake Bewacodji) could have increased the abundance of S. galilaeus juveniles and sub adults accounting together for about 98% of Lake Ahozon sub population. Probably, this water body may have been overloaded and the lake's caring capacity may have been compromised, causing decrease in fish growth rate. Nevertheless, the maximum size recorded reached 192 mm-SL. Inversely, the detritivore C. nigrodigitatus (mean size: 127.10 mm-SL in Ahozon; 100.90 mm-SL in Bewacodji), the omnivore H. niloticus (mean size: 332.24 mm-SL in Ahozon) and the intermediate carnivore C. gariepinus (mean size: 506 mm-SL in Ahozon; 153.48 mm-SL in Bewacodji) exhibited relatively high size, probably because of their relatively large ecological niche compared to S. galilaeus, a specialist planktonovore [3] [51] .

In the man-made lake of Ahozon, degradation sources were reduced and limited to sporadic agriculture using pesticides and fertilizers, and hence, should not be a major threat for the water body. The lake's users should be sensitized to practice ecological sound agriculture without pesticides and fertilizers and using only compost. In the artificial lake of Bewacodji, in addition to the utilization of pesticides and fertilizers for agriculture purpose, two other major threats were 1) the daily dumping of huge domestic wastes in the lake, causing organic pollution and 2) the proliferation of floating vegetation (Nymphea sp), covering the lake and negatively modifying the water quality by reducing the dissolved oxygen, the pH and the primary production, and hence, negatively affected the fish community structure. Furthermore, these floating plants impeded navigation and access to the lake. The same trends of floating vegetation (Ecchornia crassipes, Pistia stratiostes, Nymphea sp) proliferation have been reported for many freshwater lakes (Toho, Cele, Azili, Hlan etc.) and rivers (Oueme, Sô, Mono, Couffo, Okpara, Niger, etc.) in Benin, and has become a major degradation source for the whole aquatic ecosystem and a major threat for the biological diversity [15] .

4.3. Species Conservation and Fisheries Management Implications

The current ecological study on the man-made lakes of Ahozon and Bewacodji provide valuable information and database on habitat conditions, fish composition, richness, abundance, diversity, spatial and seasonal distributional patterns of the fishes. Also, disturbances and degradation issues were evaluated. The water quality of Lake Ahozon indicated that this habitat is favorable for a high primary production and guaranties a relatively high growth and survival of the fish resources. A holistic and sustainable management scheme to be implemented should include:

1) Aquacultural valorization: this valorization should emphasis on the introduction of target and highly economic and commercial species that could utilize the unexploited ecological niche of the lakes. In this study, in addition to the detritivore C. nigrodigitatus introduced by the fishermen, the African bonytongue, H. niloticus, an omnivore and the nile cichlid, O. niloticus, a planktinovore has been introduced to increase the fish yield and production. Heterotis is among the most valuable fish species and a single adult bonytongue cost nearly $25. A follow-up will permit to evaluate their

establishment in the future. Also, a fish culture technique, such as the "floating cage" could be installed on the artificial lake as medium environment to raise fish. Likewise, predator species such as Parachanna obscura should be introduced to regulate tilapia species population.

2) Ecological sound agriculture: The adjacent wetlands of the lakes could be utilized to grow legumes, salads, carrots, red papers, tomatoes, etc. but, without the utilization of pesticides and fertilizers capable of causing pollution.

3) Ecological sound ecotourism: This ecotourism may include aquarium, sport fisheries, aesthetical creations, and navigation.

4) Education and training program: This requires collaboration with agriculture school and especially, those involved in fish culture, fisheries and biodiversity conservation.

5) Scientific research: The man-made lakes could be used for scientific research by the University laboratory as the lakes harbor a huge aquatic biological diversity, from micro vegetation to superior animals.

6) Habitat protection and species conservation program: This program may include the protection of the lake fringes by planting grass and trees and biodiversity conservation.

7) Environment monitoring program: This monitoring should be implemented as follow-up to periodically assess the ecological status of the lakes. The monitoring could emphasize mainly on the periodic evaluation of habitat conditions, the hydrology and the living aquatic resources.

8) Particularly, in Lake Bewacodji, the grassroots and the lake users should be sensitized to stop with dumping of domestic wastes in this lake. Also, floating vegetation (Nymphea sp), should be taken out to restore the water quality and the fish community structure.

5. CONCLUSION

This investigation gives valuable information on the ecology and the fish community structure of two man-made lakes, Lake Ahozon and Lake Bewacodji of Southern Benin. Overall, the environmental conditions of Lake Ahozon were globally favorable for the primary production, the growth and the survival of the dwelling fish resources. Inversely, the man-made lake of Bewacodji exhibited a poor water quality because of the acid pH and the low dissolved oxygen concentrations caused by the floating plants and the dumping of domestic wastes. In general, species richness and species diversity indices were low. However, the two fish communities were consistently dominated by a tilapine cichlid, Sarotherodon galilaeus which constituted an important economically and commercially fish resources in these two habitats. A sustainable and successful exploitation of both lakes requires an integrated and holistic management scheme including habitat protection and species conservation, aquacultural valorization, ecological sound agriculture, ecological sound ecotourism, education and training, scientific research and environment monitoring. Also, this research should be extended to numerous man-made lakes in Benin to explore potentials for fisheries, aquaculture and biodiversity conservation.

ACKNOWLEDGEMENTS

Financial assistance and logistic were provided by the "Laboratoire d'Ecologie et de Management des Ecosystèmes Aquatiques, Département de Zoologie, Faculté des Sciences et techniques, Université d'Abomey- Calavi". We express our gratitude to Mr Doukpo Célestin, the owner of the artificial lakes to allow us to conduct this investigation on both man-made lakes. We are also grateful to Mr Houessinon M. Geoffroy, Houndjetin Louis and Djihouessi Bernold for their assistance in fish sampling and laboratory works. We thank the numerous reviewers for their thorough peer-review of this manuscript.

REFERENCES

1. Welcomme, R.L. (1979) Fisheries Ecology of Floodplain Rivers. Longman, New York.
2. Lowe-McConnell, R.H. (1987) Ecological Studies in Tropical Fish Communities. Cambridge University Press, Cambridge. http://dx.doi.org/ 10.1017/ CBO9780511721892
3. Adite, A. and Winemiller, K.O. (1997) Trophic Ecology and Ecomorphology of Fish Assemblages in Coastal Lakes of Benin, West Africa. Ecoscience, 4, 6-23.
4. Lalèyè, P., Niyonkuru, C., Moreau, J. and Teugels, G.G. (2003) Spatial and Seasonal Distribution of the Ichthyofauna of Lake Nokoué, Benin, West Africa. African Journal of Aquatic Sciences, 28, 151-161. http://dx.doi.org/10.2989/16085910309503779
5. Adite, A., Winemiller, K.O. and Fiogbé, E.D. (2006) Population Structure and Reproduction of the African Bony- tongue Heterotis niloticus in the S? River-Floodplain System (West Africa): Implications for Management. Ecology of Freshwater Fish, 15, 30-39. http://dx.doi.org/10.1111/j.1600-0633.2005.00119.x
6. Mustapha, M.K. (2010) Heterotis niloticus (Cuvier, 1829), a Threatened Fish Species in Oyun Reservoir, Offa, Nigeria: The Need for Its Conservation. Asian Journal of Experimental Biological Sciences, 1, 1-7.
7. Smith, R.L. (1992) Elements of Ecology. 3rd Edition, HarperCollins Publishers Inc., New York.
8. Fagnon, M.S., Chikou, A., Youssao, I. and Laleye, P. (2013) Caractérisation morphologique des populations de Sarotherodon melanotheron (Pisces, Cichlidae)

en eaux douces et saumatres au Sud Bénin. International Journal of Biological and Chemical Sciences, 7, 619-630. http://dx.doi.org/10.4314/ijbcs.v7i2.18
9. Hurtado, L.A., Carrera, E., Adite, A. and Winemiller, K.O. (2013) Genetic Differentiation of a Primitive Teleost, the African Bonytongue Heterotis niloticus, among River Basins and within a Floodplain River System in Benin, West Africa. Journal of Fish Biology, 83, 682-690. http://dx.doi.org/10.1111/jfb.12198
10. Lopez-Fernandez, H., Winemiller, K.O., Adite, A., Arrington, D.A. and Layman, C.A. (2003) Freshwater Fish Diversity in Benin, West Africa, and Challenge for Its Conservation. Proceedings University of Abomey-Calavi Conférence, African Fish and Fisheries, Cotonou, 85-86.
11. Lawson, E.O. (2010) Morphometric Measurements and Meristic Counts in Mudskipper (Periophthalmus papilio) from Mangrove Swamps of Lagos lagoon, Nigeria. Journal of Applied Biosciences, 34, 2166-2172.
12. Adite, A., Toko, I.I. and Gbankoto, A. (2003) Fish Assemblages in the Degraded Mangrove Ecosystems of the Coastal Zone, Benin, West Africa: Implications for Ecosystem Restoration and Resources Conservation. Journal of Environ- mental Protection, 4, 1461-1475. http://dx.doi.org/10.4236/jep.2013.412168
13. Welcomme, R.L. (1971) A Description of Certain Indigenous Fishing Methods from Southern Dahomey. African Journal of Tropical Hydrobiology and Fisheries, 1, 129-140.
14. Louis, M., Bouchon, C. and Bouchon-Navaro, Y. (1995) Spatial and Temporal Variations of Mangrove Fish Assemblages in Martinique (French West Indies). Hydrobiologia, 295, 275-284. http://dx.doi.org/10.1007/BF00029134
15. Adite, A., Abou, Y., Sossoukpe, E., Gbaguidi, G. and Fiogbe, E.D. (2013) Meristic and Morphological Characterization of the Freshwater Prawn, Macrobrachium macrobrachion (Herklots, 1851) from the Mono River-Coastal Lagoom System, Southern Benin (West Africa): Implications for Species Conservation. International Journal of Biodiversity and Conservation, 5, 704-714
16. Pliya, J. (1980) La pêche dans le Sud-ouest du Bénin: étude de géographie appliquée sur la pêche continentale et maritime. Agence de Coopération culturelle et technique, Paris.
17. Gbaguidi, G.A.M.H. (2012) Caractérisation morphométrique et écologie trophique de deux espèces de crevettes d'eaux douces Macrobrachium macrobrachion (Herklots, 1851) et Macrobrachium vollenhovenii (Herklots, 1857) du système fluvio-lagunaire Mono-Lagune C?tière (Sud-Bénin). Master's Thesis, Université d'Abomey-Calavi, Abomey Calavi.
18. Okpeicha, S.O. (2011) Biodiversité et exploitation des poissons du barrage de la SUCOBE dans la commune de Savè au Bénin. Master's Thesis, Faculté des Sciences Techniques/UAC, Abomey-Calavi.
19. Micha, J.C. (1973) Etude des populations piscicoles de l'Ouganda et tentation de sélection et d'adaptation de quelques espèces à l'étang de pisciculture. Centre Technique Forestier Tropicale, Nogent-sur-Marne,110 p.
20. Moreau, J. (1982) Exposé synoptique des données biologiques sur Heterotis niloticus (Cuvier, 1929). Food and Agriculture Organization Synopsis de pêches, 131, 1-45
21. Levêque, C., Paugy, D. and Teugels, G.G. (1990) The Fresh and Brackish Water Fishes of West Africa. Tome 1, Editions ORSTOM/MRAC, Paris.
22. Reed, W., Burchard, J., Hopson, A.J., Jenness, J. and Yaro, I. (1967) Fish and Fisheries of Northern Nigeria. Ministry of Agriculture Northern Nigeria.
23. Leveque, C. (1997) Biodiversity Dynamics and Conservation: The Freshwater Fish of Tropical Africa. Cambridge University Press, Cambridge.
24. GCLME (2006) Guinea Current Large Marine Ecosystem Report. GEF/UNIDO/UNDP/UNEP/ US-NOAA. http://Projects.csg.uwaterloo.ca/inweh/content/1188
25. Capo-Chichi, Y.J. (2006) Monographie de la commune de Grand-popo. Cabinet "Afrique Conseil", 46

26. Adite, A. and Van Thielen, R. (1995) Ecology and Fish Catches in Natural Lakes of Benin, West Africa. Environ- mental Biology of Fishes, 43, 381-391. http://dx.doi.org/10.1007/BF00001173
27. Akoegninou, A., Oyede, L.M. and Toffi, M. (1993) La mangrove du Benin: Environnement physique, végétation, et essais de gestion. Technical Report, Faculté des Sciences et Techniques, Université d'Abomey-calavi, Benin.
28. Djihouessi, B.M. (2015) Ecologie et valorization piscicole des lacs artificiels du Sud-Bénin (Bénin, Afrique du de l'Eouest). Master's Thesis, FAST/UAC, Abomey-Calavi.
29. Needham, G.J. and Needham, P.R. (1962) A Guide to the Study of Freshwater Biology. Holden-Day, Inc., San Francisco.
30. Winemiller, K.O. (1992) Fish Assemblages across a Complex, Tropical Freshwater/Marine Ecotone. Environmental Biology of Fishes, 3, 29-50. http://dx.doi.org/10.1007/BF00004783
31. Levêque, C., Paugy, D. and Teugels, G.G. (1990) The Fresh and Brackish Water Fishes of West Africa. Tome 2, Editions ORSTOM/MRAC, Paris.
32. Van Thielen, R., Hounkpe, C., Agon, G. and Dagba, L. (1987) Guide de détermination des Poissons et Crustacés des Lagunes et Lacs du Bas-Bénin. Direction des Pêches, Cotonou.
33. McConnell, R.H.L. (1975) Fish Communities in Tropical Freshwaters. Longman, London.
34. Morgan, G.A., Grieggo, O.V. and Gloekner, G.W. (2001) SPSS for Windows: An Introduction to Use and Interpretation in Research. Lawrence Erlbaum Associates, Publishers, Mahwah.
35. Margalef, R. (1968) Perspective in Ecological Theory. University of Chicago Press, Chicago.
36. Shannon, C. and Weaver, E. (1963) The Mathematical Theory of Communication. University of Illinois Press, Urbana.
37. Halliday, I.A. and Young, W.R. (1996) Density, Biomass and Species Composition of Fish in a Subtropical Rhizophora stylosa Mangrove Forest. Marine and Freshwater Research, 47, 609-615. http://dx.doi.org/10.1071/MF9960609
38. Adité, A., Winemiller, K.O. and Fiogbe, E.D. (2005) Ontogenetic, Seasonal; and Spatial Variation in the Diet of Heterotis niloticus (Osteoglossiformes: Osteoglossidae) in the S? River and Lake Hlan, Benin, West Africa. Environmental Biology of Fishes, 73, 367-378. http://dx.doi.org/10.1007/s10641-004-5563-9
39. Moyle, P.B. and Cech, J.J. (1988) Fish: An Introduction to Ichthyology. 2nd Edition, Prentice-Hall, Inc., Englewood Cliffs.
40. Skelton, P.H. (1993) A Complete Guide to the Freshwater Fishes of Southern Africa. Southern Book Publishers, Birmingham.
41. Gbaguidi, A.S. and Pfeifer, V. (1988) Statistiques des pêches continentales; année 1987. Services Etudes et Statistiques. Projet Pêche Lagunaire, Deutshe Gesellschaft Fur Technische Zusammenarbeit (GTZ) GMBH et Direction des Pêches-Cotonou, Bénin.
42. Murphy, B.R. and Willis, D.W. (1996) Fisheries Techniques. 2nd Edition, American Fisheries Society, Bethesda.
43. Krebs, C.J. (1989) Ecological Methodology. Harper & Row Publishers, New York.
44. Morton, R.M. (1990) Community Structure, Density, and Standing Crop of Fishes in a Subtropical Australian Mangrove Area. Marine Biology, 105, 385-394. http://dx.doi.org/10.1007/BF01316309
45. Williamson, I. and King Mather, C.B. (1994) A Comparison of Fish Communities in Unmodified and Modified Inshore Habitats of Raby Bay, Queensland. Estuarine, Coastal and Shelf Sciences, 39, 401-411. http://dx.doi.org/10.1006/ecss.1994.1072
46. Fryer, G. and Iles, T.D. (1972) The Cichlid Fishes of the Great Lakes of Africa: Their Biology and Evolution. Oliver & Boyd, Edinburgh.
47. Snoeks, J. (2000) How Well Known Is the Ichyodiversity of the Large East African Lakes? Advances in Ecological Research, 31, 17-38. http://dx.doi.org/10.1016/S0065-2504(00)31005-4

48. Mbega, J.D. (2004) Biodiversité des poisons du bassin inférieur de l'Ogooué (Gabon). Thèse de docteur en sciences, Facultés Universitaires Notre dame de la paix, Namur.
49. Adepo-Gourène, B. and Gourène, G. (2008) Différenciation morphologique des populations naturelles d'une sous- espèce de Tilapia Sarotherodon melanotheron melanotheron Rüppell, 1852 (Teleostei; Cichlidae) de c?te d'Ivoire. Science & Nature, 5, 15-27. http://dx.doi.org/10.4314/scinat.v5i1.42148
50. Tossavi, E.C. (2011) Evolution de la biodiversité et de l'exploitation des poissons du lac Toho (Sud-Benin): Implication pour la gestion durable des ressources halieutiques. Master's Thesis, Faculté des Sciences et Techniques, Université d'Abomey-Calavi, Abomey Calavi.
51. Bowen, S.H. and Allanson, B.R. (1982) Behavioral and Trophic Plasticity of Juvenile Tilapia mossambica in Utilization of the Unstable Littoral Habitat. Environmental Biology of Fishes, 7, 357-362. http://dx.doi.org/10.1007/BF00005570

CHAPTER 7

Restoration of a Temperate Reef: Effects on the Fish Community

Josianne Gatt Støttrup[1*], Claus Stenberg[1], Karsten Dahl[2], Louise Dahl Kristensen[1], Katherine Richardson[3]

[1]*National Institute of Aquatic Resources, Technical University of Denmark, Charlottenlund, Denmark*

[2]*Department of Bioscience, Aarhus University, Roskilde, Denmark*

[3]*Center for Macroecology, Evolution and Climate, Natural History Museum of Denmark, University of Copenhagen, Copenhagen, Denmark*

ABSTRACT

The extraction of large boulders from coastal reefs for construction of harbours and coastal protection has led to habitat degradation for local fish populations through the destruction of cavernous reefs and changes in macroalgal cover resulting from a loss of substrate. The temperate reef at Læsø Trindel in Kattegat, Denmark, has now been re-established with the aim of restoring the reef's historical structure and function. The effects of the restoration on the local fish commu- nity are reported here. Fishing surveys using gillnets and fyke nets were conducted before the restoration (2007) and four years after the restoration of the reef (2012). Species of the family Labridae, which have a high affinity for rocky reefs, dominated both before and after the restora- tion. Commercially important species such as cod Gadus morhua, and saithe Pollachius virens, oc- curred infrequently in the catches in 2007 but were significantly more abundant in the catches in 2012. Cods were especially attracted to the shallow part of the reef that was restored by adding stones. For some species, such as ballan wrasse Labrus bergylta, and cod, the proportion of larger individuals increased after the restoration. The findings highlight the importance of reef habitats for fish communities and the need for their protection.

Keywords: Reef Restoration, Impact Analysis, Labridae, Gadidae

1. INTRODUCTION

Marine reefs are important fish habitats providing complex structures that provide refuge for fish and hard substrate for benthic fauna and macroalgal forests that provide refuge and feeding sites for fish [1] -[3]. Reefs are listed in the EU Habitats Directive (1170 Reefs; Council Directive 92/43/EEC) and, for this reason, marine reefs in Danish waters have been designated as protected areas and are part of the EU-wide Natura2000 network indicating the acknowledged importance of this habitat type. Most of our knowledge on reef habitats is derived from studies on tropical reefs, although in the recent decade more focus has been directed towards the quantita- tive significance for fish communities of reef habitats in temperate areas. Most monitoring for fish assessment is limited to relatively smooth bottom areas due to the design of the survey gear [4]. Monitoring of fish communi- ties on complex habitats such as temperate reefs and biogenic reefs are often limited to specific ecological stud- ies of limited duration [5] [6]. Significantly higher catch rates of cod Gadus morhua on rough than on smooth bottom were suggested to be the main source for underestimation of the stock size of the North Sea cod [7]. Thus, limited temporal knowledge is available on the fish diversity and abundance in relation to temperate reefs and the quantitative ecological role of temperate reefs in our region.

Extensive mineral extraction of stones and boulders in coastal areas of Denmark [8] has not only led to de- struction of cavernous reefs and removal of hard bottom but also removal of biogenic structures associated with these reefs and which are a main feature of temperate reefs [9]. The removal of larger boulders increases the av- erage depth which may result in reduced benthic plant growth due to reduced light penetration with depth. A reduction of benthic plant growth reduces habitat complexity and reduces type and diversity of refugia for juve- nile fish. The removal of the top stabilizing layer of larger boulders may also result in destabilization of the re- maining reef, where smaller boulders or stones may be upturned in storms or strong current events with subse- quent loss of perennial macrophytes and potential colonization of opportunistic macroalgal species. In Denmark, mineral extraction of large stones has now ceased but the consequences of the historical removal of material and destruction of these habitat types for fish populations are largely un-documented. No archives are available on the magnitude or precise geographic locations of the extractions.

The reef at Læsø Trindel, north-east of the island Læsø in Kattegat, Denmark (Figure 1(a)) was one of the many reef areas where mineral extractions took place during a period in the last century [8]. Archival maps showed the shallowest part of the reef to be 1.25 m below the surface in 1831. The depth increased to 2.2 m in 1930 and to ~4 m in the 1970s [10]. No information is available on how many boulders were mined from this reef complex. The macroalgal vegetation at this site was included in the National Marine Monitoring Program in 1991 and the results of the monitoring showed that the status of the reef was not in a Good Environmental State mainly due to the high dominance of opportunistic species at the expense of perennial species, compared to other sites in the monitoring program [11].

Species of the Labridae family are likely to be most affected by loss of reef habitats because of their high af- finity to these habitat types and their complete dependence on this substrate for recruitment [12]. In fact, these authors suggested that a good indicator for the status of this habitat type might be the abundance of Labridae species or proportion of larger fish of Labridae as they depend on this substrate for reproduction. These species are, however, of no commercial value and are, therefore, not monitored. Thus, changes in their abundance or distribution may go undetected. In rocky habitats, species of Labridae also serve as prey for commercially important piscivors such as cod, mackerel Scombrus scombrus, saithe Pollachius virens, and whiting Merlangius merlangus that rely on teleosts as a major food source [13]. Thus, loss or degradation of rocky habitats may re- sult in changes in trophic dynamics and impact trophic integrity in and around these habitat types. In recent decades, habitat restoration has expanded from terrestrial areas to the aquatic environment. In the aquatic envi- ronment, focus has been on freshwater systems (e.g. [14]). Restoration efforts in the marine environment are di- verse, most of which have focused on restoring coral reefs [15], or large structural elements such as planting vegetation [16], restoring mangroves [17] and oyster reefs [18]. Unlike impacts from other human activity such as eutrophication and fishing, where the removal of the cause of the impact may lead to system recovery, reef habitats cannot be expected to recover without human intervention. The restoration of the marine temperate reef at Læsø Trindel (Kattegat, Denmark) represents one such large-scale intervention in European waters to restore a degraded natural reef. To our knowledge, this study represents the first temperate reef restoration in European waters. Although we have no details on the original form of the reef, we can use "before" and "after" restoration studies of biodiversity distributions on the reef to infer how reef damage affects ecosystems. In this study, the damaged reef was restored to its former function in 2008 to serve as substrate for larger kelp-forming algae.

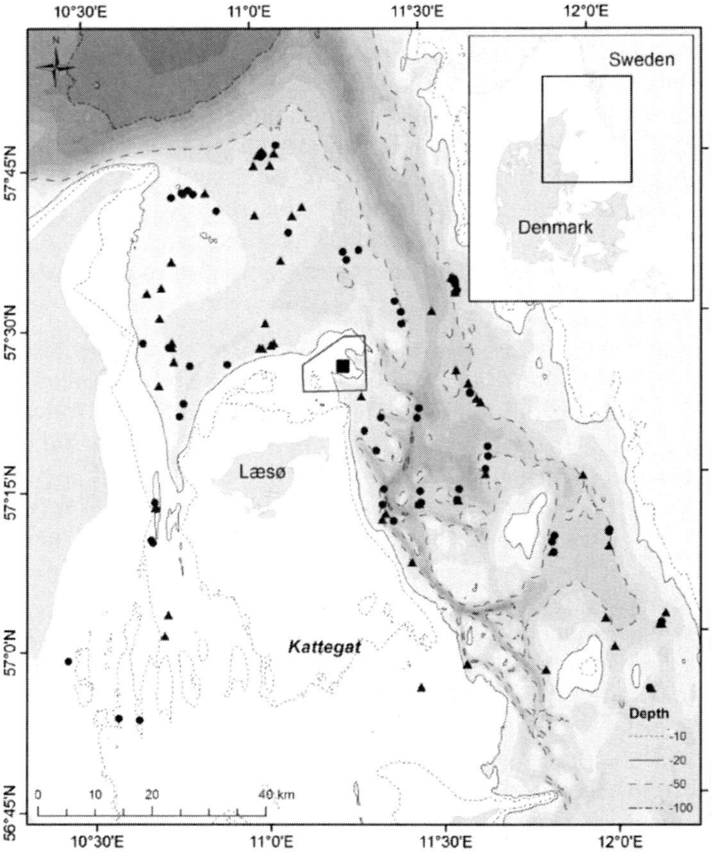

(a)

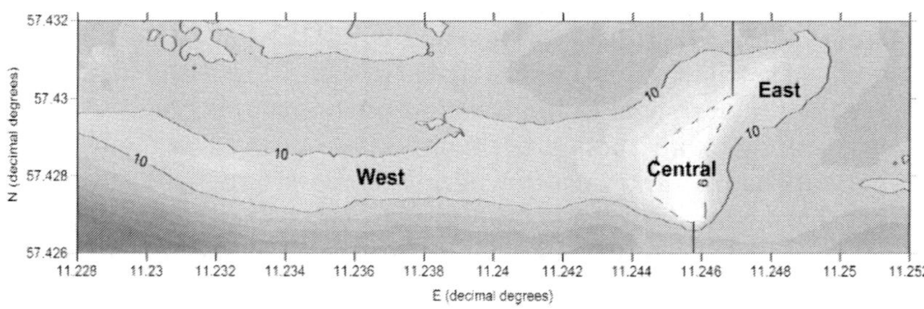

(b)

Figure 1. (a), (b) Map of the location of the reef at Læsø Trindel (black square, **Figure 1**(a)) within the NATURA 2000 site in the Kattegat (black border). Black circles (2005-2007; BEFORE) and triangles (2010-2012; AFTER) mark the positions of the stations from the research trawl surveys within 50 km of the reef site. **Figure 1**(b) shows the actual reef studied with 6 and 10 m depth contours and the location of the three sampling areas.

The aim of this study was to examine whether effects of restoring the boulder reef could be identified for the associated fish community. Effects on bottom fauna and flora of the reef restoration are presented elsewhere [10]. We hypothesized that the reef after restoration would provide more refugia and higher prey abundance than prior to restoration and that this would favor a higher abundance of reef fishes, a higher species richness and broader size distribution of fish.

2. MATERIALS AND METHODS

2.1. Study Site and Reef Restoration

The reef at Læsø Trindel is located about 12 km north-east of the island of Læsø in the Kattegat (Figure 1(a)). The reef was stabilized with a layer of large boulders ranging in individual weight from 1600 kg to 3300 kg and parts of the central reef area were restored to its original depth during 2008 (Figure 1(b); [19]). A total of ~100,000 tons of boulders originating from Norway were deployed, covering an area of ~27,600 m^2 distributed in the depth range of 4 - 10 m.

2.2. Sampling

The monitoring program was based on a "Before-After" approach. Due to economic constraints only one sam- pling was conducted before deployment of the reef in June 2007, and one sampling four years after deployment in June 2012. The sampling was a random stratified design. The area on Læsø Trindel was stratified into the ar- eas; the central shallow part of the reef at 2 - 6 m depth, the western central part at 6 - 10 m depth; the eastern central part also at 6 - 10 m depth, and the surrounding area shallower than 10 m (Figure 1(b)). Although the depth contours had changed in the area of the restored reef [19], the area stratifications were maintained within the three areas. Fish abundance was surveyed with multi-meshed gillnets [20] and fyke nets. The mesh sizes in the multi-meshed gillnets were 11, 14, 19, 24, 31, 41, 53 and 70 mm. Each panel was 1.5 m high and 6 m in length, except for the 53 and 70 mm panels which were, respectively, 12 m and 52 m long. Each gillnet had a random combination of panels, separated from each other by 1.8 m (float and sink line). The gillnets were deployed in the afternoon or evening and retrieved the following morning resulting in ~12 h fishing time. The fyke nets had a mesh size of 18 mm and were 42 cm in height with a 6.5 m leader. Five fyke nets were mounted together in a row. These were deployed in the afternoon and fished ~48 h. In 2007, 4 replicates were made in each sampling site with gillnets and 7 - 9 replicates with fyke nets. In 2012, 4 - 11 replicates were made with gillnets and 2 - 4 replicates with fyke nets. Catches were identified to species and total length of each fish measured to the nearest lower 0.5 cm and weighed. For gillnets, catch per unit effort (CPUE) was standardized as catch in numbers per species or group per gillnet length in all the combined mesh size panels in the gillnet deployment. For fyke nets, CPUE was standardized as total catch in numbers per species or group for each deployment of the combined five fyke nets.

As we had no control area for monitoring, due to economic constraints, we chose to compare the development in cod abundance BEFORE and AFTER on Læsø Trindel with CPUE data from research trawl surveys in the neighboring area (Figure 1(a)). The CPUE provides an index where it is possible to observe positive or negative changes in abundance for each gear type. The trawl surveys are conducted by DTU Aqua in spring and autumn each year. The data from these surveys is also used in stock assessment of cod in the Kattegat in ICES where additional information on the surveys can be found (e.g. [21]). We included data within a distance of 50 km to Læsø Trindel and CPUE was divided into cod smaller or larger than 30 cm total length, and two periods repre- senting the periods BEFORE (year 2005-2007) and AFTER (year 2010-2012). The stations for the trawl sam- pling are shown in Figure 1(a).

2.3. Data Analyses and Statistics

The effect of the reef restoration on fish abundance was analyzed for each group or species of fish by analyzing for the effect of BEFORE (year 2007)/AFTER (year 2012) in ANOVAs using the GLM procedure in SAS software 9.4. Copyright, SAS Institute Inc. Model residuals were tested for normal distribution by the Ander- son-Darling test in the UNIVARIATE procedure in SAS. Fish were grouped into four categories "Gadidae", "Labridae", "Pleuornectiformes" and "Other" for the remaining fish species. Catch numbers were +1 log trans- formed prior to the statistical tests. Post-Hoc comparisons were performed with Tukey's Studentized Range (HSD) test.

Species diversity was calculated using Shannon's H index

$$H = \sum pi^{*} \ln pi$$

and Shannon's equitability index E

$$E_H = H/\ln S$$

where p_i is the proportion of species i to the total number of all species and S is the total number of species.

Changes in abundance of cod of different size classes (<20 cm, 20 - 30 cm, >30 cm) in the area around Læsø Trindel in the BEFORE/AFTER period were examined by an ANOVA on the negative binomial distributed CPUE data from the DTU Aqua research trawl surveys by the procedure COUNTREG in SAS. A similar analysis was conducted for changes in abundance of cod caught in the gillnets in the three reef sampling areas. Test statistics for the residual test are not shown unless the test was statistically significant. A significance level (p) of 0.05 was applied to all tests.

3. RESULTS

The total number of species caught in 2007 in the combined catches of both gear types was 33 while a total of 30 species was caught in 2012. The number of species caught in the gillnets was 27 and was similar BEFORE and AFTER the restoration, whereas the number of species caught in the fyke nets was lower following restora- tion: 25 BEFORE decreasing to 21 AFTER (Table 1). Before the restoration, goldsinny wrasse Ctenolabrus rupestris, corkwing Symphodus melops, and small-mouthed wrasse Centrolabrus exoletus, comprised 78% of the gillnet catches and ballan wrasse Labrus bergylta, corkwing and sole Solea solea, 64% of the fyke net catches. After the restoration, the wrasses still dominated the catches but constituted 68% of the total catch. In the fyke nets, the Labridae were no longer dominant in 2012 and the catch was more evenly distributed among the groups. Thus, both Shannon's diversity index and equitability index tended to be slightly higher after the restoration, re- flecting the more even distribution of the individual species in the community AFTER the restoration than BEFORE.

The gadoids caught in the gillnets constituted mainly of cod, saithe and some whiting and Pollack Pollachius pollachius (Table 2). Abundance of gadoids was generally low in all parts of the reef BEFORE the restoration but increased significantly AFTER for all areas (Figure 2(a)) (ANOVA, $p < 0.001$). Abundance increased particularly in the central part of the restored reef at 2 - 6 m depth. Here, abundance was significantly higher (Tukey, $p < 0.05$) than over the western and eastern deeper (6 - 10 m) parts of the reef. The eastern and western parts of the reef were not significantly different from each other with respect to abundance neither BEFORE nor AFTER (Tukey, $p > 0.05$).

Table 1. Number of species caught in gillnets and fyke nets in 2007 and 2012. For each year and gear the frequency of catches of the different groups of fish species relative to total catch, Shannon's diversity index H and equitability index E_H calculated using catch per unit effort.

	Gillnets		Fyke nets	
	2007	2012	2007	2012
n species	27.00	27.00	25.00	21.00
Freq. Gadidae	0.02	0.13	0.001	0.20
Freq. Labridae	0.87	0.68	0.654	0.25
Freq. Pleuornectiformes	0.06	0.09	0.235	0.30
Freq. Other	0.05	0.10	0.110	0.26
H	1.81	1.96	2.09	2.48
E_H	0.55	0.59	0.65	0.82

Table 2. Catch per unit effort for each fish species caught in gillnetsand fyke nets.

		Gillnets		Fyke nets	
Latin name	English name	2007	2012	2007	2012
Gadidae					
Gadus morhua	Cod	4.50	13.40	0.22	11.00
Pollachius virens	Saithe	0.40	3.52	0.00	0.67
Pollachius pollachius	Pollack	0.40	0.76	0.00	0.00
Merlangius merlangus	Whiting	0.30	0.04	0.00	0.00
Labridae					
Labrus bergylta	Ballan wrasse	19.80	2.32	20.11	0.00
Labrus bimaculatus	Cuckoo wrasse	4.50	0.16	0.11	0.00
Ctenolabrus rupestris	Goldsinny wrasse	29.50	59.80	8.11	8.44
Symphodus melops	Corkwing	81.00	23.24	61.89	6.33
Centrolabrus exoletus	Small-mouthed wrasse	98.70	7.04	15.33	0.11
Pleuronectiformes					
Limanda limanda	Common dab	9.60	6.52	11.22	6.22
Psetta maxima	Turbot	1.30	0.04	0.11	0.00
Pleuronectes platessa	Plaice	0.90	3.24	1.56	0.89
Microstomus kitt	Lemon sole	0.50	0.44	2.89	3.78
Platichthys flesus	Flounder	0.00	0.04	0.11	0.00
Scophthalmus rhombus	Brill	0.80	0.24	0.89	0.33
Lepidorhombus whiffiagonis	Megrim	0.00	0.00	0.11	0.00
Solea solea	Sole	3.20	1.28	21.00	6.33
Arnoglossus laterna	Scaldfish	0.00	0.04	0.00	0.00
Zeugopterus punctatus	Topknot	0.10	0.00	0.00	0.11
Other					

Icelus bicornis	Twohorn sculpin	2.30	0.00	1.00	0.00
Trachinus draco	Greater weever fish	4.30	3.40	0.22	0.22
Callionymus lyra	Common dragonet	1.50	3.80	0.78	1.78
Ciliata mustella	Five-bearded rockling	0.00	0.04	0.00	0.00
Enchelyopus cimbrius	Four-bearded rockling	0.00	0.00	1.56	1.56
Belone belone	Garfish	0.00	0.12	0.00	0.11
Entelurus aequoreus	Snake pipefish	0.00	0.00	0.33	0.00
Cyclopterus lumpus	Lumpfish	0.10	0.00	0.00	0.00
Clupea harengus	Herring	0.90	0.00	0.00	0.00
Spinachia spinachia	Ten-spined stickleback	0.10	0.60	1.11	0.11
Pholis gunnellus	Butter fish	0.10	0.36	0.56	0.44
Hyperoplus lanceolatus	Greater sandeel	0.10	0.00	0.00	0.00
Myoxocephalus scorpius	Sculpin	2.80	3.16	3.00	2.89
Taurulus bubalis	Sea scorpion	0.00	1.60	0.67	2.56
Anguilla anguilla	Eel	0.00	0.00	4.78	1.44
Zoarces viviparus	Eelpout	0.90	0.08	3.67	4.11

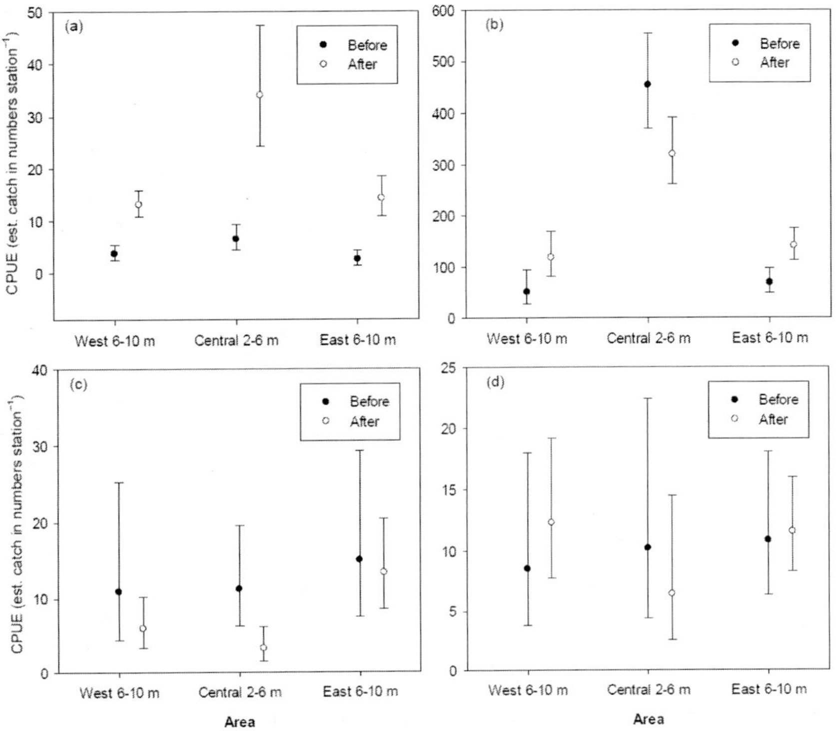

Figure 2. Catch per unit effort (CPUE) of fish caught with multi-meshed gillnets on Læsø Trindel. Estimated numbers + SE. BEFORE = 2007, AFTER = 2012. (a) Gadidae; (b) Labridae; (c) Pleuronectiformes; (d) Other.

Five species of Labridae were caught in gillnets in this study (Table 2). The highest abundances of these spe- cies were observed in the central shallow (2 - 6 m) part of the reef both BEFORE and AFTER the restoration (Figure 2(b)). However, in the BEFORE situation, the difference between the areas was more pronounced, with significantly higher abundance in the central shallow (2 - 6 m) compared to the two deeper areas (Tukey, $p > 0.005$). The abundance of Labridae in the central shallow part of the reef decreased after the restoration, al- though the change was not significant (ANOVA, $p = 0.0585$). In the deeper part (6 - 10 m), the abundances of representatives from this group increased in both the deeper western part (ANOVA, $p = 0.0419$) and the eastern part ($p = 0.0058$). While goldsinny wrasse increased in abundance AFTER the reef restoration, the other four species of Labridae decreased in abundance following restoration of the reef (Table 2).

Nine species of flatfish (Pleuronectiformes) were caught in the reef area with gillnets, most commonly dab Limanda limanda, and sole (Table 2). Abundance of these species decreased slightly after the restoration of the reef. The changes were not significantly different in the deeper western and eastern parts of the reef, but were significantly lower in the central shallow part ($p = 0.0344$) (Figure 2(c)). In contrast to all the other flatfish spe- cies, plaice Pleuronectes platessa tended to increase in abundance after the reef restoration (Table 2).

The remaining fish species caught in gillnets were pooled into a group called "Other" (see list of species in Table 2). There were no differences in abundance of this "Other" group in any of the three areas of the reef from before and after the restoration (Figure 2(d)).

Cod in fyke nets was the most frequently caught species (Table 2). The abundance of cod increased from 2007 to 2012 in all areas over the reef, although most notably in the shallow central part of the reef (West 6 - 10 m: $p = 0.0069$; East 6 - 10 m: $p = 0.0032$; Central 2 - 6 m: $p = 0.0048$) (Figure 3(a)). Another gadoid species, the saithe, caught both BEFORE and AFTER the restoration was in the size range 18 - 26 cm and increased in abundance after the restoration (Table 2). In 2007, corkwing, ballan wrasse and small-mouthed wrasse were caught most frequently in fyke nets (Table 2). After the restoration, catches of all three species declined but the decline of wrasses caught in fyke nets was only significant in the eastern deeper part of the reef (East 6 - 10 m: $p = 0.0415$), while catches of goldsinny wrasse remained at the same level as BEFORE (Figure 3(b)). Flatfish catches tended to decline but this decline was not significant in any of the reef areas. The largest declines in catches were observed for dab and sole, which were the two most abundant flatfish in the 2007 catches (Figure 3(c); Table 2). Fish species from the "Other" group showed slightly higher catches after the restoration over the deeper eastern and western parts of the reef and lower in the central shallow part but the changes were not sig- nificant (Figure 3(d)).

The analyses of the size distribution of all fish caught in gillnets over the reef showed that numbers of fish >30 cm tended to increase after the restoration (Figure 4). The size increase was mainly due to a higher occurrence of larger cod, which aggregated around the shallow part of the reef after the restoration (Figure 5(a)). The proportion of cod >30 cm increased from 0.16 to 0.24 AFTER the restoration of the reef and the difference was statistically significant (ANOVA, $p = 0.0004$). The abundance of juvenile cod also increased significantly after the restoration (<20 cm, ANOVA, $p < 0.0001$; 20 - 30 cm, ANOVA, $p = 0.0001$). Although abundance of ballan wrasse decreased after the restoration, those that were present AFTER the restoration were larger than BEFORE (Figure 5(b)). Goldsinny wrasse increased in abundance after the restoration with a slight shift towards larger individuals (Figure 5(c)).

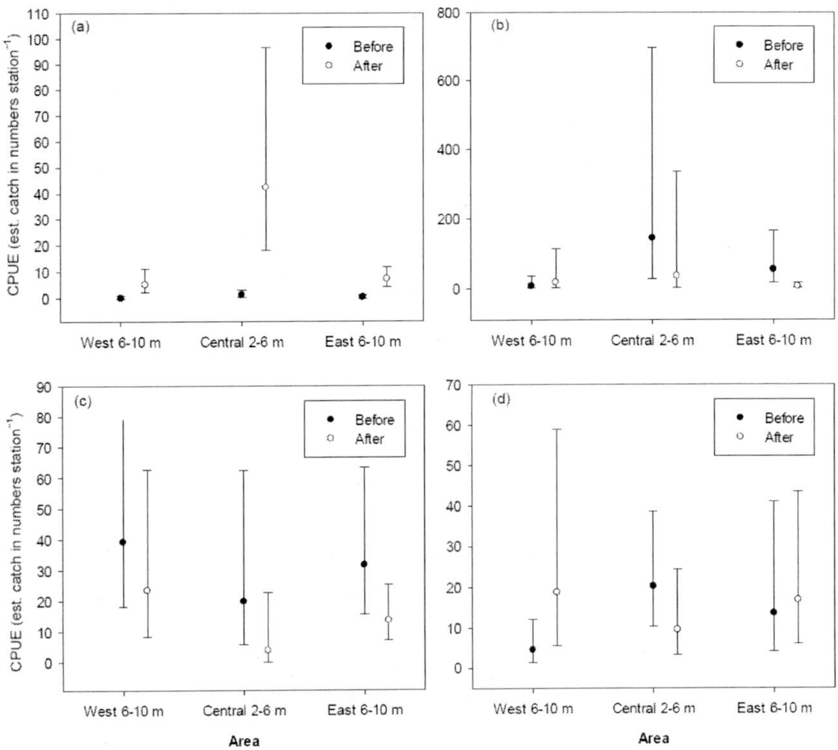

Figure 3. Catch per unit effort (CPUE) of fish caught with fykenets on Læsø Trindel. Estimated numbers + SE. BEFORE = 2007, AFTER = 2012. (a) Gadidae; (b) Labridae; (c) Pleuronectiformes; (d) Other.

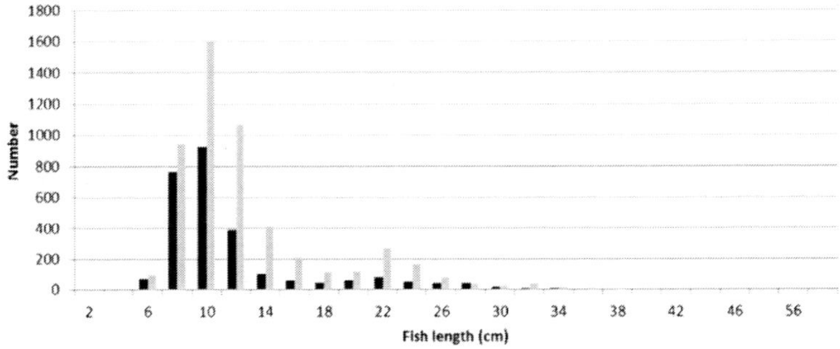

Figure 4. Length distribution of all fish caught in the multimesh gillnets BEFORE (black bars) and AFTER (light grey bars) the reef restoration.

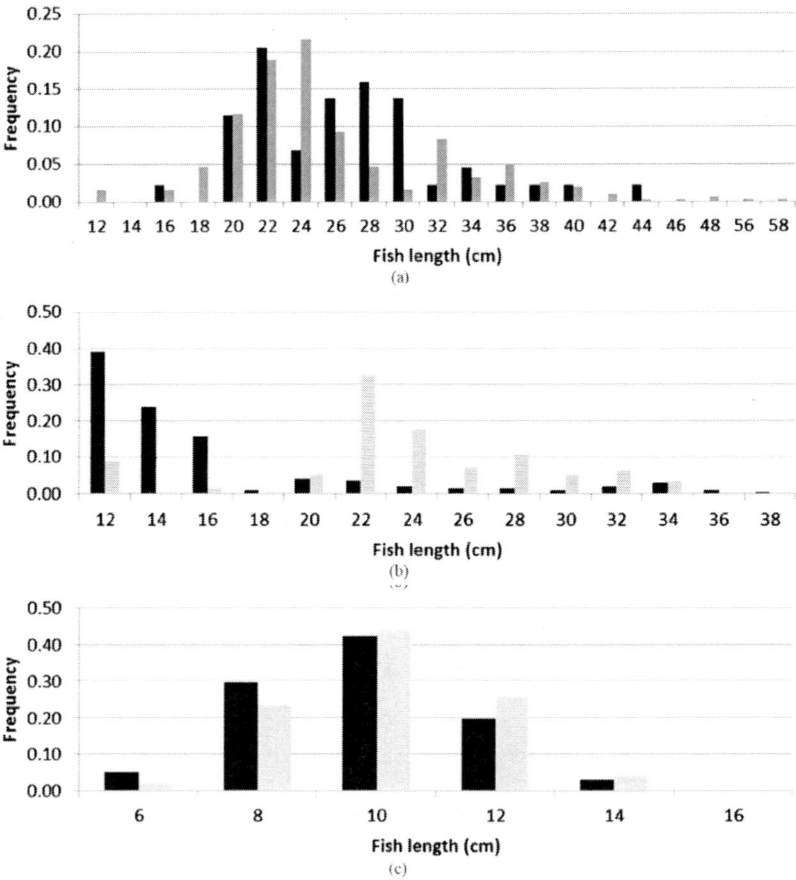

Figure 5. (a)-(c) Length frequency of cod (a), ballan wrasse (b) and goldsinny wrasse (c) caught in the multimesh gillnets BEFORE (black bars) and AFTER (light grey bars) the reef restoration.

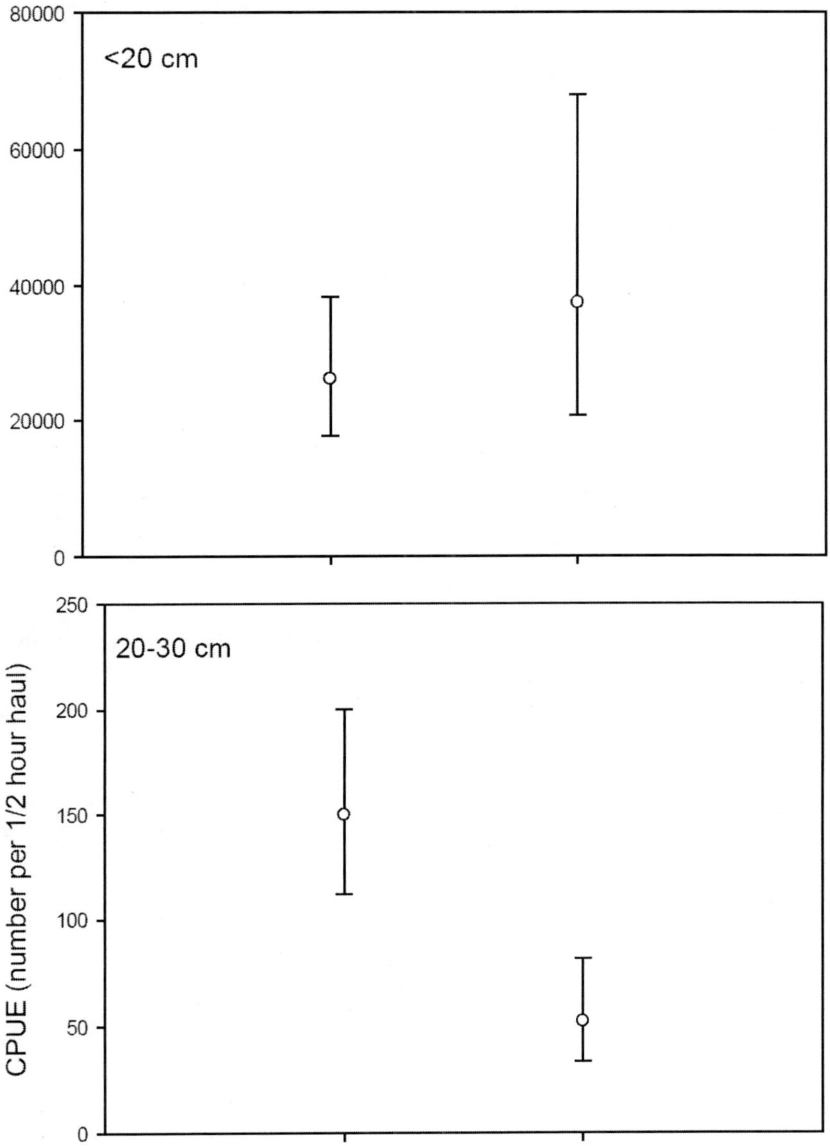
80000
60000
40000
20000
0
<20 cm
250
200
150
100
50
0
20-30 cm
CPUE (number per 1/2 hour haul)

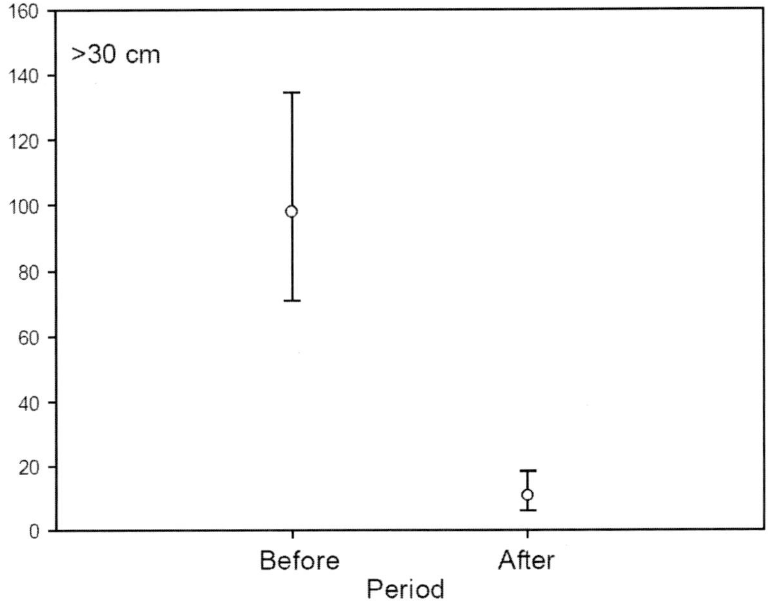

Figure 6. Catch per unit effort (CPUE) of cod in research trawl surveys within 50 km distance from Læsø Trindel. Estimated numbers + SE. BEFORE = years 2005-2007, AFTER = years 2010-2012.

Information from the research trawl surveys in the neighboring area showed that the abundance of cod < 20 cm did not change significantly between the BEFORE and AFTER period (ANOVA, $p > 0.55$) while cod from 20 - 30 cm (ANOVA, $p < 0.02$) and cod >30 cm (ANOVA, $p < 0.001$) significantly declined from BEFORE to AFTER (Figure 6). Thus, the development with respect to the abundance and population structure for cod noted over the reef following restoration did not mirror the general development in the region as a whole.

4. DISCUSSION

This study demonstrates a change in fish diversity, the distribution of species and population size characteristics for some fish species from BEFORE and AFTER reef restoration. Results from this study presented elsewhere [10] [19] documented that the reef restoration led to a re-colonization of reef regions with macroalgae. Thus, we suggest that the changes noted in the fish populations here are likely related to the restoration of the reef habitat with macroalgal vegetation.

Much of our understanding of the importance of spatial heterogeneity and structure for recruitment of reef fishes derives from studies conducted on coral reefs. However, temperate reefs differ from coral reefs in that macroalgae dominate the physical structure of temperate reefs [9]. Thus, variability in abundance and type of macroalgae between and within reefs drive the dynamics of recruitment in temperate reef fishes [2] [6]. Destruction or degradation of temperate reef habitats leading to loss of large macroalgae and dominance of opportunis- tic algae results in loss of spatial heterogeneity and structure that appear to be important for recruitment in reef fishes as well as other demersal species that utilise complex hard bottom habitat.

The total number of species (32) caught with the multi-meshed gillnets over the two sampling years in this study was slightly lower than the number (38) reported by [5]. These authors used multi-meshed gillnets in a rocky-bottom coastal habitat on the west coast of Sweden, north-east of the study site of this study. The fyke nets caught a further four species including the European eel Anguilla anguilla, bringing the total number of species caught to 36. This indicates that the combination of multi-meshed gillnets and fyke nets used in this study was sufficient to sample a wide range of the species occurring in the area. Both gears used are selective. The high size selectivity of gillnets was taken into consideration with the multi-mesh gillnets, where a broad range of mesh sizes were used to ensure a wide size range of fish. As the aim of this work was to compare the fish assemblages BEFORE and AFTER the restoration, it was assumed that the bias in species capture and size range would be similar. However, the increased complexity of the restored reef, due to both a higher variety in relief but also the presence of larger algae, may have reduced the fishing efficiency of the nets, and the catches in the 2012 surveys may thus be underestimates.

The result that more complex habitats lead to higher species diversity and abundance than less complex ones is not unexpected and has been shown earlier for tropical and temperate marine habitats [22] [23]. In this study, however, we did not alter the habitat from a smooth bottom to a complex bottom, but rather attempted to in- crease the complexity by restoring the larger boulders, creating more relief and providing a physically stable substrate for the development of macroalgae [10]. The (Ash-Free Dry Weight) AFDWm^{-2} of the macroalgal biomass increased by more than twofold at 5 - 6 m depth and this was especially due to an increase in the brown algae (Phaeophyta) (Karsten Dahl, upublished data). Apparently, the increased complexity of the restored reef was not sufficient to result in a significant increase in fish species richness. In the general Kattegat area, a de- cline in sea bottom temperature of about 0.75°C between 2007 and

2011 was associated with a decline in species richness of 3 - 4 species per degree [24]. The decline in number of species in the fyke nets could, therefore, be due to this temperature decrease, but this does not explain the similar number of species in the gillnet catches in 2007 and 2012. The species decline in the fyke nets was driven by fewer flatfish species and the lack of ballan wrasse in the 2012 catches and may be due to a less favorable habitat for these species following reef restora- tion.

In this study, although species richness declined in the fyke net catches, Shannon's diversity index increased. The dominance of Labridae in the catches before the restoration decreased after the restoration resulting in a more even distribution of species especially in the fyke net catches. Abundance of the resident fish species, goldsinny wrasse tended to increase. Goldsinny wrasse is a resident species with high affinity to rocky substrate [25] [26]. Kelp forests, similar to that developed in the restored reef [10] provide ample feeding opportunities for this species [27]. Since refuge availability seems to be the main limiting factor determining the abundance of goldsinny wrasse [28], an increase in abundance of this species was expected. The larger labrid, the ballan wrasse, decreased in abundance AFTER but a higher proportion of larger fish inhabited the reef after the resto- ration than BEFORE. The ballan wrasse is a sedentary, territorial species with slow growth and it is a protogy- nous hermaphrodite [29]. The presence of larger sizes of fish is important in sex-changing fish species to main- tain reproductive potential and population size [30]. Thus, the presence of larger specimens is important to se- cure sufficient sex ratios for effective mating. Due to the restricted home range of the species, even a small area with improved habitat and protected from fisheries would be sufficient for this and other Labridae species. The results suggest that the restored reef provided better opportunities for the larger fish predators and improved habitat for recruitment of wrasses, providing protection and potentially increasing the reproductive potential due to the increase in fish size.

The distribution of commercially important gadoids depends on the presence of heterogeneous substrate types and associated biogenic growth [31] [32]. Vegetated habitats, i.e. shallow rocky areas with algae and soft bottom sediment with seagrass beds, are important nursery areas for young juvenile cod [33]. In this study, the gadoids cod and saithe increased in abundance by a factor of 3 - 6 after the restoration. Primarily the larger juveniles (>30 and >20 cm in length for cod and saithe, respectively) increased in abundance. The increase in abundance of juveniles in this study was not reflected in the bottom trawl surveys for the Kattegat stock component [21] [34], which show a negative trend. Furthermore, the analysis on the bottom trawl data performed in this study showed a marked decline in the abundance of larger juvenile cod in the area surrounding the reef (within 50 km). These data were used in lieu of a lack of reference site for sampling before and after restoration and provide an opportunity to compare the development in abundances of different sizes of cod in the period before and after the reef restoration.

Notably, the shallow part of the reef (2 - 6 m) attracted the highest abundance of cod AFTER the restoration. This is in line with the observations of [5], where the highest abundance of fish was found in shallow water (0 - 3 m and 3 - 6 m) kelp habitats and significantly lower abundance in the deeper (6 - 10 m) rocky habitat. The study of [5] was conducted in the Swedish archipelago north-east of the site of the reef restoration in this study. Juvenile cod are especially susceptible to limitations in demersal habitat due to their density-dependent mortality [35], suggesting a high vulnerability to the loss of complex habitats and the need to preserve these habitats for maintaining or rebuilding severely depleted stocks [36].

Fish densities are positively related to vegetation biomass [9] and the increased macroalgal biomass [10] was expected to increase fish densities in the restored area of the reef. The macroalgal dominance in temperate reefs may lead to high variability in fish assemblages and broader use of a reef habitat due to linkages between habitat attributes and life-stage strategies or behavioral responses in reef associated fishes [6]. Kelp forests, such as the one in the process of forming on the restored reef area [10], are known to be important feeding grounds for many fish species including cod [27].

Habitat degradation with loss of forest-forming macroalgae can be compared to kelp-harvest activity, except in the former case this impact is of a more permanent nature because before the reef stability or structure has been restored, macroalgae cannot re-establish. In newly-harvested kelp areas in Norway, the number of juvenile gadoids was 92% lower than in un-harvested areas [32]. Lower abundances of gadoids persisted one year after the harvest. This is not unexpected considering gadoids utilize kelp forests as feeding and nursery areas and for shelter from larger predators [27]. Cod seek refuge in macroalgae to avoid predation in the presence of actively foraging predators [31], so juvenile cod tend to be segregated from larger cod [37]. In this study, the restoration of the reef resulted in higher abundances of juvenile cod (<20 cm) as well as a higher proportion of larger (>30 cm) cod in the catches. The increased complexity of the restored reef habitat may, thus, have provided shelter and food for the smaller cod and better predation opportunities for the larger cod.

The increased proportion of larger cod, increased abundance of saithe, and increased proportion of larger specimens of the larger labrid, the ballan wrasse, in the restored reef area may reflect greater prey availability on the restored reef. The increased frequency of larger cod in the catches on the restored reef was in contrast to the decline of larger cod in the catches from the bottom trawl surveys in the area surrounding the restored reef. The generally higher frequencies of larger cod in our catches further reflect the affinity of larger cod to rocky sub- strate. The Kattegat cod stock is a unique population the Kattegat and has, since 2000, been considered outside safe biological limits by ICES [21]. Commercial fishing continues to take place and, accordingly, the reduction in reproductive capacity of this stock is considered to be due to reduced stock size rather than habitat loss [38]. The fishery in the area is dominated by trawls [21] ; thus the restored reef where trawling is virtually impossible, could act as a refuge from fishing for the larger cod and contribute to the reproductive performance of the local population and, ultimately, recruitment.

Restoration is a rapidly developing field of research requiring clear objectives, appropriate definition of suc- cess criteria, and development of effective methodology to measure the success [39]. Most assessments focus on ecological attributes and although socio-economic attributes also should be targeted, it is important to measure the success of restoration to further this research field [39]. The restoration of the temperate reef in this study focused on ecological attributes; the results on the physical and social attributes will be addressed elsewhere. In this study we showed that the reef restoration brought about changes in the fish community. Dominance of wrasses was maintained after the restoration, but was less obvious than BEFORE the restoration due to the in- creased abundance of several other species resulting in a more even distribution of species. This suggests a higher variety of refuge and suitable micro-habitat types AFTER the restoration than BEFORE. Commercially important gadoid species, cod and saithe, increased 3 - 6-fold in abundance after the restoration. The restored shallow part of the reef seemed to particularly attract cod and goldsinny wrasse, although not significantly for the latter species. A higher proportion or larger specimens of cod and ballan wrasse after the restoration indicated improved foraging opportunity for the larger fish and an increase in the reproductive potential for these species. The concurrent increase in abundance of smaller cod (<20 cm) and higher proportion of larger cod (>30 cm) indicated also an increase in refuge availability.

ACKNOWLEDGEMENTS

Josianne Gatt Støttrup, Claus Stenberg, Karsten Dahl, Louise Dahl Kristensen, Katherine Richardson (Blue Reef)". LIFE06 NAT/DK/000159 and by the National Program for Coastal Recreational Fisheries "marinfiskepleje".

REFERENCES

1. Graham, M.H. (2004) Effects of Local Deforestation on the Diversity and Structure of Southern California Giant Kelp Forest Food Webs. Ecosystems, 7, 341-357. http://dx.doi.org/10.1007/s10021-003-0245-6
2. Christie, H., Jørgensen, N.M. and Norderhaug, K.M. (2007) Bushy or Smooth, High or Low; Importance of Habitat Architecture and Vertical Position for Distribution of Fauna on Kelp. Journal of Sea Research, 58, 198-208. http://dx.doi.org/ 10.1016 /j.seares.2007.03.006
3. Hunter, W.R. and Sayer, M.D.J. (2009) The Comparative Effects of Habitat Complexity on Faunal Assemblages of Northern Temperate Artificial and Natural Reefs. ICES Journal of Marine Science, 66, 691-698. http://dx.doi.org/10.1093/icesjms/fsp058
4. ICES (2007) Report of the International Bottom Trawl Survey Working Group (IBTSWG). ICES CM 2007/RMC, 05:195 pp.
5. Stål, J., Pihl, L. and Wennhage, H. (2007) Food Utilization by Coastal Fish Assemblages in Rocky and Soft Bottoms on the Swedish West Coast: Inference for Identification of Essential Fish Habitats. Estuarine, Coastal and Shelf Science, 71, 593-607. http://dx.doi.org/10.1016/j.ecss.2006.09.008
6. Pérez-Matus, A. and Shima, J.S. (2010) Disentangling the Effects of Macroalgae on the Abundance of Temperate Reef Fishes. Journal of Experimental Marine Biology and Ecology, 388, 1-10. http://dx.doi.org/10.1016/j.jembe.2010.03.013
7. Wieland, K., Pedersen, E.M.F., Olesen, H.J. and Beyer, J.E. (2009) Effect of Bottom Type on Catch Rates of North Sea Cod (Gadus morhua) in Surveys with Commercial Fishing Vessels. Fisheries Research, 96, 244-251. http://dx.doi.org/ 10.1016/j .fishres.2008.11.010
8. Dahl, K., Lundsteen, S. and Helmig, S. (2003) Stenrev, Havbundens Oaser. Gads Forlag, Copenhagen, Denmark. (In Danish)
9. Carr, M.H. (1994) Effects of Macroalgal Dynamics on Recruitment of a Temperate Reef Fish. Ecology, 75, 1320-1333. http://dx.doi.org/10.2307/1937457
10. Stenberg, C., Støttrup, J.G., Dahl, K., Lundsteen, S., Göke, C. and Andersen, O.N. (2013) Ecological Benefits from Restoring a Marine Cavernous Boulder Reef in Kattegat, Denmark. Final Report to the European Commission Regarding LIFE06 NAT/DK000159 Blue Reef, 47.
11. Dahl, K., Stenberg, C., Lundsteen, S., Støttrup, J., Dolmer, P. and Tendal, O.S. (2009) Ecology of Læsø Trindel—A Reef Impacted by Extraction of Boulders. National Environmental Research Institute, Technical Report No. 757.
12. Almada, V.C., Henriques, M. and Gonçalves, E.J. (1999) Ecology and Behaviour of Reef Fishes in the Temperate North-Eastern Atlantic and Adjacent Waters. In: Almada, V.C., Oliveira, R.F. and Gonçalves, E.J., Eds., Behaviour and Conservation of Littoral Fishes, ISPA, Lisbon, 33-68.
13. Wennhage, H. and Pihl, L. (2002) Fish Feeding Guilds in Shallow Rocky and Soft Bottom Areas on the Swedish West Coast. Journal of Fish Biology, 61, 207-228. http://dx.doi.org/10.1111/j.1095-8649.2002.tb01772.x
14. Swales, S. (1994) Habitat Restoration Methods—A Synthesis. In: Cowx, I.G., Ed., Rehabilitation of Freshwater Fisheries, Fishing News Books, Oxford, 133-137.
15. Rinkevich, B. (2014) Rebuilding Coral Reefs: Does Active Reef Restoration Lead to Sustainable Reefs? Current Opinion in Environmental Sustainability, 7, 28-36. http://dx.doi.org/10.1016/j.cosust.2013.11.018
16. Chapman, M.G. (1999) Improving Sampling Designs for Measuring Restoration in Aquatic Habitats. Journal of Aquatic Ecosystem Stress and Recovery, 6, 235-251. http://dx.doi.org/10.1023/A:1009987403481

17. Barbier, E. (2006) Natural Barriers to Natural Disasters: Replanting Mangroves after the Tsunami. Frontiers in Ecology and the Environment, 4, 124-131. http://dx.doi.org/10.1890/1540-9295(2006)004[0124:NBTNDR]2.0.CO;2
18. Kennedy, V.S., Breitburg, D.L., Christman, M.C., Luckenbach, M.W., Paynter, K., Kramer, J., Sellner, K.G., Dew- Baxter, J., Keller, C. and Mann, R. (2011) Lessons Learned from Efforts to Restore Oyster Populations in Maryland and Virginia, 1990 to 2007. Journal of Shellfish Research, 30, 719-731. http://dx.doi.org/10.2983/035.030.0312
19. Anonymous (2013) Post Bathymetry and Stability of Reef Structures. Technical Final Report LIFE06 NAT/DK/ 000159-Blue Reef, Annex 15, 5. http://naturstyrelsen.dk/naturbeskyttelse/naturprojekter/blue-reef/undersoegelser-investigations/
20. Eigaard, O.R., Støttrup, J.G. and Hovgård, H. (2000) Udvikling af standard garnserie til brug ved rutinemæssig bestandsanalyse af flad-og rundfisk i marine lavvandede områder. DFU-Report nr. 78-00. (In Danish)
21. ICES (2012) Report of the Baltic Fisheries Assessment Working Group 2012 (WGBFAS). ICES CM. 2012/ACOM: 10, 12-19 April 2012, ICES Headquarters, Copenhagen, 859 p.
22. Jones, G.P. (1988) Ecology of Rocky Reef Fish of North-Eastern New Zealand: A Review. New Zealand Journal of Marine and Freshwater Research, 22, 445-462. http://dx.doi.org/10.1080/00288330.1988.9516315
23. Gratwicke, B. and Speight, M.R. (2005) The Relationship between Fish Species Richness, Abundance and Habitat Complexity in a Range of Shallow Tropical Marine Habitats. Journal of Fish Biology, 66, 650-667. http://dx.doi.org/10.1111/j.0022-1112.2005.00629.x
24. Bryndum, K.M. (2013) Fish Species Composition in the Inner Danish Waters. An Analysis of the Fish Assemblage in Light of Changes in Sea Temperature. Master's Thesis, Center for Macroecology, Evolution and Climate, University of Copenhagen, Copenhagen.
25. Sayer, M.D.J., Gibson, R.N. and Atkinson, R.J.A. (1993) Distribution and Density of Populations of Goldsinny Wrasse (Ctenolabrus rupestris) on the West Coast of Scotland. Journal of Fish Biology, 43, 157-167. http://dx.doi.org/10.1111/j.1095-8649.1993.tb01185.x
26. Gjøsæter, J. (2002) Distribution and Density of Goldsinny Wrasse (Ctenolabrus rupestris) (Labridae) in the Risør and Arendal Areas along the Norwegian Skagerrak Coast. Sarsia, 87, 75-82. http://dx.doi.org/10.1080/003648202753631758
27. Norderhaug, K.M., Christie, H., Fosså, J.H. and Fredriksen, S. (2005) Fish-Macrofauna Interactions in a Kelp (Laminaria hyperborea) Forest. Journal of the Marine Biological Association of the United Kingdom, 85, 1279-1286. http://dx.doi.org/10.1017/S0025315405012439
28. Costello, M.J. (1991) Review of the Biology of Wrasse (Labridae: Pisces) in Northern Europe. Progress in Underwater Science, 16, 29-51.
29. Villegas-Ríos, D., Alos, J. and March, D. (2013) Home Range and Diel Behavior of the Ballan Wrasse, Labrus bergylta, Determined by Acoustic Telemetry. Journal of Sea Research, 80, 61-71. http://dx.doi.org/10.1016/j.seares.2013.02.009
30. Alonzo, S.H. and Mangel, M. (2004) The Effects of Size-Selective Fisheries on the Stock Dynamics of and Sperm Limitation in Sex-Changing Fish. Fisheries Bulletin, 102, 1-13.
31. Gotceitas, V., Fraser, S. and Brown, J.A. (1995) Habitat Use by Juvenile Atlantic Cod (Gadus morhua) in the Presence of an Actively Foraging and Nonforaging Predator. Marine Biology, 123, 421-430. http://dx.doi.org/10.1007/BF00349220
32. Lorentsen, S.H., Sjøtun, K. and Grémillet, D. (2010) Multi-Trophic Consequences of Kelp Harvest. Biological Conservation, 143, 2054-2062. http://dx.doi.org /10.1016/j.biocon.2010.05.013
33. Stål, J., Paulsen, S., Pihl, L., Rönnbäck, P., Söderqvist, T. and Wennhage, H. (2008) Coastal Habitat Support to Fish and Fisheries in Sweden: Integrating Ecosystem

Functions into Fisheries Management. Ocean & Coastal Management, 51, 594-600. http://dx.doi.org/10.1016/j.ocecoaman.2008.06.006
34. Vinther, M. and Eero, M. (2013) Quantifying Relative Fishing Impact on Fish Populations Based on Spatio-Temporal Overlap of Fishing Effort and Stock Density. ICES Journal of Marine Science, 70, 618-627. http://dx.doi.org/ 10.1093 /icesjms/fst001
35. Meyers, R.A. and Cadigan, N.G. (1993) Density-Dependent Juvenile Mortality in Marine Demersal Fish. Canadian Journal of Fisheries and Aquatic Sciences, 50, 1576-1590. http://dx.doi.org/10.1139/f93-179
36. Gregory, R.S. and Anderson, J.T. (1997) Substrate Selection and Use of Protective Cover by Juvenile Atlantic Cod Gadus morhua in Inshore Waters of Newfoundland. Marine Ecology Progress Series, 146, 9-20. http://dx.doi.org/10.3354/meps146009
37. Fraser, S., Gotceitas, V. and Brown, J.A. (1995) Interactions between Age-Classes of Atlantic Cod and Their Distribution among Bottom Substrates. Canadian Journal of Fisheries and Aquatic Sciences, 53, 305-314. http://dx.doi.org/10.1139/f95-183
38. Svedäng, H., Stål, J., Sterner, T. and Cardinale, M. (2010) Subpopulation Structure in Cod (Gadus morhua) Puts Strain on the Management Toolbox. Reviews in Fisheries Science, 18, 139-150. http://dx.doi.org/10.1080/10641260903511420
39. Wortley, L., Hero, J.M. and Howes, M. (2013) Evaluating Ecological Restoration Success: A Review of the Literature. Restoration Ecology, 21, 537-543. http://dx.doi .org/10.1111/rec.12028

CHAPTER 8

Population Ecology of the Sea Lamprey (*Petromyzon Marinus*) as an Invasive Species in the Laurentian Great Lakes and an Imperiled Species in Europe

Michael J. Hansen, Charles P. Madenjian, Jeffrey W. Slade, Todd B. Steeves, Pedro R. Almeida, Bernardo R. Quintella

Hammond Bay Biological Station, Great Lakes Science Center, U.S. Geological Survey, 11188 Ray Road, Millersburg, MI 49759, USA

Great Lakes Science Center, U.S. Geological Survey, 1451 Green Road, Ann Arbor, MI 48105, USA

Sea Lamprey Control Centre, Fisheries and Oceans Canada, 1219 Queen St. East, Sault Ste. Marie, ON P6A 2E5, Canada

Departamento de Biologia, Escola de Cie^ncias e Tecnologia, MARE – Centro de Cie^ncias do Mar e do Ambiente, Universidade de E´ vora, Largo dos Colegiais, 7004-516 E´ vora, Portugal

Departamento de Biologia Animal, Faculdade de Cie^ncias, MARE – Centro de Cie^ncias do Mar e do Ambiente, Universidade de Lisboa, Campo Grande, 1749-016 Lisbon, Portugal

ABSTRACT

The sea lamprey *Petromyzon marinus* (Linnaeus) is both an invasive non-native species in the Laurentian Great Lakes of North America and an imperiled species in much of its native range in North America and Europe. To compare and contrast how understanding of population ecology is useful for control programs in the Great Lakes and restoration programs in Europe, we review current understanding of the population ecology of the sea lamprey in its native and introduced range. Some attributes of sea lamprey population ecology are particularly useful for both control programs in the Great Lakes and restoration programs in the native range. First, traps within fish ladders are beneficial for removing sea lampreys in Great Lakes streams and passing sea lampreys in the native range. Second, attractants and repellants are suitable for luring sea lampreys into traps for control in the Great Lakes and guiding sea lamprey passage for conservation in the native range. Third, assessment methods used for targeting sea lamprey control in the Great Lakes are useful for targeting habitat protection in the native range. Last, assessment methods used to quantify numbers of all life stages of sea lampreys would be appropriate for measuring success of control in the Great Lakes and success of conservation in the native range.

Keywords: Sea lamprey, Population ecology, Management, Conservation

Jeffrey W. Slade was retired from U.S. Fish and Wildlife Service, Ludington Biological Station, 229 S. Jebavy Drive, Ludington, Michigan 49431, USA.

INTRODUCTION

The sea lamprey *Petromyzon marinus* (Linnaeus) is both an invasive exotic species in the Laurentian Great Lakes of North America and an imperiled species in much of its native range along the north Atlantic coasts of North America and Europe (Fig. 1). In the Laurentian Great Lakes, the sea lamprey evidently invaded the Great Lakes from the Atlantic Ocean (Christie and Goddard 2003; Eshenroder 2014), and were first found in Lake Ontario in 1835 (although this date has been disputed by Eshenroder 2014), Lake Erie in 1921, Lake Michigan in 1936, Lake Huron in 1937, and Lake Superior in 1938 (Applegate 1950; Lawrie 1970; Smith 1979; Smith and Tibbles 1980; Smith 1985). By the 1950s, sea lampreys were abundant in all Great Lakes, where they imposed high mortality on nearly all teleost species, but especially the lake trout *Salvelinus namaycush* (Hansen 1999). Control of sea lamprey populations began in the 1950s, initially with mechanical and electrical barriers to upstream migration, and later with a selective pesticide, 3-trifluoromethyl-4-nitrophenol (TFM; Smith and Tibbles 1980). Suppression of sea lamprey populations continues to rely on TFM, but was expanded to also include use of an integrated program of physical (barriers and traps) and biological (sterile-male releases) control methods (Christie and Goddard 2003), although sterile-male releases were suspended until further research could be completed to confirm its efficacy.

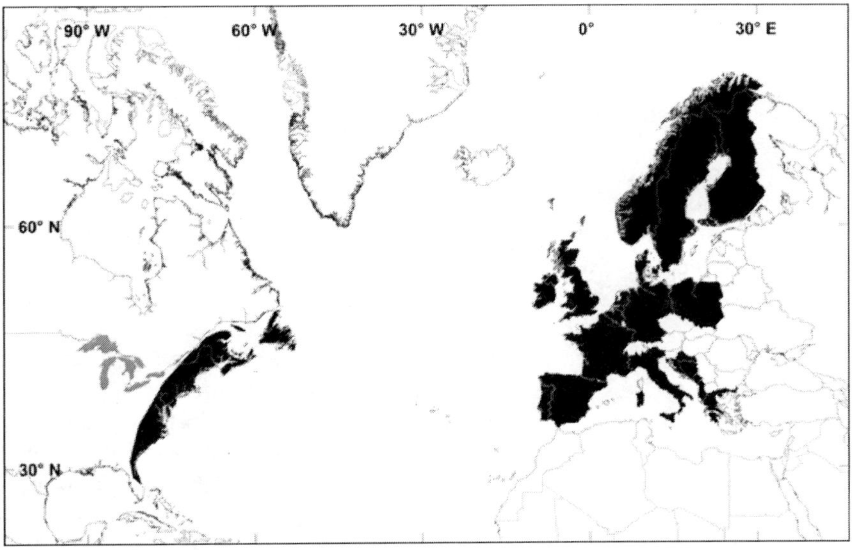

Figure. 1 Worldwide distribution of native (*black shading* North Atlantic Ocean) and non-native (*gray shading* Laurentian Great Lakes, North America) sea lamprey populations

In its native range, the sea lamprey is considered threatened in France, Spain, and Portugal, European countries where the main populations are found, although the species is considered of Least Concern according to the International Union for Conservation of Nature (IUCN) Red List of Threatened Species, and the European Red List of Freshwater Fishes (Mateus et al. 2012). The sea lamprey is highly valued as a food fish where populations are large enough to be exploited (Quintella 2006), so commercial overfishing is a serious threat for the species in areas such as the Iberian Peninsula (Mateus et al. 2012) and elsewhere in Europe (Maitland et al. 2015). In general, however, the sea lamprey has declined over the last 25 years in Europe from a combination of (1) habitat loss associated with dam construction, (2) degradation of water quality from mining, industrial, and urban development, (3) direct loss of habitat by sand extraction and dredging, (4) overfishing, and (5) changes in water quality (temperature) and quantity associated with climate change (Mateus et al. 2012). To enable population recovery, lost or damaged habitat must be restored, while sustainably managing commercial fisheries (Mateus et al. 2012).

The purpose of this review is to synthesize and relate the state of knowledge concerning the population ecology of sea lamprey between Continents, to increase understanding of how knowledge of population ecology bears on control programs in the Great Lakes and restoration programs in Europe. First, current understanding of the life cycle is reviewed, including adult life stage, reproduction, larval life stage, metamorphosis, juvenile life stage, feeding, and effects on host species, pointing out when present differences between North American versus European populations and/or landlocked versus anadromous form. Second, implications for future status of the species are reviewed, including how global climate change will affect sea lamprey population ecology in the Great Lakes and Europe, which attributes of sea lamprey population ecology can be used to control populations in the Great Lakes, which attributes of sea lamprey population ecology can be used to restore and conserve populations in Europe, which attributes of sea lamprey population ecology are in need of further study for management of the species worldwide, and how understanding of population ecology bears on both control programs in the Great Lakes and restoration programs in Europe. For each topic below, we attempted to include information for both non-native populations in the Laurentian Great Lakes and native populations in Europe, although gaps in available information prevented complete coverage of all topics in both the Great Lakes and Europe.

Overview

Lampreys have an unusual life cycle for a vertebrate because of a relatively long larval life stage and a relatively short adult life stage. A complete life cycle of a sea lamprey takes at least 4–5 years (range = 3–10+ years), including 12–18 months (minimum of 14 months for anadromous populations) in the juvenile and adult life stage (Fig. 2). The long larval life stage, a disadvantage for most anadromous fishes because of increased risk of predation during this vulnerable period, is more beneficial than detrimental to the sea lamprey because larvae spend that period of life burrowed in river sediments and are mostly sedentary. Osmotic, bioenergetics, and predation-exposure costs in moving between riverine and oceanic or lake ecosystems are compensated by a reduced predation on early life stages in riverine environments and access to greater trophic resources in marine or lake environments (Gross 1987). For the sea lamprey, in particular, richness of the marine or lake diet is measured not only in terms of numbers of potential host species and individuals, but also the size of parasitized species needed to sustain an adult sea lamprey.

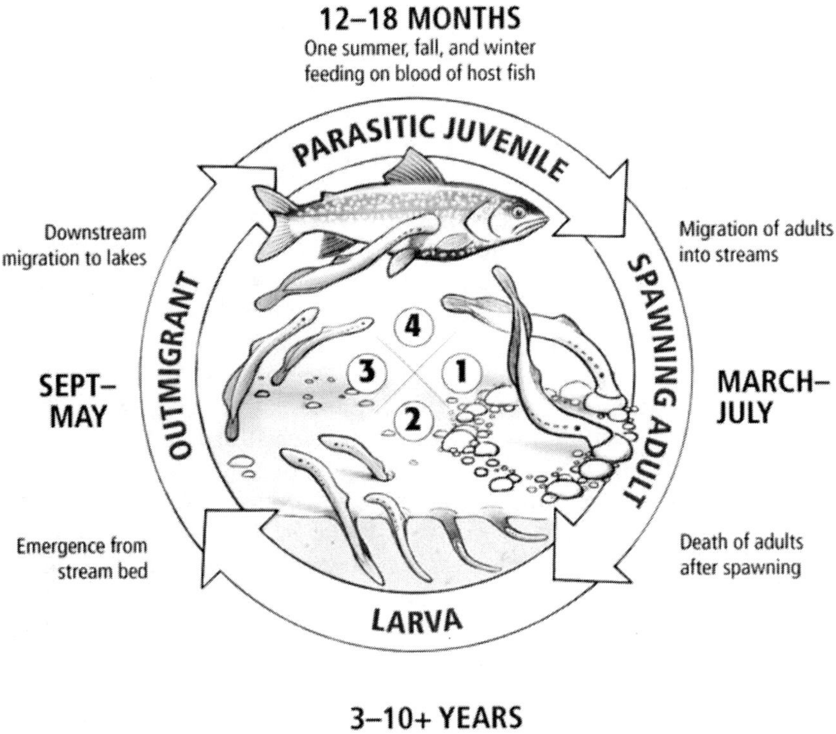

Figure. 2 Duration of and general habitat characteristics used by spawning adult, larval, outmigrating, and parasitic juvenile sea lamprey life stages in the Great Lakes (Great Lakes Fishery Commission)

Debate regarding the native or non-native origin of sea lampreys in Lake Ontario continues (Siefkes et al. 2013; Eshenroder 2014), but once established in the upper Great Lakes, sea lampreys spread rapidly to reproduce in ~500 Great Lakes tributaries (Lake Superior = 161, Lake Michigan = 126, Lake Huron = 122, Lake Erie = 23 and Lake Ontario = 66). Following their invasion, sea lampreys spread rapidly throughout the five Laurentian Great Lakes and are now found in streams of Minnesota (Eddy and Underhill 1974), Wisconsin (Becker 1983), Michigan (Applegate 1950), Illinois (Smith 1979), Indiana (Gerking 1955), Ohio (Trautman 1981), Pennsylvania (Emery 1985), New York (Smith 1985), and throughout Ontario (Adair and Sullivan 2013). As evidenced by marking on host fishes, juveniles are distributed throughout open waters of all Great Lakes, and larvae are currently distributed in streams from eastern Lake Ontario to western Lake Superior.

Adult life stage

Following completion of their 1–2 year-long marine trophic phase (Beamish 1980; Silva et al. 2013a, b), anadromous adult sea lampreys migrate upstream to river stretches where they build nests, spawn, and die (Larsen 1980; Moser et al. 2015). Passage from sea to fresh water is a stressful stage of migration, so adults use estuaries to acclimate from salt-water to fresh-water osmoregulation (Bartels and Potter 2007). The spawning migration ranges from September to March along the east coast of North America (Beamish 1980); begins in December, peaks in February–March, and ends with spawning in April-June in Southwestern Europe Portuguese rivers (Almeida et al. 2000; Oliveira et al. 2004); and begins in February, continues through May–June, and ends with spawning between the end of May and early July in the Northwestern Europe Severn River, Britain (Hardisty 1986). Upstream spawning migration is triggered by flow variation and temperature, so increased migratory activity in periods of high discharge is likely a behavior adopted by sea lampreys to overcome difficult passage stretches to reach upstream spawning areas (Almeida et al. 2002a; Andrade et al. 2007; Binder et al. 2010). Regulated increased river discharge at night (i.e. hydropeaking) seems to stimulate lamprey movement, although reduced ground speed of upstream movement has also been observed (Almeida et al. 2002a).

Adult sea lampreys do not appear to home to natal streams (Bergstedt and Seelye 1995; Waldman et al. 2008; Swink and Johnson 2014), but rather, select spawning streams through innate attraction using other sensory cues (Sorensen and Vrieze 2003; Li et al. 2003; Vrieze et al. 2010, 2011). In the Great Lakes, adults from the same stream and cohort migrate into numerous streams throughout one or more of the Great Lakes, and tagged juveniles have been recaptured as adults in streams more than 400 km from their natal stream (Swink and Johnson 2014; Johnson et al. 2014; U.S. Fish and Wildlife Service, unpublished data). Mechanistic factors driving these movements are still unknown, although adults locate spawning streams using a three-phase odor-mediated strategy that includes searching a shoreline while casting vertically, followed by stream-water-induced turning toward a stream mouth, where they ascend using rheotaxis (Vrieze et al. 2011; Meckley et al. 2014). In the Great Lakes, adult sea lampreys are highly selective in the choice of spawning streams (Morman et al. 1980), and choose streams with high larval density (Moore and Schleen 1980), signaled by bile acid-based pheromones released by larvae (Bjerselius et al. 2000). Lack of homing is also evident from genetic studies of spawning migrants returning from the Atlantic Ocean (Bryan et al. 2005; Waldman et al. 2008). However, fatty acids and morphology of sea lampreys from major Portuguese rivers indicate that dispersion in the ocean may be limited, which suggests some degree of geographical fidelity when adults return from feeding areas (Lança et al. 2014). Further, absence of genetic exchange among sea lamprey populations spawning in the western and eastern Atlantic suggests adults do not intermingle (Rodríguez-Muñoz et al. 2004).

Adult sea lampreys congregate at stream mouths from January through March each year, but the precise timing of upstream migration differs with latitude (Moser et al. 2015). In the Great Lakes, upstream migration of adults begins in March–April (Applegate 1950) when water temperature reaches ~15 °C (Binder et al. 2010). In Europe, adult begin to migrate into streams in December–January, with the peak of migration in February–April, and spawning in April–May. Migratory activity is stimulated when water temperature increases daily, and is suppressed when water temperature decreases daily (Binder et al. 2010). Migration distance depends on river size, location of suitable spawning areas, and length of river stretches downstream of impassable barriers (Hardisty 1986). In Great Lakes streams without natural or man-made barriers, adult sea lampreys migrate more than 100 km upstream. In the Iberian Peninsula, 80 % of accessible sea lamprey habitat was lost by obstruction of lower stretches of all major rivers: historical available habitat in the main stretch of the larger rivers was 516 km in the Douro, 633 km in the Tagus, 648 km in the Guadiana, 394 km in the Guadalquivir, and 680 km in the Ebro (Mateus et al. 2012). In Portugal, the present distribution of the sea lamprey is quite limited, with spawning areas located below impassable dams, with the upstream limit ranging between 27 km (Cávado River) and 150 km (Tagus River) (Almeida et al. 2002b; Mateus et al. 2012). In Britain, spawning habitat lies within 10–100 km of the tidal limit (Hardisty 1986). Sea lampreys historically migrated upstream 850 km in the Rhine River, Europe (Hardisty 1986) and 320 km in the Delaware and Susquehanna rivers, North America (Bigelow and Schroeder 1948), until construction of the Conowingo Dam near the mouth of the Susquehanna in 1928 obstructed migration (Waldman et al. 2009).

Adult sea lampreys migrating upstream have ceased feeding, the digestive system has atrophied and is non-functional, and the sea lamprey invests remaining energy in gamete production, nest construction and spawning. Lampreys are negatively phototaxic, so move upstream in fresh water primarily during dusk and darkness (Almeida et al. 2000, 2002a) and seek refuge before dawn (Andrade et al. 2007). The adaptive value of nocturnal behavior might be related to the greater protection afforded by darkness. When swimming through slow river stretches, adult sea lampreys can maintain constant activity at an average ground speed of 0.76 body lengths/s (Quintella et al. 2009), although typical swimming is at a ground speed of 0.2–0.4 body lengths/s (Andrade et al. 2007). Adults seek cool, well-oxygenated water with a unidirectional flow over rock, gravel, and sand substrate (Applegate 1950; Hardisty and Potter 1971b). Nests are typically constructed by males using their mouth to move stones, while flushing smaller particles from the nest with rapid body movements. Spawning pairs intertwine in the nest, and with a series of convulsions, release milt and eggs into the nest. Sea lampreys are semelparous, so they die shortly after spawning (Applegate 1950; Johnson et al. 2015a, b).

Mortality of adult sea lampreys prior to spawning is poorly understood, but has been observed (Applegate 1950), although dead carcasses observed in one stream may have been caused by spawning or chemo-sterilization (Hanson and Manion 1980). Natural mortality of adult sea lampreys introduced into two Lake Ontario tributaries ranged from 6 to 30 % and mortality from predation ranged from 1 to 11 % (O'Connor 2001). Mortality from predation, particularly on nest sites has been assumed to be relatively small, although predators could prevent successful spawning in streams with few adults (Applegate 1950; Morman et al. 1980). The Eurasian otter (*Lutra lutra* Linnaeus) commonly prey sea lamprey adults during spawning period with estimated predation rates around 8 % (Andrade et al. 2007; Maitland et al. 2015). Emigration of adult sea lampreys introduced into streams ranged from 8 to 49 % within a spawning season (Manion and McLain 1971; Hanson and Manion 1980; Kelso and Gardener 2000; Dolinsek et al. 2014). Depending upon their state of maturation, sea lampreys that emigrated from streams may die (Applegate 1950; Applegate and Smith 1951) or move to another stream to spawn (Dolinsek et al. 2014).

Lamprey can use their oral disc to attach to substrate and rest between bouts of swimming, a strategy referred to as "burst-and-attach" (Quintella et al. 2009). In areas of fast water velocity, a combination of intermittent burst swimming and periods of rest when attached to the substrate is characteristic behavior (Applegate 1950; Hardisty and Potter 1971b; Haro and Kynard 1997; Mesa et al. 2003; Quintella et al. 2004). This highly active swimming is the most energy-inefficient form of activity (Beamish 1978) and can only be achieved for short periods. Nevertheless, absence of a swim bladder to sustain neutral buoyancy (Hardisty and Potter 1971b) and less-efficient anguilliform propulsion used by lampreys (Webb 1978; van Ginneken et al. 2005) makes this pattern the most energetically conservative for overcoming rapid flow or man-made obstacles (Quintella et al. 2004).

Adult anadromous sea lampreys are larger in Europe than in North America, and exhibit latitudinal and temporal trends, with body size increasing from north to south, and length of spawners increasing from 1980 to 2005 (Beaulaton et al. 2008). In Europe, adults are the largest in Portugal (Beaulaton et al. 2008), where they averaged 85 cm in TL and 1.2 kg in weight (Fig. 2). In North America, adult anadromous sea lampreys averaged nearly 20-cm shorter (66 cm) in the East Machias River, Maine (Davis 1967) and nearly 15 cm shorter in the Connecticut River (71 cm), St. John (72 cm) River, and New Brunswick Rivers (Beamish and Potter 1975; Stier and Kynard 1986) than in Portugal. In the Laurentian Great Lakes, adult sea lampreys are even smaller, and averaged only 48 cm in TL (SD = 4.5 cm) over all lakes (Fig. 3), with small variation among lakes Superior (44 cm; SD = 4.2 cm), Huron (48 cm; SD = 3.9 cm), Michigan (49 cm; SD = 3.8 cm), Erie (51 cm; SD = 4.2 cm), and Ontario (49 cm; SD = 4.4 cm).

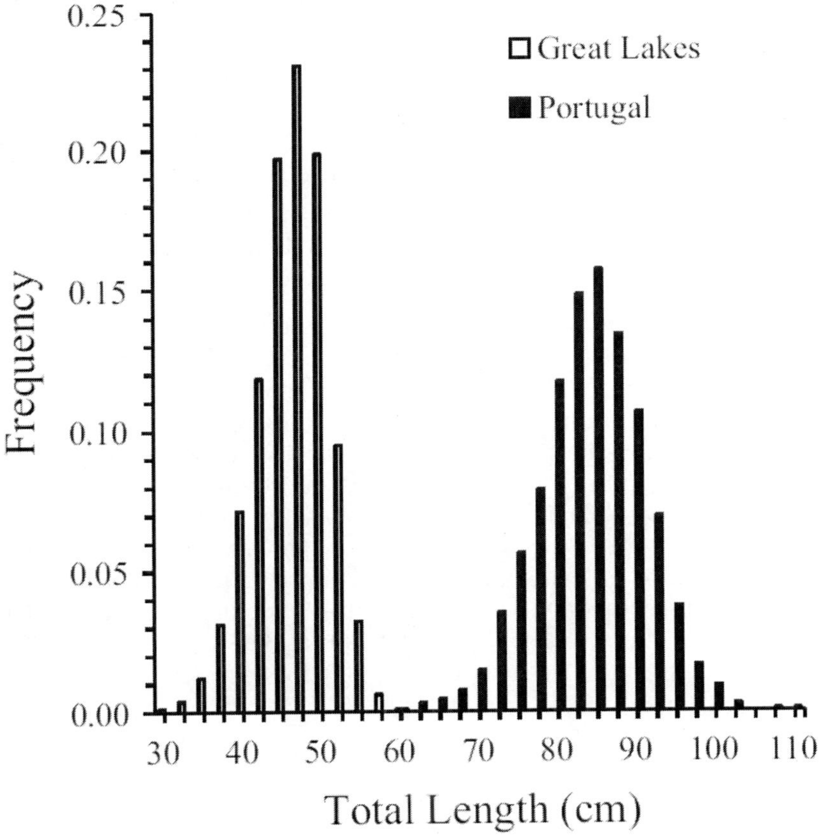

Figure. 3 Length-frequency distributions of adult sea lampreys caught during upstream spawning migrations in streams throughout the Laurentian Great Lakes basin, North America, 2000–2015 (N = 50,348; USFWS and DFO unpublished data), and in Europe (Portugal) in rivers Minho, Lima, Cávado, Douro, Vouga, Mondego, Tagus and Guadiana, 2000–2014 (N = 1580; Quintella et al. 2003)

Fecundity of landlocked sea lampreys, measured as the number of eggs per gram of body weight, ranged from 339 eggs/g for Lake Ontario populations (O'Connor 2001) to 670 eggs/g for a Lake Superior population (Manion 1972). This measure of relative fecundity depends on when gravid females are collected, because adult sea lampreys stop feeding, so females collected later in the spawning run expended more energy and body mass searching for mates and building nests. Consequently, wet weight is lower late in the spawning run, so the ratio of eggs to gram of body mass increases (Manion 1972). Absolute fecundity is proportional to sea lamprey size, so larger sea lampreys in Great Lakes with warmer water (Michigan, Erie, and Ontario) are more fecund than in Great Lakes with colder water (Superior and Huron, Sullivan and Adair 2014). Absolute fecundity is also related to diet quality, so female sea lampreys are more fecund in Lake Superior (67,000 eggs/female), with a large population of lake trout, than in Lake Huron (46,000 eggs/female) or Cayuga Lake (43,000 eggs/female), with fewer preferred prey (Manion 1972). This difference in absolute fecundity among lakes may be lower since the 1960s, because sea lamprey control continued to suppress sea lamprey populations while host populations increased (Heinrich et al. 1980).

In the Laurentian Great Lakes, lake-wide adult sea lamprey abundance has been a primary metric used to evaluate success of the bi-national sea lamprey control program since the late 1970s and early 1980s. Lake-wide estimates are generated by summing estimates of adult sea lamprey abundance from tributaries that sea lampreys use for spawning. In streams with traps, stream-specific estimates are generated using mark-recapture or measures of trap efficiency. Estimates for streams without traps are generated using a model that incorporates five independent variables (Mullett et al. 2003). Since the onset of sea lamprey control in the Great Lakes, lake-wide estimates of adult sea lamprey abundance ranged from 261,000 in 1981 to 12,000 in 1994 in Lake Superior, 169,000 in 2004 to 29,000 in 1997 in Lake Michigan, 450,000 in 1993 to 42,000 in 1997 in Lake Huron, 33,000 in 2009 to 1700 in 2002 in Lake Erie, and 297,000 in 1982 to 23,000 in 1994 in Lake Ontario.

In Europe, abundance of the anadromous sea lamprey has rarely been measured or monitored, although abundance and exploitation of the population in the Garonne River basin in France, a river with one of the largest populations and fisheries for the species in Europe, have been estimated (Beaulaton et al. 2008). The average catch of sea lampreys from the Garonne basin was 72 t (~67,000 individuals) during 1985–2003. Abundance, estimated using a generalized linear model (GLM) of catch per unit of effort (CPUE) during 1943–2000, peaked at 10–15 kg/day (weighted average) in 1957–1965. From 1973 through the 1990s, abundance was stable at 35–40 % of the maximum in 1957–1965. Since the end of the 1990s, abundance increased, and by 2000, reached abundance levels last observed in the 1960s. Portuguese rivers also support large populations of sea lampreys, but lack of good records from annual catches makes accurate estimation of population abundance difficult for most rivers. In the River Minho, northern Portugal, official records of landings dating back to 1914 show an increase in the number of sea lampreys caught by professional fisherman, with a peak in 2009 of about 60,000 specimens (Mota 2014). For the River Mondego, Portugal, a professional fisheries survey in 2014 enabled an estimate of annual harvest of 30,000 individuals and a total estimate of 100,000 individuals that entered the river during the spawning migration (ICES 2014).

REPRODUCTION

Physical factors essential for successful reproduction include steady, unidirectional water flow, and suitable sand and gravel (0.9–5.1 cm diameter) substrates, water velocity (0.5–1.5 m/s), depth (13–170 cm), and temperature (10.0–26.1 °C) (Manion and Hanson 1980). Nest building begins when water temperature warms to ~15 °C (Applegate 1950). Nest construction is usually initiated by males, but females may initiate nest construction near the end of the spawning season when numerically dominant. Nest construction takes 1–3 days, with activity increasing after females join males (Applegate 1950). Mature spermiated males release a potent sex pheromone that induces preference and searching behavior in ovulated females ascending to upstream spawning areas (Li et al. 2002). On average, spawning lasts ~2–5 s and is repeated every 4–5 min (see Johnson et al. 2015a, b for a thorough review of spawning behavior). Spawning activity is typically monogamous (Applegate 1950; Manion and McLain 1971; Hanson and Manion 1978, 1980), although polyandry may increase when the adult sex ratio is skewed away from 1:1 (Hanson and Manion 1978, 1980). Genetic evidence suggests that both polyandry and polygyny are widespread (Scribner and Jones 2002). In the upper Great Lakes, each female produces an average of ~60,000 eggs, and fertilization and survival of eggs in the nest can be as great as 90 % (Manion and Hanson 1980). Egg development depends on temperature (Piavis 1961) and hatching success typically averages only ~6.3 % (Manion 1968). After eggs incubate for about 2 weeks, pro-larvae emerge from the nest at night over a period of several weeks and range 5–12 mm in length (Applegate 1950; Derosier et al. 2007).

Larval life stage

After hatching, blind, poor-swimming larvae are carried downstream from nests to depositional areas of sand, silt, and detritus, where they burrow into soft sediments to feed on suspended organic matter (Sutton and Bowen 1994; Dawson et al. 2015). Dispersal from nests is highly variable and is influenced by larval density and water temperature (Derosier et al. 2007). Age-0 larvae can remain within a few hundred meters of the nest in their first year of life (Manion and McLain 1971), or they may move downstream after emerging to reduce density-dependent effects on recruitment (Derosier et al. 2007). Although larval sea lampreys seldom leave their burrows, some larvae move downstream spontaneously in response to hydrologic conditions or at night (Hardisty and Potter 1971a, b). Downstream movement is stimulated more in streams with higher gradient, particularly during periods of high water discharge (Quintella et al. 2005), which may lead to the observation of larger (older) larvae in the lower reaches of some streams where spawning habitat is limited (Hardisty and Potter 1971a, b; Quintella et al. 2003). Redistribution of larvae over short distances may result in burrowing in more suitable habitat (Hardisty and Potter 1971a, b; Yap and Bowen 2003). Depth of the burrow is directly correlated with lamprey size (Hardisty and Potter 1971a). Habitat selection, indexed as larval density, is correlated with water velocity and substrate hardness (Thomas 1962). Age-1-and-older larvae are also known to migrate from silty substrates to locations with coarser substrates in summer, presumably as temperature increases and oxygen concentration declines in depositional areas (Sullivan 2003). Larvae are sensitive to light, and withdraw into burrows during daytime. Larvae remain burrowed for 3–5 years on average, and filter-feed on seston, diatoms, and biofilm (Sutton and Bowen 1994; Quintella et al. 2003). Larvae are able to move upstream against relatively slow currents (<0.2 m/s) (Quintella et al. 2005).

Overlap in habitat among lamprey species may lead to interspecific competition that affects survival and growth of sea lamprey larvae (Beamish and Lowartz 1996). For example, Northern brook lamprey (*Ichthyomyzon fossor*) select habitats that leads to higher-quality diet (Yap and Bowen 2003), and consequently, they may compete with sea lamprey larvae for resources. Year-class strength is driven by both density-independent and density-dependent forces, so environmental variation plays a large role in determining year-class strength that is not explained by adult abundance (Jones et al. 2003; Dawson 2007; Dawson and Jones 2009). Density-independent recruitment variation leads to occasionally strong year classes of larvae even when adult abundance is low (Jones et al. 2003). Strong year classes can be produced by adult density as low as 1.0 female/100 m^2 of larval habitat, but evidently not at adult density sizes below 0.2 females/100 m^2 (Dawson and Jones 2009). Therefore, control programs could aim to reduce adult female density to fewer than 0.2 females/100 m^2 of suitable nursery habitat for recruits, and conversely, recovery programs could seek to increase adult female density to more than 0.2 females/100 m^2 of nursery habitat for recruits.

Typical larval habitat is protected from major fluctuations in water levels or stream flow, where current velocity is slow. Such conditions are often found in eddies or backwaters at bends in a river, where soft silt and sand accumulate to provide suitable substrate for burrowing larvae (Table 1). Such habitat is often partially shaded, so diatoms often encrust the interface between silt and water, thereby contributing to stability of such microenvironments (Hardisty 1979). Most importantly, existence of suitable conditions for larval colonization depends on stream gradients that determine overall current velocity, the size of deposited substrate particles, and accumulation of organic debris (Hardisty and Potter 1971a). Suitable river substrate is essential for development of larval lampreys, to enable burrow construction and to maintain water flow (Hardisty 1979; Kainua and Valtonen 1980; Malmqvist 1980; Morman et al. 1980; Potter 1980; Young et al. 1990a, b; Kelso and Todd 1993; Beamish and Jebbink 1994; Ojutkangas et al. 1995; Beamish and Lowartz 1996; Sugiyama and Goto 2002; Goodwin et al. 2008). Larvae depend on unidirectional water flow through their branchial chamber, to provide detritus food and to exchange respiratory gases and metabolic wastes (Hardisty and Potter 1971a). Small larvae (20–60 mm TL) prefer small-grained substrate (silt-sand), medium-sized larvae (60–140 mm TL) prefer medium-grained substrate (gravel-silt-sand), and large larvae (140–200 mm TL) prefer coarse-grained sediments (gravelly-sand and sand) (Almeida and Quintella 2002; Sullivan 2003; Table 4). Small larvae are often found in fine-grained sediments, likely because soft sediments allow young larvae with reduced swimming capacity to propel their head and branchial region into the substrate (Quintella et al. 2007). In contrast, large larvae colonize a wider range of sediments because their ability to burrow is considerably higher (Quintella et al. 2007), and because they have had a greater amount of time to colonize multiple habitats (Sullivan et al. 2008). Because selection of sediment is size-dependent, differences in preference for distinct sediment types within the same age group may have resulted from larval redistribution at the end of each annual growing season, perhaps as a strategy to avoid high density in areas colonized by younger individuals to reduce intraspecific competition for space and food (Almeida and Quintella 2002). Larval distribution is also associated with slow current, although sediment particle size is strongly determined by current velocity, thereby confounding the relative importance of current and substrate particle size in determining larval distribution (Young et al. 1990b; Almeida and Quintella 2002).

Table 1 Habitat variables important for larval sea lamprey at different spatial scales (adapted from Dawson et al. 2015)

Variables	Study type	Source
Substrate (medium-fine sand)	Laboratory	Lee (1989)
Substrate (sand)	Field	Young et al. (1990a)
Substrate (silt-sand)	Field	Young et al. (1990b)
Substrate (sand)	Field	Almeida and Quintella (2002)
Substrate/distance from stream mouth/slope of the lake bottom	Field	Fodale et al. (2003)[a]
Substrate (sand/fine organic matter)	Field	Slade et al. (2003)
Geomorphic features (river slope – radius of curvature)	Field	Neeson et al. (2008)
Substrate (fine-medium sand)/water depth (>2 m)/current (slow)/macrophyte roots	Field	Taverny et al. (2012)

[a] Lentic habitat

Small-scale studies of larval lamprey habitat have been useful for developing a general understanding of the biology of lampreys. However, conservation and management of lamprey populations requires the ability to evaluate and predict spatial patterns in larval abundance at several scales (Torgersen and Close 2004). The evaluation of ecological patterns and processes at multiple scales may reveal causal factors that are important at one scale, but are less important or have an opposite effect at other scales (Torgersen and Close 2004). Studies of lamprey ecology in streams and rivers have addressed the interplay of macro- and micro-environmental factors as influences on larval distribution (Baxter 1957; Hardisty and Potter 1971a, b). Broad-scale distribution patterns of larval lampreys, including the juxtaposition of adult spawning habitat upstream of larval habitat, is related to variation in channel gradient within and among streams (Baxter 1957; Young et al. 1990a). For example, larval sea lamprey density varies longitudinally with channel gradient, and the influence of channel gradient on larval density increases with spatial scale (Torgersen and Close 2004).

Larval abundance is directly linked to environmental variables, but biological factors, such as the spawning distribution of adults, also plays an important role in larval distribution (Torgersen and Close 2004). For example, larval distribution along a river is strongly associated with spawning areas, with larval density inversely related to distance downstream from spawning areas (Morman et al. 1980; Almeida and Quintella 2002; Quintella et al. 2003; Derosier et al. 2007). Similarly, adults must have access to spawning habitat, so the presence of migration barriers influences larval abundance and distribution (Goodwin et al. 2008). Last, metamorphosing lampreys are often found in the same sites where larvae of all sizes are found (Potter 1980; Quintella et al. 2003). During the initial stage of metamorphosis, transforming juveniles are relatively more sedentary than larvae (Quintella et al. 2005), whereas later on, juveniles burrow less, so are found hiding between pebbles, and under aquatic vegetation, rocks, and other structures (Dawson et al. 2015).

Prior to burrowing, larval mortality from predation is likely high (Potter 1980), although estimates of age-specific larval survival are limited by a lack of age estimates. Larval survival was 96 % between age 1 and age 4 (Morman 1987), although this estimate was likely too high because caged larvae were protected from mortality (Johnson et al. 2014). In contrast, results from modeling studies have yielded estimates of much lower larval survival in streams of Lake Erie (39.5 %, Irwin et al. 2012), Lake Michigan (45 % Jones et al. 2009), and Lake Ontario (51.8 % Irwin et al. 2012). In the St. Marys River, estimates of survival have ranged from 35–49 % (Haeseker et al. 2003) to 66–91 % (Jones et al. 2012). Variability among estimates may reflect environmental conditions or uncertainty associated with using multiple parameters to simulate the response of sea lamprey populations to management actions with management strategy evaluation (MSE) models. Regardless, the wide range of survival estimates is consistent with a large range in variation of density-independent survival (Jones et al. 2003).

Larval sea lampreys hatch at ~9 mm in length, and growth depends on biotic and abiotic factors like population density and water temperature (Table 2; Morman 1987; Murdoch et al. 1992). In a Lake Ontario tributary, the first year class to infest a stream following lampricide treatment grew faster than subsequent year classes (Weise and Pajos 1998), which suggests growth was density dependent (Dawson et al. 2015). In the River Mondego, Portugal, larval sea lampreys averaged 59 mm in TL (range = 53–74 mm) at age 0+ (<9 months post-hatch), 95 mm in TL (range = 58–144 mm) at age 1+ (9–21 months post-hatch), 140 mm in TL (range = 93–179 mm) at age 2+ (21–33 months post-hatch), 166 mm in TL (range = 142–190 mm) at age 3+ (33–45 months post-hatch), and 184 mm in TL (range = 179–188 mm) at age 4+ (>45 months post-hatch; Quintella et al. 2003). Larval growth is nearly 0 mm in some years, and some individuals may stay 1 year in a river to accumulate fat needed to metamorphose (i.e. sometimes termed the "retarded growth phase"). Larvae from the River Mondego, Portugal, had shorter larval stage duration (Quintella et al. 2003) than in more northerly river basins, such as anadromous populations in Canada (range = 6–8 years; Beamish and Potter 1975) and Great Britain (average = 5 years; Hardisty 1969a, b), likely because higher productivity associated with warmer water enhances feeding efficiency and growth (Morman 1987). As expected for a poikilothermic organism, growth of sea lamprey larvae is correlated with water temperature (Potts et al. 2015), and varies among geographic regions with different climatic conditions, with faster growth in more favorable climates (Potter 1980; Dawson et al. 2015).

Table 2 Minimum, maximum, and mean lengths of sea lampreys at larval (minimum and maximum only), recently metamorphosed juvenile, feeding juvenile (minimum and maximum only), and spawning adult life stages in the Laurentian Great Lakes (landlocked form; USFWS and DFO unpublished data) and North American and European rivers draining to the Atlantic Ocean (anadromous form; Quintella et al. 2003; doi:10.1006/jfbi.2000.1465)

Life stage	**Total length (mm)**		**Mean**
	Minimum	**Maximum**	
Laurentian Great Lakes—landlocked			
Larvae	9	196	
Recently Metamorphosed Juvenile	100	196	135
Feeding Juvenile	100	639	
Spawning Adult	49	624	477
Europe—anadromous			
Larvae	9	200	
Recently Metamorphosed Juvenile	105	188	149
Feeding Juvenile	–	–	
Spawning Adult	607	1113	853

Age estimation of sea lamprey larvae, particularly in controlled populations, initially relied on analysis of length-frequency data (Table 3) but can be supplemented by knowing the number of years since known recruitment (Weise and Pajos 1998; Hansen et al. 2003; Dawson et al. 2015). However, growth rates vary greatly within and among cohorts within stream populations, so age cannot be accurately assigned from length-frequency histograms (Dawson et al. 2009), especially for older age classes (Hardisty and Potter 1971a; Hardisty 1979). Statoliths, like otoliths in teleosts, are calcium carbonate structures that accumulate annuli for use in age estimation of sea lampreys (Volk 1986; Hollett 1998; Henson et al. 2003). Although statolith-based age estimates may be biased for sea lamprey larvae (Dawson et al. 2009), accurate estimates of age can be obtained by combining length-frequency information with a sample of bias-corrected statolith-based age in a statistical model of larval growth. Statolith-based age estimation is not reliable for populations that do not form cohesive statoliths (Barker et al. 1997) or for larvae that resorb statoliths (Lochet et al. 2013). In addition, age estimates vary greatly among readers (Dawson et al. 2009).

Table 3 Duration of the sea lamprey larval life stage (methods listed are those used to estimate age and growth; adapted from Dawson et al. 2015)

Larval stage (years)	Method	Source
3.4–3.9	Length–frequency	Applegate (1950)
5	Length–frequency	Hardisty (1969)
6	Length–frequency	Lowe et al. (1973)
6–8	Length–frequency	Beamish and Potter (1975)
5–12	Known recruitment date	Manion and Smith (1978)
5	Length–frequency	Hardisty (1979)
3–7	Length–frequency	Purvis (1979)
5	Cage	Morman (1987)
2	Length–frequency/statolith	Morkert et al. (1998)
3–4	statolith	Griffiths et al. (2001)
4	Length–frequency/statolith	Quintella et al. (2003)

Although analysis of length–frequency distributions and statolith readings are the only method for estimating larval sea lamprey age, growth rate can be estimated with other methods, such as in situ cages (c.f. Malmqvist 1983; Morman 1987; Zerrenner 2004), laboratory growth experiments (c.f. Murdoch et al. 1992; Rodriguez-Muñoz et al. 2003), release of adult spawners into naturally inaccessible streams or stream reaches (Dawson et al. 2009), and surveys within lampricide-treated streams in years following treatment (Griffiths et al. 2001). Use of in situ cages and laboratory methods to measure larval growth rate have been useful for assessing effects of factors such as density (Malmqvist 1983; Mallatt 1983; Morman 1987; Murdoch et al. 1992; Rodriguez-Muñoz et al. 2003; Zerrenner 2004) and temperature (Mallatt 1983; Holmes 1990; Rodríguez-Muñoz et al. 2001). Most such studies confirmed that larval growth was inversely related to larval density (Malmqvist 1983; Morman 1987; Murdoch et al. 1992; Rodriguez-Muñoz et al. 2003), although one study of larvae held for 1 year in 0.16-m^2 circular cages suggested larval growth was independent of density (Zerrenner 2004). In the Great Lakes, larval sea lampreys are present in streams where winter water temperatures near 0 °C (USFWS and DFO unpublished data). Growth was highest in streams with mean annual water temperature of ~8 °C, discharge of 0.5–2.0 m^3/s, and >300 µS conductivity (Griffiths et al. 2001) and larvae reach a maximum length of 196 mm in total length (Table 1). In Europe, 8 °C is close to the minimum temperature where the sea lamprey is present, and growth is highest in systems with higher average temperatures, with some larvae attaining 190 mm in total length (Quintella et al. 2003).

Gonadogenesis in larval sea lampreys usually begins when larvae are 40–60 mm TL and ages 1–2 and gonadal differentiation is usually complete when larvae are 90–100 mm TL and ages 3–4 (Hardisty 1969a; Docker 1992; Wicks et al. 1998). Biotic and abiotic factors are thought to contribute to sex determination in lamprey species, such as larval density, temperature, pH, and when physiological resources are diverted into somatic growth, although mechanisms are not well established (Hardisty 1965a, b; Barker et al. 1998; Neave 2004; Dawson et al. 2015). In Great Lakes sea lamprey populations, selection pressure from pesticide application is implicated in sex determination (Smith 1971; Heinrich et al. 1980; Wicks et al. 1998). For example, males predominated in high pre-control sea lamprey populations (Smith 1971), with 70 % males in Lake Superior and 68 % males in lakes Huron and Michigan, whereas sex ratios shifted toward females after years of treatment, with only 28 % males in Lake Superior and 21 % males in Lake Michigan (Heinrich et al. 1980). A similar shift in the sex ratio in Lake Huron began before treatment reduced larval density, and was as low as 31 % males by 1975, likely because of environmental conditions and extremely low abundance of lake trout in the 1950s (Purvis 1979; Heinrich et al. 1980). Contemporary sex ratios have typically been 50–70 % males in the Great Lakes (Fig. 4), although the increase in the proportion of males in Lake Erie in 2007 and 2008 corresponds to record-high abundance of adult lampreys in those years, despite consecutive years treating sea lamprey producing streams in 2008–2009 and 2009–2010. In French rivers, the sex ratio favors females for both exploited and unexploited rivers (Beaulaton et al. 2008). For North American anadromous sea lamprey populations, males typically outnumber females (Beamish 1980). Estimates of adult sex ratio are fraught with uncertainty, depending on sampling time and methods (see Johnson et al. 2015a, b).

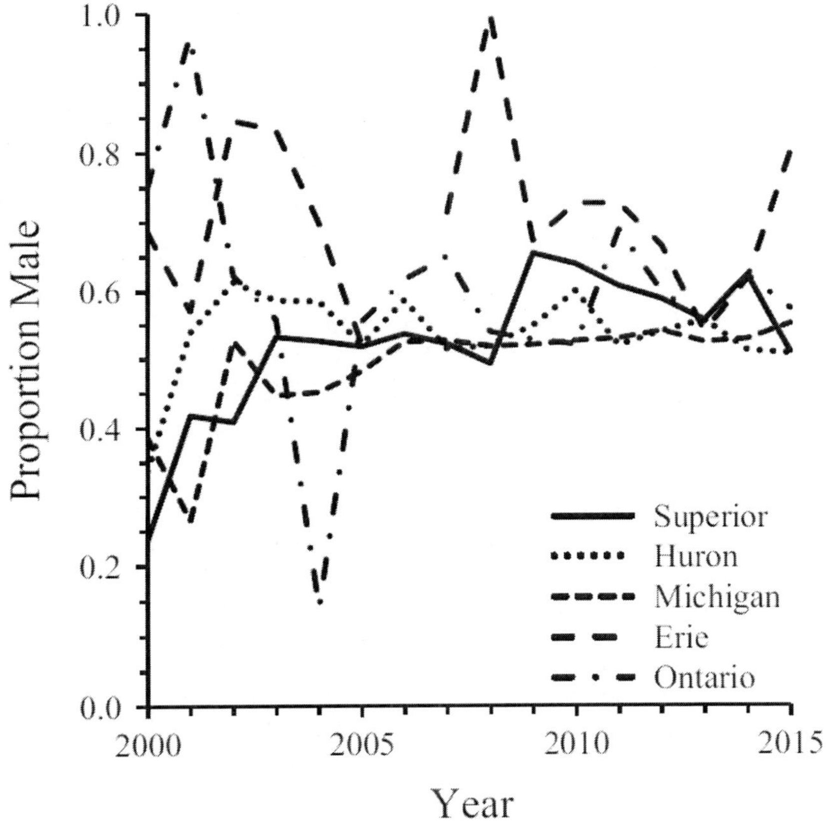

Figure. 4 Proportion of adult male sea lampreys caught during upstream spawning migrations in streams throughout the Laurentian Great Lakes basin, North America, 2000–2015 (N = 50,348; USFWS and DFO unpublished data)

Abundance of larval sea lampreys was estimated in most Great Lakes tributaries during 1995–2013 using quantitative assessment methods, which were generally greater than those estimated by mark-recapture (Slade et al. 2003; Hansen and Jones 2008; Dawson et al. 2015). Stream-specific estimates of abundance ranged from less than 50 to more than 6-million larvae. Larval abundance has also been estimated in small to mid-sized tributaries using mark-recapture techniques, with estimates ranging from 3,000 to nearly 2-million larvae (Steeves 2002; Hansen and Jones 2008). Estimates of maximum larval abundance in Great Lakes tributaries currently infested with larval sea lampreys suggest that these streams are capable of producing nearly 69 million larvae, with 2.7 million in Lake Superior, 15 million in Lake Michigan, 22 million in Lake Huron, 0.5 million in Lake Erie, and 4.5 million in Lake Ontario.

Metamorphosis

In spring of the year of metamorphosis, larvae are usually 3–5 years old, and are at least 100-mm long in the Great Lakes and 120-mm long in Europe (average = 130–140 mm; Table 1; Dawson et al. 2015). Larvae metamorphose into juveniles by development of an oral disk, appearance of teeth, eruption of eyes, enlargement of fins, and changes in pigmentation (Hardisty and Potter 1971b), all of which enable transformed individuals to change from sedentary larvae filter-feeding in streams to free-swimming predators in marine or lake environments (Manion and Stauffer 1970; Youson et al. 1977; Potter and Beamish 1977; Youson 1980; Hardisty 2006). Metamorphosis depends on length and weight, and particularly, on accumulation of lipids as energy stores, because the sea lamprey does not feed during metamorphosis (Holmes and Youson 1994; Youson 2003; Treble et al. 2008). Metamorphosis generally commences in July with changes to internal organs and body proportions, and concludes by September (Potter and Beamish 1977). Larval sea lampreys were once thought to transfer somatic growth strictly to mass and accumulation of lipids in the year of metamorphosis (Potter 1980), but recent data suggest that larvae also change in total length prior to metamorphosis (Hollett 1998; Treble et al. 2008). Age at metamorphosis is primarily related to growth and accumulation of lipid stores in larvae (Youson 1980; Treble et al. 2008), and inversely related to latitude, with larvae in southern streams growing faster than in northern streams (Hansen et al. 2003; Treble et al. 2008; Quintella et al. 2003). In cold untreated streams, larvae may not metamorphose until age ten or more, as in the St. Marys River (Haeseker et al. 2003) or the Big Garlic River (Manion and McLain 1971; Manion and Smith 1978). Sea lampreys then migrate from a stream, typically during high water events in autumn or spring (Applegate 1950), and begin the parasitic stage of life in an ocean or lake environment.

Juvenile life stage

Downstream migration of juveniles is triggered by increasing stream flow (Potter 1980; Bird et al. 1994; Hardisty 2006) that coincides with peak stream flows in autumn and spring in the Great Lakes (Applegate 1950; Potter 1980). Numbers of downstream migrating juveniles are usually greater in autumn than spring in the Great Lakes (Applegate and Brynildson 1952; Applegate 1961; Hodges 1972; Potter and Huggins 1973; Potter 1980; Hanson and Swink 1989; Swink and Johnson 2014). Downstream migration begins as early as September and ends as late as mid- to late-April or early-May, and is predominantly at night, with out-migrating juveniles burrowing into substrates or hiding beneath cover during the day (Applegate 1950, 1961; Hanson and Swink 1989; Swink and Johnson 2014). For anadromous populations in Europe, downstream migration is from late autumn through early winter (Taverny and Elie 2009), although movement is between October and May, with a peak in March, in the Galician region of northwestern Spain (Silva et al. 2013a, b). In Western Europe, milder weather may lead to a more continuous and gradual downstream migration than in the Great Lakes region of North America (Silva et al. 2013a, b). Timing of downstream migration was markedly bimodal in North American sea lamprey populations (landlocked and anadromous populations entering Canadian Atlantic Rivers) in autumn and spring (Applegate 1950; Hardisty 2006). This bimodal distribution, typical for North American populations, is not followed by European anadromous populations, which typically migrate with progressively increasing number of individuals moving toward a peak in March, although peak movement varies annually (Silva et al. 2013b). The spatial distribution of out-migrating juveniles in the water column is believed to be greatest in the thalweg near the surface where current velocity is greatest.

Survival of juvenile sea lampreys is poorly understood, but is likely related to factors associated with conditions in natal streams when juveniles out-migrate and availability of hosts in lake or ocean environments near natal streams when juveniles begin to feed (Young et al. 1990a). In lakes Michigan and Huron, mark-recapture studies indicated that survival of out-migrating juveniles to migratory adult stage was highly variable and was as high as 90 % (Sullivan and Adair 2010). Survival did not differ between autumn and spring juveniles of three parasitic cohorts that were tagged and released while out-migrating and recovered as adults 12–18 months later (Swink and Johnson 2014). Therefore, the opportunity for fall migrants to feed over winter did not improve survival.

Little is known about the marine life-history phase of anadromous sea lampreys (Beamish 1980; Halliday 1991), whereas studies of the landlocked form to estimate the duration of the parasitic phase using diverse methods suggest a juvenile parasitic stage of ~1 year (Table 4) or about 18–20 months between completion of metamorphosis and reproduction. Similarly, a mark-recapture study suggests a period of 18–20 months between completion of metamorphosis and reproduction in the Galician River, northwest Spain, where 408 sea lampreys were captured and tagged with external T-bar anchor tags at the onset of feeding (Silva et al. 2013a, b).

Table 4 Duration of the sea lamprey juvenile and adult life stage (methods listed are those use to determine duration)

Juvenile and adult stage[a] (years)	Ecotype/population	Method	Source
1–1.7	Landlocked	Captures at Great Lakes	Applegate (1950)
2–2.5	Anadromous/North America	Captures at sea	Beamish (1980)
1.5	Anadromous/North America	Captures at sea	Halliday (1991)
1	Landlocked	Captures at Great Lakes/Mark-recapture	Bergstedt and Swink (1995)
1.5–1.7	Anadromous/Europe	Mark-recapture	Silva et al. (2013a, b)

[a] Period between completion of metamorphosis (completion of stage 7 and prior to outmigration) and reproduction

Growth of juvenile sea lampreys increases with the onset of sexual maturity (Bergstedt and Swink 1995), where mass increases linearly from June through September, but more sharply in October (Madenjian et al. 2003). Growth of juvenile sea lampreys is also greater at higher temperatures, as is mortality of host fish following sea lamprey attack (Swink 2003; Farmer et al. 1977). However, the observed increase in growth during October was not associated with an increase in water temperature, but rather, may have been due to an increased likelihood of attaching to a host and actively feeding on host blood during October (Madenjian et al. 2003). Models of juvenile sea lamprey growth in Lake Superior indicate that sea lampreys can reach spawning size within 12 months even in the coldest temperature regimes (Moody et al. 2011). The effect of climate change is expected to increase the range of thermal habitat, and subsequently, the growth rate and attained size of sea lampreys (Cline et al. 2013). Juvenile sea lamprey grew 227–268 g in weight in one summer, from outmigration in autumn and spring 1998–2000 (initial weight = 5.33–6.01 g) to recapture in spring 1 year later, in the Black Mallard Creek, Michigan, a tributary to Lake Huron (Swink and Johnson 2014). In the Galician River, Spain, one lamprey grew from 218 mm in total length and 20 g in weight when tagged to 895 mm in total length and 1218 g in weight 13.5 months later when recaptured during its spawning migration into the same river (Silva et al. 2013a, b).

The period between final metamorphosis (October–November) and downstream migration can extend up to 3–4 months in European rivers, so sea lampreys may start feeding in freshwater streams after metamorphosis is complete (Potter and Beamish 1977; Beamish 1980; Silva et al. 2013a, b). For example, small juvenile sea lampreys were attached to resident brown trout *Salmo trutta* in a Spanish river (Silva et al. 2013a, b), which corroborates observations of North American populations for which 10–30 % of metamorphosed juveniles started feeding in rivers before migrating downstream (Davis 1967; Potter and Beamish 1977). The golden grey mullet *Liza aurata* (Risso, 1810), is an important and very abundant prey species in lower parts of European estuaries (Silva et al. 2013a, b). In rivers Mondego and Tagus, Portugal, thin-lipped grey mullets (*Liza ramada*, Risso, 1827) are also commonly observed with wounds from sea lamprey juveniles (Almeida, unpublished data). Large schools of mullets in European estuaries (Almeida 1996) may constitute an easy prey during this last phase of adaptation to the marine environment (Almeida and Quintella 2013). Juvenile lampreys have limited swimming capacities (Dauble et al. 2006), so predation on an intermediate species that is moving to the sea may increase survival of young sea lampreys during initial parasitic feeding in predator-rich marine environments.

Abundance of out-migrating juvenile sea lampreys estimated using mark-recapture methods ranged widely in Lake Huron (536,000–1,100,000) and in Lake Michigan (325,000–813,000; Sullivan and Adair 2010; Young et al. 2003). Young et al. (2003) concluded that these mark-recapture methods likely overestimated abundance of out-migrating sea lampreys. For example, mark-recapture estimates of the abundance of five cohorts of out-migrating juveniles (639,000–803,000) were larger on average than those of five cohorts of feeding juveniles (515,000–2,342,000), likely because tagging and handling mortality was higher for feeding juveniles than for out-migrating juveniles (Bergstedt et al. 2003). In the absence of control, these estimates of juvenile sea lamprey abundance for controlled populations would have been much greater.

Feeding

A bioenergetics model for the sea lamprey by Kitchell and Breck (1980) was developed from laboratory studies of standard metabolism (Beamish 1973), blood consumption (Farmer et al. 1975), egestion and excretion (Farmer et al. 1975), and the effect of water temperature on sea lamprey growth (Farmer et al. 1977). Sea lampreys used in laboratory experiments by Beamish (1973) showed very little, if any, movement within the respiration chamber, and therefore these respiration rate measurements were categorized as standard metabolic rates (SMRs). However, the energy budget of sea lampreys from the Laurentian Great Lakes would not balance based on respiration that only included SMR, so Kitchell and Breck (1980) hypothesized that sea lampreys must exhibit some activity and multiplied SMR by an activity multiplier (ACT = 1.5) to balance the energy budget. Recent underwater video of sea lampreys attached to lake trout in Lake Champlain confirms that sea lampreys swim even when attached to hosts (E. Marsden, University of Vermont, personal communication; Madenjian et al. 2013).

Size-based interactions between individual sea lampreys and individual lake trout, the preferred host in the Laurentian Great Lakes, appeared to be important factors affecting mortality inflicted by sea lampreys on lake trout, so individual-based models (IBMs) have been applied to Great Lakes sea lamprey populations (MacKay 1992; Madenjian et al. 2003). Feeding models for the sea lamprey were based on laboratory studies in which sea lamprey growth was measured during the time of attachment to a host, and then fitted to observed growth to estimate sea lamprey feeding rate with the sea lamprey bioenergetics model (Cochran and Kitchell 1989; Cochran et al. 1999). Bioenergetics and feeding models were incorporated into IBMs to estimate blood consumption and sea lamprey-induced mortality on lake trout by a sea lamprey population. Results indicated that sea lamprey growth, blood consumption by sea lampreys, and sea lamprey-induced mortality on lake trout peaked in October and November. The incidence of fresh wounds by sea lamprey on lake trout should therefore peak in October–November. Application of the sea lamprey bioenergetics model to sea lamprey in Lake Ontario confirmed that growth and blood consumption by sea lampreys peaked in autumn, which suggests that sea lamprey-induced mortality on hosts should peak in autumn (Kitchell and Breck 1980).

Recent comparisons of contaminant concentrations between male and female adult sea lampreys from northern Lake Huron suggested that males expended energy at a faster rate than females, because of higher swimming activity and possibly higher SMR (Madenjian et al. 2013, 2014). Mercury (Hg) and polychlorinated biphenyls (PCBs) have both been used as tracers of food consumption by fish, and male sea lampreys were 16–17 % higher in Hg and PCB concentrations than female sea lampreys, which suggests males fed at a 16–17 % higher rate than females. Higher swimming activity by adult male sea lampreys than adult female sea lampreys has been documented in the laboratory (Madenjian et al. 2013).

Effects on host species

Interactions between sea lampreys and their hosts in the Great Lakes are well-described by a Type-2 functional response in the attack rate and a development response in the growth rate (Bence et al. 2003). A Type-2 functional response assumes that the sea lamprey attack rate on a host increases with host density to a maximum attack rate at which the sea lamprey population is satiated by prey density. Similarly, a developmental response assumes that sea lamprey growth increases with host density to a maximum growth rate at which sea lamprey growth is satiated by host density. Sea lamprey populations in the Great Lakes do not exhibit numerical responses in either attack or growth rates, but rather, survival from metamorphosis to the summer–autumn period is assumed to be constant ($S = 0.5$) for parasitic-phase animals (Bence et al. 2003). However, survival of young parasites, as well as older parasites, is variable and partially depends on host density and also perhaps on water temperature (Christie and Kolenosky 1980; Eshenroder et al. 1995; Swink 1995).

Host selection is influenced by host size, with sea lampreys preferring to attack larger hosts (Cochran and Kitchell 1989; Bence et al. 2003; Swink 2003). Moreover, the lake trout is the preferred host of sea lampreys in the Great Lakes (Christie and Kolenosky 1980; Johnson and Anderson 1980; Bence et al. 2003; Morse et al. 2003; Harvey et al. 2008), although ciscoes (*Coregonus* spp.) are preferred by small (<15 g) sea lampreys (Johnson and Anderson 1980; Harvey et al. 2008). Sea lampreys may change their selectivity for hosts, perhaps in response to changing host abundance (Bence et al. 2003). For example, sea lamprey wounding on lake whitefish (*Coregonus clupeaformis*) was lower than wounding on lake trout when lake trout were abundant, whereas wounding on lake whitefish increased when lake trout were scarce. Similarly, wounding of large Chinook salmon (*Oncorhynchus tshawytscha*) was relatively high when lake trout abundance was low in northern Lake Huron. Based on stable-isotope signatures of sea lampreys and their prey in six regions of Lake Superior, sea lampreys fed predominantly on lake trout across the lake, but fed heavily on *Coregonus* spp. and *Catostomus* spp. in Black Bay, where these alternate hosts were more abundant than lake trout (Harvey et al. 2008).

Lethality of a sea lamprey attack on host species is a key determinant of sea lamprey-induced mortality on a host species' population (Bence et al. 2003), but available estimates vary widely and those based on tank studies may have been biased by confinement stress. Estimates of lethality of a sea lamprey attack on lake trout vary from nearly 0 % (negligible lethality) to 82 %, and mortality from a sea lamprey attack was higher (64 %) for small lake trout (469–557 mm TL) than for medium (44 %; 559–643 mm TL) and large (43 %; 660–799 mm TL) lake trout in laboratory tanks (Swink 2003). However, stress from confinement in relatively small (151 L in volume) tanks may have contributed additional mortality in these laboratory experiments. Lake sturgeon (*Acipenser fulvescens*) similar in size to young adult lake trout survived sea lamprey attacks at a higher rate (70 %) in larger (>1000 L) tanks (Patrick et al. 2009), perhaps because lake sturgeon have tougher skin than lake trout. Further, 66–74 % of adult lake trout survived sea lamprey attacks based on field data from Lake Champlain, Lake Ontario, and Lake Huron (Madenjian et al. 2008a). In contrast, the lethality of a sea lamprey attack on age-3 to age-5 lake whitefish was 75 %, based on a mark-recapture study in Lake Huron (Spangler et al. 1980), lethality of sea lamprey attacks on lake whitefish should be substantially higher than on lake trout because adult lake trout are substantially larger than adult lake whitefish. Similarly, lethality of sea lamprey attacks on adult burbot (*Lota lota*) should be higher than on adult lake trout.

Sea lampreys are capable of contributing substantially to declining abundance of host populations in aquatic ecosystems they invade. For example, predation by sea lamprey, along with overfishing, were responsible for collapses of lake trout populations in lakes Superior, Michigan, and Huron in the 1940s and 1950s (Hansen 1999). Lake trout abundance was already declining in each of these three lakes from overfishing in the early 1900s, but sea lamprey invasion in the 1930s accelerated rates of declining lake trout abundances in each lake in the 1940s and 1950s (Hansen 1999). In the 1960s, sea lamprey control in conjunction with intensive stocking enabled a buildup of lake trout populations in the 1970s and 1980s in the upper Great Lakes, and widespread natural reproduction by lake trout in Lake Superior. Further, sea lamprey predation contributed to declining abundance of lake whitefish in the upper Great Lakes in the 1950s (Lawrie and Rahrer 1972; Berst and Spangler 1973; Wells and McLain 1973), after which sea lamprey control contributed to recovery of lake whitefish populations (Madenjian et al. 2008b). Sea lamprey predation has also been suspected of having some effect on abundance of coregonines other than lake whitefish in the upper Great Lakes (Lawrie and Raher 1972; Berst and Spangler 1973; Wells and McLain 1973). Last, sea lamprey predation contributed to declining burbot abundance in the upper Great Lakes in the 1950s (Berst and Spangler 1973; Wells and McLain 1973; Gorman and Sitar 2013), after which sea lamprey control enabled recovery of burbot populations in the Laurentian Great Lakes (Madenjian et al. 2008b; Stapanian et al. 2008).

Even with reduced sea lamprey populations due to control activities, sea lamprey predation is still a large source of mortality on host populations in the Laurentian Great Lakes. For example, sea lamprey predation was the predominant source of mortality experienced by the lake trout population in Lake Huron during 1984–1993 (Sitar et al. 1999). More recently, sea lamprey predation is still believed to be an important source of mortality on lake whitefish populations in northern Lake Huron (M. Ebener, Chippewa Ottawa Resource Authority, personal communication). Of course, accuracy of these estimates of sea lamprey-induced mortality depend on accuracy of estimates of sea lamprey attack lethality. In addition to predation effects, sea lamprey attacks on hosts can also lead to sub-lethal effects. For example, lake sturgeon that survived a sea lamprey attack suffered acute anemia (Sepúlveda et al. 2012). Similarly, immune function was reduced and lipid stores were depleted in lake trout surviving sea lamprey attacks (Smith 2013). Long-term influences of such sub-lethal effects on host population dynamics have not been quantified.

Future status

How will climate change affect sea lamprey population ecology in the Laurentian Great Lakes and Europe?

Global climate change has been identified as a driver of change in thermal habitat of fish species that would increase biomass of sea lampreys in Lake Superior (Moody et al. 2011). Forecasts of increased fecundity and sea lamprey induced mortality are expected to negatively impact native fish species, particularly siscowet lake trout in Lake Superior (Cline et al. 2013). Global climate change is expected to increase negative effects of sea lamprey-induced mortality on host fish populations in Lake Superior, one of the most rapidly warming lakes on Earth (Kitchell et al. 2014). As Lake Superior continues to warm, sea lamprey growth is predicted to increase, thereby producing larger sea lampreys, greater blood consumption by sea lampreys, and increased lethality of sea lamprey attacks on hosts (Cline et al. 2014; Kitchell et al. 2014). Further, climate change may cause longer feeding seasons, increased growth, and larger sea lampreys (Cline et al. 2013), with increased fecundity, particularly in Lake Superior (Moody et al. 2011), which may lead to increased sea lamprey abundance. Climate change is also predicted to increase the magnitude of flood events that could lead to reduced effectiveness of low-head barriers used to block upstream spawning migrations of sea lampreys in the Great Lakes (Rahel and Olden 2008). The impact of climate change on the geographic distribution of sea lampreys in the Great Lakes is unknown, but changes in water levels, discharge, and temperature are likely to result in variation in streams selected for spawning, with varying impacts on larval recruitment and growth. In total, global climate change will likely require additional effort to maintain the current level of control of sea lampreys in the Great Lakes.

Global climate change is expected to greatly alter the geographic distribution of the sea lamprey in Europe (Lassalle et al. 2008). Specifically, the presence of sea lampreys in basins bordering the east coast of the Adriatic Sea, most Italian basins, and in the Iberian Peninsula (including Spain and Portugal) is expected to decrease in the future, based on predictions of precipitation and water temperature in tributaries. In the northern part of the species' range, conditions would likely remain suitable and waters in Iceland could even become suitable for the sea lamprey (Lassalle et al. 2008; Maitland et al. 2015). In Southwestern Europe, the sea lamprey is economically valuable, and supports commercial fisheries in major river systems, but is vulnerable to reductions in suitable essential habitats. Conservation measures to rehabilitate sea lamprey habitat and to manage for sustained commercial exploitation have been proposed (Almeida and Quintella 2002; Andrade et al. 2007). Further, conservation plans for sea lamprey should include projections of the geographic distribution of the sea lamprey in response to global climate change, with priority given to restoring suitable basins in Portugal (Lassalle et al. 2008).

Which attributes of sea lamprey population ecology can be used to control populations in the Laurentian Great Lakes?

Adult

Migratory habits of adult sea lampreys are currently being exploited by using barriers to block upstream migration and traps to capture and remove upstream migrating adults in Great Lakes tributaries (McLaughlin et al. 2007). Nonetheless, increased knowledge of adult sea lamprey movement and barrier design will be crucial to effective control (McLaughlin et al. 2007). Further, current barrier technology limits passage of non-jumping fishes, so new barrier designs must enable fish passage (McLaughlin et al. 2007). Trapping of adult sea lampreys is presently not used for control, but could be important if trapping efficacy is improved and the number of spawning adults is reduced enough to minimize larval recruitment (McLaughlin et al. 2007). Therefore, development of new trap designs that exploit ecological and behavioral aspects of adult migration, such as staging behavior at river mouths or movements within streams, while minimizing by-catch of other species, would benefit future control (Bravener and McLaughlin 2013). Use of attractants or repellents may increase trap efficacy, thereby reducing adult abundance and larval recruitment, or reducing infested areas of streams by making them less desirable for migrating or spawning adults (Johnson et al. 2015). Enhanced barrier design and more effective trapping can also reduce reliance on chemical use (McLaughlin et al. 2007).

Larval

The sedentary life habits that last 2–5 years appears to be the critical ecological attribute for control at this life stage. Current control depends on reducing recruitment of juvenile sea lampreys to the Great Lakes by killing larvae during their sedentary life stage with strategic applications of lampricides every 2–4 years, depending upon larval recruitment and growth. Further exploitation of this life stage should be explored, by developing new lampricides to further reduce or eliminate larval populations, or autocidal technologies (Thresher 2008) to manipulate genes that slow growth, determine sex, prohibit metamorphosis, or increase larval mortality by disrupting feeding behavior (McCauley et al. 2015). Last, because larvae spend multiple years in the same stream environment, they acquire regionally unique chemical signatures that identify their general natal origin (Hand et al. 2008). Refining methods to determine the specific stream of origin of larvae that survive or are not exposed to lampricide applications would be beneficial for targeting future control.

Juvenile

Success of sea lamprey control in the Great Lakes depends on the ability to apply pesticides to natal streams prior to sea lamprey metamorphosis and migration. Research has attempted to identify factors controlling metamorphosis (Docker et al. 2003), such as endocrinology (Youson 1994; Youson et al. 1994; Youson 1999), environmental conditions such as temperature (Purvis 1980; Youson 2003), conspecific density (Purvis 1980; Morman 1987; Treble et al. 2008), and gender distortion (Thresher et al. 2014). Recent sequencing of the sea lamprey genome may provide control opportunities if genes that regulate metamorphosis can be exploited (Smith et al. 2013; McCauley et al. 2015). Further, downstream migration of juveniles is a stage in the sea lamprey life history that is not presently exploited for control, but holds promise for the future (Johnson and Miehls 2014). Because out-migration of juveniles typically occurs at night during protracted periods of high discharge, efforts to trap this life stage are costly and have met with limited success. However, this life stage of sea lamprey ecology represents the last opportunity to remove this pest before it inflicts damage on host fishes. Development of new trapping technologies or guidance systems to lead out-migrating juveniles into traps or areas where survival can be reduced may result in more effective methods for exploiting this stage of the sea lamprey life history.

Future research on sea lamprey bioenergetics and feeding should focus on sea lamprey feeding rates, feeding models, and sex-specific bioenergetics models. Weight-specific feeding rates of sea lampreys in summer did not significantly change as sea lamprey weight increased from 20 to 160 g (Farmer et al. (1975), but weight-specific consumption rates may substantially decrease as lamprey body weight increases from 160 to 400 g and may also be influenced by seasonal cues (Madenjian et al. 2003). Attachment time is also suspected to be influenced by sea lamprey size, host size, and water temperature. Sex-specific bioenergetics models will require SMR and swimming activity to be determined for male and female juvenile (parasitic-phase) sea lampreys at various sizes, water temperatures, and times of the year, and would, for example, be useful for modeling how climate change would affect population fecundity and predation effects on host species.

Which attributes of sea lamprey population ecology can be used to restore and conserve populations in Europe?

An absence of genetic differences among sea lamprey populations within western Atlantic (Waldman et al. 2008) and eastern Atlantic rivers suggests regional panmixia with reproductively isolated European and North American sea lamprey populations (Rodríguez-Muñoz et al. 2004; Bryan et al. 2005; Genner et al. 2012). Nevertheless, recent studies in Portugal suggested the existence of a population structure composed of three different stocks (North/, Tagus and Guadiana) possibly promoted by sea bed topography (Lança et al. 2014). The hypothesis of a limited dispersion capacity in the ocean raises questions about management and conservation of this species in Europe, because oceanic regions and specific watersheds likely play an important role in the conservation of sea lamprey populations in southwestern Europe. Sea bed topography could play a major role in dispersion of sea lampreys from natal streams, thereby causing dispersion to differ between European and North American Atlantic coasts. Such differences are particularly important in areas where the species presents conservation challenges, because success of any management plan implemented to sustain the species' fisheries depends on sound understanding of the species life cycle (Almeida and Quintella 2013). More research on this subject is warranted, to guide future restoration and management of the species in Europe.

Because of its decline across Europe, the sea lamprey was given some legal protection (Lelek 1987; Renaud 1997; Mateus et al. 2012) under Annex II of the European Union Habitats Directive, Appendix III of the Bern Convention, and in the OSPAR (Oslo-Paris) convention list (Convention for the Protection of the Marine Environment of the North- east Atlantic). Aquatic pollution, and habitat fragmentation and reduction caused by construction of large dams, weirs, and other man-made barriers, are among the greatest threats to European sea lamprey population (Gardner et al. 2012; Mateus et al. 2012; Hogg et al. 2013; Rooney et al. 2015; Maitland et al. 2015). Since the late twentieth century, sea lamprey populations have increased slightly following improvement of water quality in some European rivers (Beaulaton et al. 2008). The sea lamprey is presently assessed as Least Concern in the European Red List of Freshwater Fishes (Freyhof and Brooks 2011), although it is considered threatened in Red List categories of several European countries (Mateus et al. 2012). Recovery of sea lamprey populations in Europe will depend mostly on restoration of freshwater habitat and on promoting sustainable management of commercial exploitation in both estuarine and freshwater environments. Lack of reliable records of commercial harvest from each river basin where this species occurs is a major drawback that introduces subjectivity and risk to fishery management.

Pheromones are crucial for the sea lampreys' life-history that could potentially be useful as management tools for both control programs in the

Great Lakes and conservation programs in Europe. The sea lamprey produces at least two types of pheromones. Stream-dwelling larvae release a migratory bile acid-based pheromone that guides adult sea lampreys to watersheds that were successful as spawning and nursery areas for larvae (Bjerselius et al. 2000). Mature spermiated male sea lampreys also release a potent sex pheromone that induces preference and searching behavior by ovulated female lampreys ascending to upstream spawning areas (Li et al. 2002). The bile acid pheromone released by larvae may be used to attract adults to rivers where habitat was restored, thereby increasing speed of recolonization. Male sex pheromone may be used to increase the efficiency of fish passage devices by increasing attractiveness of a fish passage entrance.

Which attributes of sea lamprey population ecology are useful for both control programs in the Great Lakes and restoration programs in Europe?

Many of the same features of sea lamprey population ecology that are useful for control programs in the Great Lakes of North America are also useful for restoration programs in Europe. For example, migration by adult sea lampreys into streams to spawn has been exploited since the late 1940s as a population control strategy by blocking upstream migration with traps and barriers (McLaughlin et al. 2007). Non-target effects of traps and barriers on species other than the sea lamprey have been the focus of research and engineering aimed to enable passage of non-target species while simultaneously blocking passage of sea lampreys (McLaughlin et al. 2007). Not surprisingly, dams on rivers in Europe are presently one of the primary factors limiting sea lamprey access to spawning areas upstream (Mateus et al. 2012). We propose that understanding of how to block adult sea lamprey passage in streams for control purposes in the Great Lakes is also useful for allowing passage of adult sea lampreys for restoration purposes in Europe and elsewhere where the species is imperiled (Moser et al. 2015). For example, traps within fish ladders to sort and remove adult sea lampreys in Great Lakes streams would be useful designs for traps within fish ladders to catch and transport adult sea lampreys to European streams free from dams (McLaughlin et al. 2007). Similarly, use of attractants and repellants to lure sea lampreys into physical or ecological traps for their control in the Great Lakes would be equally useful for luring sea lampreys into traps and fishways for their conservation (e.g. to guide their passage around or through barriers) where they are imperiled in their native range (Li et al. 2007).

The long period of life spent in streams by larval and juvenile sea lampreys that makes the species vulnerable to pesticide control in the Great Lakes also provides a focus for habitat protection, mitigation, and enhancement where the species is imperiled in Europe and elsewhere. Since 1958, sea lamprey control in the Great Lakes targeted the stream-dwelling non-parasitic larval phase of the sea lamprey life history, because larvae are relatively sedentary and live for several years in streams (McDonald and Kolar 2007). Pesticide control of sea lamprey populations in the Great Lakes was subsequently directed by assessments of the abundance and distribution of larval sea lampreys within and among streams, including assessment of the suitability of instream habitat (Hansen et al. 2003; Jones 2007). These same assessment methods for targeting sea lamprey control in the Great Lakes are equally useful for targeting habitat protection, mitigation, and enhancement of sea lamprey populations where they are imperiled in their native range. Similarly, assessment methods for juvenile and adult sea lampreys that aim to enhance effectiveness of control in the Great Lakes (e.g. migration timing, mating systems, and factors influencing spawning success, Jones 2007) would be useful for enhancing effectiveness of conservation measures where the species is imperiled in its native range. Assessment methods presently used for measuring success of sea lamprey control in the Great Lakes (e.g. quantifying numbers of all life stages, Jones 2007) would be equally useful for measuring success of sea lamprey conservation elsewhere in its native range.

Which attributes of sea lamprey population ecology are in need of further study for management of the species worldwide?

Much has been learned about sea lamprey population dynamics, but some areas of sea lamprey population dynamics require further study, especially in relation to abundance, survival, recruitment, and feeding. For example, abundance of the sea lamprey in Europe has rarely been measured or monitored (except for the population in the Garonne River, France; Beaulaton et al. 2008), but is crucial for understanding the population status of the species. Methods by which juvenile and adult sea lamprey abundance is monitored in the Laurentian Great Lakes may be useful models for European population monitoring (e.g. Hansen et al. 2003; Jones 2007). Similarly, little is known about the marine phase of anadromous sea lampreys, especially habitat and host preferences, dynamics and mechanisms of movement, although much has now been learned about how adults locate freshwater spawning streams and mates with streams using pheromones (e.g. Bjerselius et al. 2000; Li et al. 2002; Silva et al. 2013a). Research into the effects of pheromones in the different spatial and chemical scales of the marine environment may provide management tools applicable to the regional stocks of anadromous sea lampreys. Telemetry (acoustic and archival) may be useful for increasing knowledge of movement by feeding-phase sea lampreys in the Atlantic Ocean and Great Lakes (e.g. Johnson et al. 2015a, b). Movement of out-migrating juveniles is only generally understood, so is an area of fruitful future study (e.g. Johnson and Miehls 2014). Recruitment is highly dynamic, because of the interplay of strong density-independent and density-dependent forces, but must be studied more broadly in both native and introduced populations, to better understand general patterns of recruitment dynamics. As a general subject of study, dynamics of survival and mortality of sea lampreys is poorly understood for all life stages, including adults, larvae, recently metamorphosed juveniles, and feeding juveniles. Feeding dynamics of sea lamprey have benefited from bioenergetics studies, but more research is needed on feeding rates, feeding models, and sex-specific bioenergetics models (e.g. Madenjian et al. 2003). Last, long-term sub-lethal effects of feeding-phase sea lampreys on host populations should be studied, to improve understanding of effects of invasive non-native sea lampreys in the Laurentian Great Lakes.

ACKNOWLEDGMENTS

We thank Erin Dunlop, Ontario Ministry of Natural Resources, and four anonymous reviewers for their many helpful comments on the manuscript. Use of trade, product, or firm names is for descriptive purposes and does not imply endorsement by the U.S. Government. The findings and conclusions in this article are those of the authors and do not necessarily represent the views of the U.S. Fish and Wildlife Service. This article is Contribution 2031 of the U.S. Geological Survey, Great Lakes Science Center.

REFERENCES

1. Adair R, Sullivan P (2013) Sea lamprey control in the Great Lakes 2012. Great Lakes Fishery Commission, Annual Report, May 29–30, 2013, Ann Arbor, Michigan
2. Almeida PR (1996) Biologia e Ecologia de Liza ramada (Risso, 1826) e Chelon labrosus (Risso, 1826) (Pisces, Mugilidae) no Estuário do Mira (Portugal). Inter-relações com o Ecossistema Estuarino. PhD dissertation, University of Lisbon, Portugal
3. Almeida PR, Quintella BR (2002) Larval habitat of the sea lamprey (*Petromyzon marinus* L.) in the river Mondego (Portugal). In: Collares-Pereira MJ, Coelho MM, Cowx IG (eds) Freshwater fish conservation: options for the future. Blackwell Science, Oxford, pp 121–130
4. Almeida PR, Quintella BR (2013) Sea Lamprey migration: a millenial journey. In: Ueda H, Tsukamoto K (eds) Physiology and ecology of fish migration. CRC Press, Boca Raton, pp 105–131
5. Almeida PR, Silva HT, Quintella BR (2000) The migratory behaviour of the sea lamprey *Petromyzon marinus* L., observed by acoustic telemetry in the River Mondego (Portugal). In: Moore A, Russel I (eds) Advances in fish telemetry. CEFAS, Lowestoft, pp 99–108

6. Almeida PR, Quintella BR, Dias NM (2002a) Movement of radio-tagged anadromous sea lamprey during the spawning migration in the River Mondego (Portugal). Hydrobiologia 483:1–8
7. Almeida PR, Quintella BR, Dias NM, Andrade N (2002b) The anadromous sea lamprey in Portugal: biology and conservation perspectives. In: Moser M, Bayer J, MacKinlay D (eds) The biology of lampreys, symposium proceedings, International Congress on the Biology of Fish, American Fisheries Society, 21–25 July, Vancouver, British Columbia, pp 49–58
8. Andrade NO, Quintella BR, Ferreira J, Pinela S, Póvoa I, Sílvia P, Almeida PR (2007) Sea lamprey (*Petromyzon marinus* L.) spawning migration in the Vouga river basin (Portugal): poaching impact, preferential resting sites and spawning grounds. Hydrobiologia 582:121–132
9. Applegate VC (1950) Natural history of the sea lamprey (*Petromyzon marinus*) in Michigan. US Fish Wildl Serv Spec Sci Rep Fish 55, Washington, DC
10. Applegate VC (1961) Downstream movement of lampreys and fishes in the Carp Lake River, Michigan. US Fish Wildl Serv Spec Sci Rep Fish 387, Washington, DC
11. Applegate VC, Brynildson C (1952) Downstream movement of recently transformed sea lampreys, *Petromyzon marinus*, in the Carp Lake River, Michigan. Trans Am Fish Soc 81:275–290
12. Applegate VC, Smith BR (1951) Movement and dispersion of a blocked spawning run of sea lampreys in the Great Lakes. Trans N Am Wildl Conf 16:243–351
13. Barker LA, Morrison BJ, Wicks BJ, Beamish FWH (1997) Age discrimination and statolith diversity in sea lamprey from streams with varying alkalinity. Trans Am Fish Soc 126:1021–1026
14. Barker LA, Morrison BJ, Wicks BJ, Beamish FWH (1998) Potential fecundity of landlocked sea lamprey larvae, *Petromyzon marinus*, with typical and atypical gonads. Copeia 1998:1070–1075
15. Bartels H, Potter IC (2007) Cellular composition and ultrastructure of the gill epithelium of larval and adult lampreys: implications for osmoregulation in fresh and seawater. J Exp Biol 207:3447–3462
16. Baxter EW (1957) Lamprey distribution in streams and rivers. Nature 180:1145
17. Beamish FWH (1973) Oxygen consumption of adult *Petromyzon marinus* in relation to body weight and temperature. J Fish Res Board Can 30:1367–1370
18. Beamish FWH (1978) Swimming capacity. In: Hoar WS, Randall DJ (eds) Fish physiology, vol VII. Academic Press, New York, pp 101–187
19. Beamish FWH (1980) Biology of the North American anadromous sea lamprey, *Petromyzon marinus*. Can J Fish Aquat Sci 37:1924–1943
20. Beamish FWH, Jebbink J (1994) Abundance of lamprey larvae and physical habitat. Environ Biol Fish 39:209–214
21. Beamish FWH, Lowartz S (1996) Larval habitat of American brook lamprey. Can J Fish Aquat Sci 53:693–700
22. Beamish FWH, Potter IC (1975) The biology of the anadromous sea lamprey (*Petromyzon marinus*) in New Brunswick. J Zool 177:57–72
23. Beaulaton L, Taverny C, Castelnaud G (2008) Fishing, abundance and life history traits of the anadromous sea lamprey (*Petromyzon marinus*) in Europe. Fish Res 92:90–101
24. Becker GC (1983) Fishes of Wisconsin. University of Wisconsin Press, Madison
25. Bence JR, Bergstedt RA, Christie GC, Cochran PA, Ebener MP, Koonce JF, Rutter MA, Swink WD (2003) Sea lamprey (*Petromyzon marinus*) parasite–host interactions in the Great Lakes. J Great Lakes Res 29(Suppl 1):253–282
26. Bergstedt RA, Seelye JG (1995) Evidence for lack of homing by sea lampreys. Trans Am Fish Soc 124:235–239
27. Bergstedt RA, Swink WD (1995) Seasonal growth and duration of the parasitic life stage of the landlocked sea lamprey (*Petromyzon marinus*). Can J Fish Aquat Sci 52:1257–1264

28. Bergstedt RA, McDonald RB, Mullett KM, Wright GM, Swink WD, Burnham KP (2003) Mark-recapture population estimates of parasitic sea lampreys (*Petromyzon marinus)* in Lake Huron. J Great Lakes Res 29(Suppl 1):226–239
29. Berst AH, Spangler GR (1973) Lake Huron—The ecology of the fish community and man's effects on it. Great Lakes Fishery Commission, Technical Report 21, Ann Arbor, Michigan
30. Bigelow HB, Schroeder WC (1948) Fishes of the western North Atlantic, part 1, cyclostomes. Mem Sears Found Mar Res 1:29–58
31. Binder TR, McLaughlin RL, McDonald DG (2010) Relative importance of water temperature, water level, and lunar cycle to migratory activity in spawning phase sea lampreys in Lake Ontario. Trans Am Fish Soc 139:700–712
32. Bird DJ, Potter IC, Hardisty MW, Baker BI (1994) Morphology, body size and behaviour of recently-metamorphosed sea lampreys, *Petromyzon marinus*, from the lower River Severn, and their relevance to the onset of parasitic feeding. J Fish Biol 44:67–74
33. Bjerselius R, Li W, Teeter JH, Seelye JG, Johnsen PB, Maniak PJ, Grant GC, Polkinghorne CN, Sorensen PW (2000) Direct behavioral evidence that unique bile acids released by larval sea lamprey function as a migratory pheromone. Can J Fish Aquat Sci 57:557–569
34. Bravener GA, McLaughlin RL (2013) A behavioural framework for trapping success and its application to invasive sea lamprey. Can J Fish Aquat Sci 70:1438–1446
35. Bryan MB, Zalinski D, Filcek KB, Libants S, Li W, Scribner KT (2005) Patterns of invasion and colonization of the sea lamprey (*Petromyzon marinus*) in North America as revealed by microsatellite genotypes. Mol Ecol 14:3757–3773
36. Christie GL, Goddard CI (2003) Sea lamprey international symposium (SLIS II): advances in the integrated management of sea lamprey in the Great Lakes. J Great Lakes Res 29(Suppl 1):1–14
37. Christie WJ, Kolenosky DP (1980) Parasitic phase of the sea lamprey (*Petromyzon marinus*) in Lake Ontario. Can J Fish Aquat Sci 37:2021–2038
38. Cline TJ, Bennington V, Kitchell JF (2013) Climate change expands the spatial extent and duration of preferred thermal habitat for Lake Superior fishes. PLoS ONE 8(4):e62279. doi:10.1371/journal.pone.0062279
39. Cline TJ, Kitchell JF, Bennington V, McKinley GA, Moody EK, Weidel BC (2014) Climate impacts on landlocked sea lamprey: implications for host–parasite interactions and invasive species management. Ecosphere 5:68
40. Cochran PA, Kitchell JF (1989) A model of feeding by parasitic lampreys. Can J Fish Aquat Sci 46:1845–1852
41. Cochran PA, Swink WD, Kinziger AP (1999) Testing and extension of a sea lamprey feeding model. Trans Am Fish Soc 128:403–413
42. Dauble DD, Moursund RA, Bleich MD (2006) Swimming behaviour of juvenile Pacific lamprey, *Lampetra tridentata*. Environ Biol Fish 75:167–171
43. Davis RM (1967) Parasitism by newly-transformed anadromous sea lampreys on landlocked salmon and other fishes in a coastal Maine lake. Trans Am Fish Soc 96:11–16
44. Dawson HA (2007) Recruitment dynamics of Great Lakes sea lamprey (*Petromyzon marinus*) populations and implications for integrated pest management. PhD dissertation, Michigan State University, East Lansing
45. Dawson HA, Jones ML (2009) Factors affecting recruitment dynamics of Great Lakes sea lamprey (*Petromyzon marinus*) populations. J Great Lakes Res 35:353–360
46. Dawson HA, Jones ML, Scribner KT, Gilmore SA (2009) An assessment of age determination methods for Great Lakes larval sea lampreys. N Am J Fish Manage 29:914–927
47. Dawson HA, Quintella BR, Almeida PR, Treble AJ, Jolley JC (2015) The ecology of larval and metamorphosing lampreys. In: Docker MF (ed) Lampreys: biology, conservation and control. Springer, Berlin, pp 75–137

48. Derosier AL, Jones ML, Scribner KT (2007) Dispersal of sea lamprey larvae during early life: relevance for recruitment dynamics. Environ Biol Fish 78:271–284
49. Docker MF (1992) Labile sex determination in lampreys: the effect of larval density and sex steroids on gonadal differentiation. PhD dissertation, University of Guelph
50. Dolinsek IJ, McLaughlin RL, Grant JW, O'Connor LM, Pratt TC (2014) Do natural history data predict the movement ecology of fishes in Lake Ontario streams? Can J Fish Aquat Sci 71:1171–1185
51. Eddy S, Underhill JC (1974) Northern fishes, with special reference to the Upper Mississippi Valley, 3rd edn. University of Minnesota Press, Minneapolis
52. Emery L (1985) Review of fish introduced into the Great Lakes, 1819–1974. Great Lakes Fishery Commission, Technical Report 45, Ann Arbor, Michigan
53. Eshenroder RL (2014) The role of the Champlain canal and Erie canal as putative corridors for colonization of Lake Champlain and Lake Ontario by sea lampreys. Trans Am Fish Soc 143:634–649
54. Eshenroder RL, Payne NR, Johnson JE, Bowen C II, Ebener MP (1995) Lake trout rehabilitation in Lake Huron. J Great Lakes Res 21(Supplement 1):108–127
55. Farmer GJ, Beamish FWH, Robinson GA (1975) Food consumption of the adult landlocked sea lamprey, *Petromyzon marinus*, L. Comp Biochem Physiol 50A:753–757
56. Farmer GJ, Beamish FWH, Lett PF (1977) Influence of water temperature on the growth rate of landlocked sea lamprey (*Petromyzon marinus*) and the associated rate of host mortality. J Fish Res Board Can 34:1373–1378
57. Fodale MF, Bronte CR, Bergstedt RA, Cuddy DW, Adams JV (2003) Classification of lentic habitat for sea lamprey (*Petromyzon marinus*) larvae using a remote seabed classification device. J Great Lakes Res 29(Suppl 1):190–203
58. Freyhof J, Brooks E (2011) European red list of freshwater fishes. Publications Office of the European Union, Luxembourg
59. Gardner C, Coghlan SM Jr, Zydlewski J (2012) Distribution and abundance of anadromous sea lamprey spawners in a fragmented stream: current status and potential range expansion following barrier removal. Northeast Nat 19:99–110
60. Genner MJ, Hillman R, McHugh M, Hawkins SJ, Lucas MC (2012) Contrasting demographic histories of European and North American sea lamprey (*Petromyzon marinus*) populations inferred from mitochondrial DNA sequence variation. Mar Freshw Res 63:827–833
61. Gerking SD (1955) Key to the fishes of Indiana. Investig Indiana Lakes Streams 4:49–86
62. Goodwin CE, Dick JTA, Rogowski DL, Elwood RW (2008) Lamprey (*Lampetra fluviatilis* and *Lampetra planeri*) ammocoete habitat associations at regional, catchment and microhabitat scales in Northern Ireland. Ecol Freshw Fish 17:542–553
63. Gorman OT, Sitar SP (2013) Ups and downs of burbot and their predator lake trout in Lake Superior, 1953-2011. Trans Am Fish Soc 142:1757–1772
64. Griffiths RW, Beamish FWH, Morrison BJ, Barker LA (2001) Factors affecting larval sea lamprey growth and length at metamorphosis in lampricide-treated streams. Trans Am Fish Soc 130:289–306
65. Gross M (1987) The evolution of diadromy in fishes. Am Fish Soc Symp 1:14–25
66. Haeseker SL, Jones ML, Bence JR (2003) Estimating uncertainty in the stock-recruitment relationship for St. Marys River sea lampreys. J Great Lakes Res 29(Suppl. 1):728–741
67. Halliday RG (1991) Marine distribution of the sea lamprey (*Petromyzon marinus*) in the northwest Atlantic. Can J Fish Aquat Sci 48:832–842
68. Hand CP, Ludsin SA, Fryer BJ, Marsden EJ (2008) Statolith microchemistry as a technique for discriminating among Great Lakes sea lamprey (*Petromyzon marinus*) spawning tributaries. Can J Fish Aquat Sci 65:1153–1164
69. Hansen MJ (1999) Lake trout in the Great Lakes: basinwide stock collapse and binational restoration. In: Taylor WW, Ferreri CP (eds) Great Lakes fisheries policy

and management: a binational perspective. Michigan State University Press, East Lansing, pp 417–453
70. Hansen GJA, Jones ML (2008) A rapid assessment approach to prioritizing streams for control of Great Lakes sea lampreys: a case study in adaptive management. Can J Fish Aquat Sci 65:2471–2484
71. Hansen MJ, Adams JV, Cuddy DW, Doemel JM, Fodale MF, Larson GL, Ollila DJ, Slade JW, Steeves TB, Young RJ, Zerrenner A (2003) Optimizing larval assessment to support sea lamprey control in the Great Lakes. J Great Lakes Res 29(Suppl 1):766–782
72. Hanson LH, Manion PJ (1978) Chemosterilization of the sea lamprey *Petromyzon marinus*. Great Lakes Fishery Commission, Technical Report 29, Ann Arbor, Michigan
73. Hanson LH, Manion PJ (1980) Sterility method of pest control and its potential role in an integrated sea lamprey (*Petromyzon marinus*) control program. Can J Fish Aquat Sci 37:2108–2117
74. Hanson LH, Swink WD (1989) Downstream migration of recently metamorphosed sea lampreys in the Ocqueoc River, Michigan, before and after treatment with lampricides. N Am J Fish Manage 9:327–331
75. Hardisty MW (1965a) Sex differentiation and gonadogenesis in lampreys I, the ammocoete gonads of the brook lamprey, *Lampetra planeri*. J Zool 146:305–345
76. Hardisty MW (1965b) Sex differentiation and gonadogenesis in lampreys II, the ammocoete gonads of the landlocked sea lamprey, *Petromyzon rnarinus*. J Zool 146:346–387
77. Hardisty MW (1969a) A comparison of the gonadal development in the ammocoetes of the landlocked and anadromous forms of the sea lamprey, *Petromyzon marinus* L. J Fish Biol 2:153–166
78. Hardisty MW (1969b) Information on the growth of the ammocoete larva of the anadromous sea lamprey, *Petromyzon marinus* in British rivers. J Zool 159:139–144
79. Hardisty MW (1979) Biology of the cyclostomes. Chapman & Hall, London
80. Hardisty MW (1986) *Petromyzon marinus* Linnaeus, 1758. In: Holčík J (ed) The freshwater fishes of Europe, vol 1, part 1, Petromyzontiformes. AULA-Verlag, Wiesbaden, pp 96–116
81. Hardisty MW (2006) Lampreys: life without jaws. Forrest Text, Ceredigion
82. Hardisty MW, Potter IC (1971a) The behaviour, ecology and growth of larval lampreys. In: Hardisty MW, Potter IC (eds) The biology of lampreys, vol I. Academic Press, London, pp 85–125
83. Hardisty MW, Potter IC (1971b) The general biology of adult lampreys. In: Hardisty MW, Potter IC (eds) The biology of lampreys. Academic Press, London, pp 127–205
84. Haro A, Kynard B (1997) Video evaluation of passage efficiency of American shad and sea lamprey in a modified Ice Harbor fishway. N Am J Fish Manag 17:981–987
85. Harvey CJ, Ebener MP, White CK (2008) Spatial and ontogenetic variability of sea lamprey diets in Lake Superior. J Great Lakes Res 34:434–449
86. Heinrich JW, Weise JG, Smith BR (1980) Changes in biological characteristics of the sea lamprey (*Petromyzon marinus*) as related to lamprey abundance, prey abundance, and sea lamprey control. Can J Fish Aquat Sci 37:1861–1871
87. Henson MP, Bergstedt RA, Adams JV (2003) Comparison of spring measures of length, weight, and condition factor for predicting metamorphosis in two populations of sea lamprey (*Petromyzon marinus*) larvae. J Great Lakes Res 29(Suppl 1):204–214
88. Hodges JW (1972) Downstream migration of recently transformed sea lampreys before and after treatment of a Lake Michigan tributary with a lampricide. J Fish Res Board Can 29:1237–1240
89. Hogg R, Coghlan SM Jr, Zydlewski J (2013) Anadromous sea lampreys recolonize a Maine coastal river tributary after dam removal. Trans Am Fish Soc 142:1381–1394

90. Hollett AK (1998) Condition factor and statolith aging in assessment of metamorphosis in sea lampreys (*Petromyzon marinus*), in the Great Lakes. MS Thesis, University of Toronto
91. Holmes J (1990) Sea lamprey as an early responder to climate change in the Great Lakes basin. Trans Am Fish Soc 119:292–300
92. Holmes J, Youson JH (1994) Fall condition factor and temperature influence the incidence of metamorphosis in sea lampreys, *Petromyzon marinus*. Can J Zool 72:1134–1140
93. ICES (2014) Report of the Workshop on Lampreys and Shads (WKLS), 27–29 November 2014, Lisbon, Portugal. ICES CM 2014/SSGEF:13, 206pp
94. Irwin BJ, Li W, Bence JR, Jones ML (2012) Defining economic injury levels for sea lamprey control in the Great Lakes basin. N Am J Fish Manage 32:760–771
95. Johnson BGH, Anderson WC (1980) Predatory-phase sea lampreys (*Petromyzon marinus*) in the Great Lakes. Can J Fish Aquat Sci 37:2007–2020
96. Johnson NS, Miehls SM (2014) Guiding out-migrating juvenile sea lamprey (*Petromyzon marinus*) with pulsed direct current. River Res Appl 30:1146–1156
97. Johnson NS, Swink DS, Brenden TO, Slade JW, Steeves TB, Fodale MF, Jones ML (2014) Survival and metamorphosis of low-density population of larval sea lampreys (*Petromyzon marinus*) in streams following lampricide treatment. J Great Lakes Res 40:155–163
98. Johnson NS, Buchinger TJ, Li W (2015a) Reproductive ecology of lampreys. In: Docker MF (ed) Lampreys: biology, conservation and control. Springer, Berlin, pp 265–303
99. Johnson NS, Siefkes MJ, Wagner CM, Bravener G, Steeves T, Twohey MB, Li W (2015b) Factors influencing capture of sea lamprey in traps baited with a synthesized sex pheromone component. J Chem Ecol 41:913–923
100. Jones ML (2007) Toward improved assessment of sea lamprey population dynamics in support of cost-effective sea lamprey management. J Great Lakes Res 33 (Special Issue 2):35–47
101. Jones ML, Bergstedt RA, Twohey MB, Fodale MF, Cuddy DW, Slade JW (2003) Compensatory mechanisms in Great Lakes sea lamprey populations: implications for alternative controls. J Great Lakes Res 29(Suppl 1):113–129
102. Jones ML, Irwin BJ, Hansen GJA, Dawson HA, Treble AJ, Li W, Dai W, Bence JR (2009) An operating model for the integrated pest management of Great Lakes sea lamprey. Open Fish Sci J 2:59–73
103. Jones ML, Brenden TO, Irwin BJ (2012) Evaluating integrated pest management in the St. Marys River. Great Lakes Fishery Commission, Project Completion Report, Ann Arbor, Michigan
104. Kainua K, Valtonen T (1980) Distribution and abundance of European river lamprey (*Lampetra fluviatilis*) larvae in three rivers running into Bothnian Bay, Finland. Can J Fish Aquat Sci 37:1960–1966
105. Kelso JR, Gardner W (2000) Emigration upstream movement and habitat use by sterile and fertile sea lampreys in three Lake Superior tributaries. N Am J Fish Manage 20:144–153
106. Kelso JRM, Todd PR (1993) Instream size segregation and density of *Geotria australis* ammocoetes in two New Zealand streams. Ecol Freshw Fish 2:108–115
107. Kitchell JF, Breck JE (1980) Bioenergetics model and foraging hypothesis for sea lamprey (*Petromyzon marinus*). Can J Fish Aquat Sci 37:2159–2168
108. Kitchell JF, Cline T, Bennington V, McKinley G (2014) Climate change challenges management of invasive sea lamprey in Lake Superior. In: Keller RP, Cadotte MW, Sandiford G (eds) Invasive species in a globalized world: ecological, social, and legal perspectives on policy. University of Chicago Press, Chicago, pp 209–232
109. Lança MJ, Machado M, Mateus CS, Lourenço M, Ferreira AF, Quintella BR, Almeida PR (2014) Investigating population structure of sea lamprey (*Petromyzon marinus*, L.) in western Iberian Peninsula using morphological characters and heart fatty

acid signature analyses. PLoS ONE 9(9):e108110. doi:10.1371/journal.pone.0108110
110. Larsen LO (1980) Physiology of adult lampreys, with special regard to natural starvation, reproduction, and death after spawning. Can J Fish Aquat Sci 37:1762–1779
111. Lassalle G, Béguer M, Beaulaton L, Rochard E (2008) Diadromous fish conservation plans need to consider global warming issues: an approach using biogeographical models. Biol Conserv 141:1105–1118
112. Lawrie AH (1970) The sea lamprey in the Great Lakes. Trans Am Fish Soc 99:766–775
113. Lawrie AH, Rahrer JF (1972) Lakes Superior—A case history of the lake and its fisheries. Great Lakes Fishery Commission, Technical Report 19, Ann Arbor, Michigan
114. Lee DS (1989) Proximate determinants of larval lamprey habitat selection. PhD dissertation, Michigan State University, East Lansing
115. Lelek A (1987) The freshwater fishes of Europe. Threatened fishes of Europe, vol 9. Aula-Verlag, Wiesbaden
116. Li W, Scott AP, Siefkes MJ, Yan H, Liu Q, Yun S, Gage DA (2002) Bile acid secreted by male sea lamprey that acts as a sex pheromone. Science 296:138–141
117. Li W, Siefkes MJ, Scott AP, Teeter JH (2003) Sex pheromone communication in the sea lamprey: implications for integrated management. J Great Lakes Res 29(Suppl 1):85–94
118. Li W, Twohey M, Jones M, Wagner M (2007) Research to guide use of pheromones to control sea lamprey. J Great Lakes Res 33 (Special Issue 2):70–86
119. Lochet A, Marsden JE, Fryer BJ, Ludsin SA (2013) Instability of statolith elemental signatures revealed in newly metamorphosed sea lamprey (*Petromyzon marinus*). Can J Fish Aquat Sci 70:565–573
120. Lowe DR, Beamish FWH, Potter IC (1973) Changes in the proximate body composition of the landlocked sea lamprey *Petromyzon marinus* (L.) during larval life and metamorphosis. J Fish Biol 5:673–682
121. MacKay NA (1992) Evaluating the size effects of lampreys and their hosts: application of an individual-based model. In: DeAngelis D, Gross L (eds) Individual-based models and approaches in ecology: populations, communities, and ecosystems. Chapman and Hall, New York, pp 278–294
122. Madenjian CP, Cochran PA, Bergstedt RA (2003) Seasonal patterns in growth, blood consumption, and effects on hosts by parasitic-phase sea lampreys in the Great Lakes: an individual-based model approach. J Great Lakes Res 29(Suppl 1):332–346
123. Madenjian CP, Chipman BD, Marsden JE (2008a) New estimates of lethality of sea lamprey (*Petromyzon marinus*) attacks on lake trout (*Salvelinus namaycush*): implications for fisheries management. Can J Fish Aquat Sci 65:535–542
124. Madenjian CP, O'Gorman R, Bunnell DB, Argyle RL, Roseman EF, Warner DM, Stockwell JD, Stapanian MA (2008b) Adverse effects of alewives on Laurentian Great Lakes fish communities. N Am J Fish Manage 28:263–282
125. Madenjian CP, Johnson NS, Binder TR, Rediske RR, O'Keefe JP (2013) Polychlorinated biphenyl concentrations and activity of sea lamprey *Petromyzon marinus* vary by sex. Arch Environ Contam Toxicol 65:693–703
126. Madenjian CP, Johnson NS, Siefkes MJ, Dettmers JM, Blum JD, Johnson MW (2014) Mercury accumulation in sea lamprey (*Petromyzon marinus*) from Lake Huron. Sci Total Environ 470–471:1313–1319
127. Maitland PS, Renaud CB, Quintella BR, Close DA, Docker MF (2015) Conservation of native lampreys. In Lampreys: biology, conservation and control. Springer, Dordrecht, pp 376–428
128. Mallatt J (1983) Laboratory growth of larval lampreys (*Lampetra* (*Entosphenus*) *tridentata* Richardson) at different food concentrations and animal densities. J Fish Biol 22:293–301

129. Malmqvist B (1980) Habitat selection of larval brook lampreys (*Lampetra planeri*, Bloch) in a south Swedish stream. Oecologia 45:35–38
130. Malmqvist B (1983) Growth, dynamics, and distribution of a population of the brook lamprey *Lampetra planeri* in a South Swedish stream. Holarct Ecol 6:404–412
131. Manion PJ (1968) Production of sea lamprey larvae from nests in two Lake Superior streams. Trans Am Fish Soc 97:484–486
132. Manion PJ (1972) Fecundity of the sea lamprey (*Petromyzon marinus*) in Lake Superior. Trans Am Fish Soc 101:718–720
133. Manion PJ, Hanson LH (1980) Spawning behavior and fecundity of lampreys from the upper three Great Lakes. Can J Fish Aquat Sci 37:1635–1640
134. Manion P J, McLain AL (1971) Biology of larval sea lampreys (*Petromyzon marinus*) of the 1960 year class isolated in the Big Garlic River, Michigan 1960–65. Great Lakes Fishery Commission, Technical Report 16, Ann Arbor, Michigan
135. Manion PJ, Smith BR (1978) Biology of larval and metamorphosing sea lampreys, *Petromyzon marinus*, of the 1960 year class in the Big Garlic River, Michigan, Part II, 1966–72. Great Lakes Fishery Commission, Technical Report 30, Ann Arbor, Michigan
136. Manion PJ, Stauffer TM (1970) Metamorphosis of the landlocked sea lamprey, *Petromyzon marinus*. J Fish Res Board Can 27:1735–1746
137. Mateus CS, Rodríguez-Muñoz R, Quintella BR, Alves MJ, Almeida PR (2012) Lampreys of the Iberian Peninsula: distribution, population status and conservation. Endanger Species Res 16:183–198
138. McCauley DW, Docker MF, Whyard S, Li W (2015) Lampreys as diverse model organisms in the genomics era. Bioscience 65:1046–1056
139. McDonald DG, Kolar CS (2007) Research to guide the use of lampricides for controlling sea lamprey. J Great Lakes Res 33 (Special Issue 2):20–34
140. McLaughlin RL, Hallett A, Pratt TC (2007) Research to guide use of barriers, traps, and fishways to control sea lamprey. J Great Lakes Res 33 (Special Issue 2):7–19
141. Meckley TD, Wagner CM, Gurarie E (2014) Coastal movements of migrating sea lamprey (*Petromyzon marinus*) in response to a partial pheromone added to river water: implications for management of invasive populations. Can J Fish Aquat Sci 71(4):533–544
142. Mesa MG, Bayer JM, Seelye JG (2003) Swimming performance and physiological responses to exhaustive exercise in radio-tagged and untagged pacific lampreys. Trans Am Fish Soc 132:483–492
143. Moody EK, Weidel BC, Ahrenstorff TD, Mattes WP, Kitchell JF (2011) Evaluating the growth potential of sea lampreys (*Petromyzon marinus*) feeding on siscowet lake trout (*Salvelinus namaycush*) in Lake Superior. J Great Lakes Res 37:343–348
144. Moore HH, Schleen LP (1980) Changes in spawning runs of sea lamprey (Petromyzon marinus) in selected streams of Lake Superior after chemical control. Can J Fish Aquat Sci 37:1851–1860
145. Morkert SB, Swink WD, Seelye JG (1998) Evidence for early metamorphosis of sea lampreys in the Chippewa River, Michigan. N Am J Fish Manage 18:966–971
146. Morman RH (1987) Relationship of density to growth and metamorphosis of caged larval sea lampreys, *Petromyzon marinus* Linnaeus, in Michigan streams. J Fish Biol 30:173–181
147. Morman RH, Cuddy DW, Rugen PC (1980) Factors influencing the distribution of sea lamprey (*Petromyzon marinus*) in the Great Lakes. Can J Fish Aquat Sci 37:1811–1826
148. Morse TJ, Ebener MP, Koon EM, Morkert SB, Johnson DA, Cuddy DW, Weisser JW, Mullett KM, Genovese JH (2003) A case history of sea lamprey control in Lake Huron: 1979 to 1999. J Great Lakes Res 29(Suppl 1):599–614
149. Moser ML, Almeida PR, Kemp PS, Sorensen PW (2015) Lamprey spawning migration. In: Docker MF (ed) Lampreys: biology, conservation and control. Springer, Berlin, pp 215–263

150. Mota, MFR (2014) Biology and ecology of the allis shad, *Alosa alosa* (Linnaeus, 1758), in the Minho river. PhD dissertation. Instituto de Ciências Biomédicas Abel Salazar da Universidade do Porto

151. Mullett KM, Heinrich JW, Adams JV, Young RJ, Henson MP, McDonald RB, Fodale MF (2003) Estimating lake-wide abundance of spawning phase sea lampreys *(Petromyzon marinus)* in the Great Lakes: extrapolating from sampled streams using regression models. J Great Lakes Res 29(Suppl 1):240–252

152. Murdoch SP, Docker MF, Beamish FWH (1992) Effect of density and individual variation on growth of sea lamprey (*Petromyzon marinus*) larvae in the laboratory. Can J Zool 70:184–188

153. Neave FB (2004) The utility of meristic, morphometric, pigmentation and gonad analysis on the identification of *Ichthyomyzon* lamprey larvae. MS thesis, University of Guelph

154. Neeson TM, Koonce JF, Whiting PJ (2008) Factors affecting accuracy of stream channel slope estimates derived from geographical information systems. N Am J Fish Manage 28:722–732

155. O'Connor LM (2001) The proportion of introduced sea lampreys (*Petromyzon marinus*) that spawn in two streams tributary to Lake Ontario. MS Thesis, University of Guelph

156. Ojutkangas E, Aronen K, Laukkanen E (1995) Distribution and abundance of river lamprey (*Lampetra fluviatilis*) ammocoetes in the regulated river Perhonjoki. Regul River 10:239–245

157. Oliveira JM, Ferreira MT, Pinheiro AN, Bochechas JH (2004) A simple method for assessing minimum flows in regulated rivers: the case of sea lamprey reproduction. Aquat Cons 14:481–489

158. Patrick HK, Sutton TM, Swink WD (2009) Lethality of sea lamprey parasitism on lake sturgeon. Trans Am Fish Soc 138:1065–1075

159. Piavis GW (1961) Embryological stages in the sea lamprey and effects of temperature on development. Fish Bull 61:111–143

160. Potter IC (1980) Ecology of larval and metamorphosing lampreys. Can J Fish Aquat Sci 37:1641–1657

161. Potter IC, Beamish FWW (1977) The freshwater biology of adult anadromous sea lampreys *Petromyzon marinus*. J Zool Lond 181:113–130

162. Potter IC, Huggins RJ (1973) Observations on the morphology, behaviour and salinity tolerance of downstream migrating River lampreys (*Lampetra fluviatilis*). J Zool Lond 169:365–379

163. Potts DD, Dawson HA, Jones ML (2015) Validation of a relationship between statolith size and age of larval Great Lakes sea lamprey (*Petromyzon marinus*). Environ Biol Fish 98:1859–1869

164. Purvis HA (1979) Variation in growth, age at transformation, and sex ratio of sea lampreys re-established in chemically treated tributaries of the Upper Great Lakes. Great Lakes Fishery Commission, Technical Report 35, Ann Arbor, Michigan

165. Purvis HA (1980) Effects of temperature on metamorphosis and the age and length at metamorphosis in sea lamprey (*Petromyzon marinus*) in the Great Lakes. Can J Fish Aquat Sci 37:1827–1834

166. Quintella BR (2006) Biology and conservation of the sea lamprey (*Petromyzon marinus* L.) in Portugal. PhD thesis, University of Lisbon

167. Quintella BR, Andrade NO, Almeida PR (2003) Distribution, larval stage duration and growth of the sea lamprey ammocoetes, *Petromyzon marinus* L., in a highly modified river basin. Ecol Freshw Fish 12:1–8

168. Quintella BR, Andrade NO, Koed A, Almeida PR (2004) Behavioural patterns of sea lampreys spawning migration during difficult passage areas studied by electromyogram telemetry. J Fish Biol 65:1–12

169. Quintella BR, Andrade NO, Espanhol R, Almeida PR (2005) The use of PIT telemetry to study movements of ammocoetes and metamorphosing sea lampreys in river beds. J Fish Biol 66:97–106

170. Quintella BR, Andrade NO, Dias NM, Almeida PR (2007) Laboratory assessment of sea lamprey larvae burrowing performance. Ecol Freshw Fish 16:177–182
171. Quintella BR, Povoa I, Almeida PR (2009) Swimming behaviour of upriver migrating sea lamprey assessed by electromyogram telemetry. J Appl Ichthyol 25:46–54
172. Rahel FJ, Olden JD (2008) Assessing the effects of climate change on aquatic invasive species. Conserv Biol 22:521–533
173. Renaud CB (1997) Conservation status of Northern Hemisphere lampreys (Petromyzontidae). J Appl Ichthyol 13:143–148
174. Rodriguez-Muñoz R, Nicieza AG, Brana F (2003) Density-dependent growth of sea lamprey larvae: evidence for chemical interference. Funct Ecol 17:403–408
175. Rodríguez-Muñoz R, Nicieza AG, Braña F (2001) Effects of temperature on developmental performance, survival, and growth of sea lamprey embryos. J Fish Biol 58:475–486
176. Rodríguez-Muñoz R, Waldman JR, Grunwald C, Roy NK, Wirgin I (2004) Absence of shared mitochondrial DNA haplotypes between sea lamprey from North American and Spanish rivers. J Fish Biol 64:783–787
177. Rooney SM, Wightman G, Ó'Conchúir R, King JJ (2015) Behaviour of sea lamprey (*Petromyzon marinus* L.) at man-made obstacles during upriver spawning migration: use of telemetry to assess efficacy of weir modifications for improved passage. Biol Environ Proc R Irish Acad 115B:125–136
178. Scribner KT, Jones ML (2002) Genetic assignment of larval parentage as a means of assessing mechanisms underlying adult reproductive success and larval dispersal. Great Lakes Fishery Commission, Research Completion Report, Ann Arbor, Michigan
179. Sepúlveda MS, Patrick HK, Sutton TM (2012) A single sea lamprey attack causes acute anemia and mortality in lake sturgeon. J Aquat Anim Health 24:91–99
180. Siefkes MJ, Steeves TB, Sullivan WP, Twohey MB, Li W (2013) Sea lamprey control: past, present, and future. In: Taylor WW (ed) Great Lakes fisheries policy & management: a binational perspective. Michigan State University Press, East Lansing, pp 651–704
181. Silva S, Servia MJ, Vieira-Lanero R, Barca S, Cobo F (2013a) Life cycle of the sea lamprey *Petromyzon marinus*: duration of and growth in the marine life stage. Aquat Biol 18:59–62
182. Silva S, Servia MJ, Vieira-Lanero R, Cobo F (2013b) Downstream migration and hematophagous feeding of newly metamorphosed sea lampreys (*Petromyzon marinus* Linnaeus, 1758). Hydrobiologia 700:277–286
183. Sitar SP, Bence JR, Johnson JE, Ebener MP, Taylor WW (1999) Lake trout mortality and abundance in southern Lake Huron. N Am J Fish Manage 19:881–900
184. Slade JW, Adams JV, Christie GC, Cuddy DW, Fodale MF, Heinrich JW, Quinlan HR, Weise JG, Weisser JW, Young RJ (2003) Techniques and methods for estimating abundance of larval and metamorphosed sea lampreys in Great Lakes tributaries, 1995 to 2001. J Great Lakes Res 29(Suppl 1):137–151
185. Smith BR (1971) Sea lampreys in the Great Lakes of North America. In: Hardisty MW, Potter IC (eds) The biology of lampreys, vol 1. Academic Press, New York, pp 207–247
186. Smith PW (1979) The fishes of Illinois. University of Illinois Press, Urbana
187. Smith CL (1985) The inland fishes of New York State. New York State Department of Environmental Conservation, Albany
188. Smith SE (2013) Sublethal immune, endocrine and bioenergetics response to sea lamprey parasitism in two lake trout morphotypes from Lake Superior. MS thesis, Michigan State University, East Lansing
189. Smith BR, Tibbles JJ (1980) Sea lamprey (*Petromyzon marinus*) in Lakes Huron, Michigan, and Superior: history of invasion and control, 1936–78. Can J Fish Aquat Sci 37:1780–1801

190. Smith JJ, Kuraku S, Holt C, Sauka-Spengler T, Jiang N, Campbell MS, Yandell MD, Manousaki T, Meyer A, Bloom OE, Morgan JR, Buxbaum JD, Sachidanandam R, Sims C, Garruss AS, Cook M, Krumlauf R, Wiedemann LM, Sower SA, Decatur WA, Hall JA, Amemiya CT, Saha NR, Buckley KM, Rast JP, Das S, Hirano M, McCurley N, Guo P, Rohner N, Tabin CJ, Piccinelli P, Elgar G, Ruffier M, Aken BL, Searle SMJ, Muffato M, Pignatelli M, Herrero J, Jones M, Brown CT, Chung-Davidson Y-W, Nanlohy KG, Libants SV, Yeh C-Y, McCauley DW, Langeland JA, Pancer Z, Fritzsch B, de Jong PJ, Zhu B, Fulton LL, Theising B, Flicek P, Bronner ME, Warren WC, Clifton SW, Wilson RK, Li W (2013) Sequencing of the sea lamprey (*Petromyzon marinus*) genome provides insights into vertebrate evolution. Nat Genet 45:415–421
191. Sorensen PW, Vrieze LA (2003) Chemical ecology and potential application of the sea lamprey migratory pheromone. J Great Lakes Res 29(Suppl 1):66–84
192. Spangler GR, Robson DS, Regier HA (1980) Estimates of lamprey-induced mortality in whitefish, *Coregonus clupeaformis*. Can J Fish Aquat Sci 37:2146–2150
193. Stapanian MA, Madenjian CP, Bronte CR, Ebener MP, Lantry BF, Stockwell JD (2008) Status of burbot populations in the Laurentian Great Lakes. Am Fish Soc Symp 59:111–130
194. Steeves TB (2002) Uncertainty in estimating sea lamprey (*Petromyzon marinus*) abundance in Great Lakes tributaries. MS Thesis, Michigan State University, East Lansing
195. Stier K, Kynard BB (1986) Abundance, size and sex ratio of adult sea-run sea lampreys, *Petromyzon marinus*, in the Connecticut river. Fish Bull 84:476–480
196. Sugiyama H, Goto A (2002) Habitat selection by larvae of a fluvial lamprey, *Lethenteron reissneri* in a small stream and in an experimental aquarium. Ichthyol Res 49:62–68
197. Sullivan WP (2003) Substrate as a correlate of density and distribution of larval sea lampreys in streams. MS Thesis, University of Guelph
198. Sullivan WP, Adair R (2010) Integrated Management of Sea Lampreys in the Great Lakes 2009. Great Lakes Fishery Commission, Annual Report, Ann Arbor, Michigan
199. Sullivan WP, Adair R (2014) Integrated management of sea lampreys in the Great Lakes 2013. Great Lakes Fishery Commission, Annual Report, Ann Arbor, Michigan
200. Sullivan WP, Morrison BJ, Beamish FWH (2008) Adaptive cluster sampling: estimating density of spatially autocorrelated larvae of the sea lamprey with improved precision. J Great Lakes Res 34:86–97
201. Sutton TM, Bowen SH (1994) Significance of organic detritus in the diet of larval lampreys in the Great Lakes basin. Can J Fish Aquat Sci 51:2380–2387
202. Swink WD (1995) Growth and survival of newly parasitic sea lampreys at representative winter temperatures. Trans Am Fish Soc 124:380–386
203. Swink WD (2003) Host selection and lethality of attacks by sea lamprey (*Petromyzon marinus*) in laboratory studies. J Great Lakes Res 29(Suppl 1):307–319
204. Swink WD, Johnson NS (2014) Growth and survival of sea lampreys from metamorphosis to spawning in Lake Huron. Trans Am Fish Soc 143:380–386
205. Taverny C, Elie P (2009) Bilan des connaissances biologiques et de l'etat des habitats des lamproies migratrices dans le bassin de la Gironde—Propositions d'actions prioritaires. Rapport Final. Etude Cemagref, no. 123 Groupement de Bordeaux, Bourdeaux. France
206. Taverny C, Lassalle G, Ortusi I, Roqueplo C, Lepage M, Lambert P (2012) From shallow to deep waters: habitats used by larval lampreys (genus *Petromyzon* and *Lampetra*) over a western European basin. Ecol Freshw Fish 21:87–99
207. Thomas JD (1962) Observations on ecology of ammocetes *Petromyzon marinus L.* and *Entosphenus lamottei*. MS Thesis, University of Toronto
208. Thresher RE (2008) Autocidal technology for the control of invasive fish. Fisheries 33:114–121

209. Thresher RE, Hayes K, Bax NJ, Teem J, Benfey TJ, Gould F (2014) Genetic control of invasive fish: technological options and its role in integrated pest management. Biol Invasions 16:1201–1216
210. Torgerson CE, Close DA (2004) Influence of habitat heterogeneity on the distribution of larval Pacific lamprey (*Lampetra tridentata*) at two spatial scales. Fresh Biol 49:614–630
211. Trautman MB (1981) The fishes of Ohio. The Ohio State University Press, Columbus
212. Treble AJ, Jones ML, Steeves TB (2008) Development and evaluation of a new predictive model for metamorphosis of Great Lakes larval sea lamprey (*Petromyzon marinus*) populations. J Great Lakes Res 34:404–417
213. van Ginneken VJT, Maes GE (2005) The European eel (*Anguilla anguilla*, Linnaeus), its lifecycle, evolution and reproduction: a literature review. Rev Fish Biol Fish 15:367–398
214. Volk EC (1986) Use of calcareous otic elements (statoliths) to determine age of sea lamprey ammocoetes (*Petromyzon marinus*). Can J Fish Aquat Sci 43:718–722
215. Vrieze LA, Bjerselius R, Sorensen PW (2010) Importance of the olfactory sense to migratory sea lampreys *Petromyzon marinus* seeking riverine spawning habitat. J Fish Biol 76:949–964
216. Vrieze LA, Bergstedt RA, Sorensen PW (2011) Olfactory-mediated stream finding behavior of migratory adult sea lamprey (*Petromyzon marinus*). Can J Fish Aquat Sci 68:523–533
217. Waldman J, Grunwald C, Wirgin I (2008) Sea lamprey *Petromyzon marinus*: an exception to the rule of homing in anadromous fishes. Biol Lett 4:659–662
218. Waldman J, Daniels R, Hickerson M, Wirgin I (2009) Mitochondrial DNA analysis indicates sea lampreys are indigenous to Lake Ontario: response to comment. Trans Am Fish Soc 138:1190–1197
219. Webb PW (1978) Hydrodynamics: nonscombroid fish. In: Hoar WS, Randall DJ (eds) Fish physiology, vol 7. Academic Press, New York, pp 189–237
220. Weise JG, Pajos TA (1998) Intraspecific competition between larval sea lamprey year-classes as Salem Creek was recolonized, 1990–1994, after a lampricide application. N Am J Fish Manage 18:561–568
221. Wells L, McLain AL (1973) Lake Michigan—man's effects on native fish stocks and other biota. Great Lakes Fishery Commission, Technical Report 20, Ann Arbor, Michigan
222. Wicks BJ, Barker LA, Morrison BJ, Beamish FWH (1998) Gonadal variation in Great Lakes sea lamprey, *Petromyzon marinus*, larvae. J Great Lakes Res 24:962–968
223. Yap MR, Bowen SH (2003) Feeding by northern brook lamprey (*Ichthyomyzon fossor*) on sestonic biofilm fragments: habitat selection results in ingestion of a higher quality diet. J Great Lakes Res 29(Suppl. 1):15–25
224. Young RJ, Kelso JRM, Weise JG (1990a) Occurrence, relative abundance, and size of landlocked sea lamprey (*Petromyzon marinus*) ammocoetes in relation to stream characteristics in the Great Lakes. Can J Fish Aquat Sci 47:1773–1778
225. Young RJ, Houston KA, Weise JG, Kelso JRM (1990b) The effect of environmental variables on the population dynamics of sea lamprey, *Petromyzon marinus*. Can Tech Rep Fish Aquat Sci 1736, Sault Ste Marie, Ontario
226. Young RJ, Jones ML, Bence JR, McDonald RB, Mullett KM, Bergstedt RA, McDonald RB (2003) Estimating parasitic sea lamprey abundance in Lake Huron from heterogeneous data sources. J Great Lakes Res 29(Suppl. 1):214–225
227. Youson JH (1980) Morphology and physiology of lamprey metamorphosis. Can J Fish Aquat Sci 37:1687–1710
228. Youson JH (1994) Environmental and hormonal cues and endocrine glands during lamprey metamorphosis. In: Davey KG, Peter RE, Tobe SS (eds) Perspectives in comparative endocrinology. National Research Council of Canada, Ottawa, pp 400–408

229. Youson JH (1999) The lability and endocrinology of lamprey metamorphosis: an ancient developmental strategy. In: Roubos EW, Wendelaar Bonga SE, Vaudry H, De Loof A (eds) Recent developments in comparative endocrinology and neurobiology. Shaker, Maastrict, pp. 285–288
230. Youson JH (2003) The biology of metamorphosis in sea lampreys: endocrine, environmental and physiological cues and events and their potential application to lamprey control. J Great Lakes Res 29(Suppl. 1):26–49
231. Youson JH, Wright GM, Oot EC (1977) The timing of changes in several internal organs during metamorphosis of anadromous larval lamprey, *Petromyzon marinus* L. Can J Zool 55:469–473
232. Youson JH, Plisetskaya EM, Leatherland JF (1994) Concentrations of insulin and thyroid hormones in the serum of landlocked sea lampreys (*Petromyzon marinus*) of three larval year classes, larvae exposed to two temperature regimes, and both during and after metamorphosis. Gen Comp Endocrinol 94:294–304
233. Zerrenner A (2004) Effect of density and age on larval sea lamprey growth and survival in three Lake Champlain streams. J Freshw Ecol 19:515–519

CHAPTER 9

Fish Diversity and Fish Assemblage Structure in Seagrass Meadows at Sikao Bay, Trang Province, Thailand

Wikit Phinrub[1,2*], Bunyat Montien-Art[1], Jongkol Promya[1], Apinun Suvarnaraksha[1]

[1]*Faculty of Fisheries Technology and Aquatic Resources, Maejo University, Chiang Mai, Thailand*

[2]*Faculty of Science and Fisheries Technology, Rajamangala University of Technology, Srivijaya, Thailand*

ABSTRACT

We investigated the relationship between environmental variables as descriptors of the fish community, in terms of species diversity indices and fish assemblage patterns, in seagrass meadows at Sikao Bay, Trang Province, Thailand. Fish data and water quality parameters were collected from January to December 2012. A total of 10,596 fish specimens of 97 taxa in 48 families from four stations, Kham Bay (7°30'9.21''N, 99°18'7.67''E), Boonkong Bay (7°31'2.49''N, 99°17'40.09''E), Ban Pak Klong (7°36'17.67''N, 99°16'32.89''E) and Sai Cape (7°38'30.19''N, 99°14'49.85''E), were collected using gillnets of three different mesh sizes. The three most diverse families were Leiognathidae, Carangidae and Tetraodontidae, respectively. The dominant species were Atherinomorus duodecimalis, Sillago sihama and Pelates quadrilineatus. Specimens were highly abundant in July and less so in January and species richness was high in July and less so in March. Fish assemblages were classified into two patterns and the average of the Shannon index was 2.7. The environmental parameters in each month were analyzed by one-way ANOVA which did not show significant difference (P > 0.05) of pH, orthophosphate, wind speed and rainfall. Fish diversity and assemblage, and environmental parameters, were categorized into four groups. These could be promoted to local fisheries so that conservation programs cpuld be set up to ensure the ecological sustainability of seagrass meadows.

Keywords: Fish, Diversity, Canonical Correspondence Analysis, Seagrass, Sikao Bay

1. INTRODUCTION

Seagrass meadows are considered a valuable component of coastal ecosystems, as they are an important habitat for numerous fish, invertebrates and other animals, perform different ecological functions and act as environmental reserves. They supply a permanent habitat to fulfil the life cycles of aquatic animals, especially as a temporary nursery area for juvenile stages [1] [2] , and as feeding areas for all life stages [3] [4] . Seagrass meadows also provide benefits to humans in terms of protein source security, as well as occupation for coastal populations [5] [6] . Seagrass beds have high primary productivity; the habitat provides suitable protective cover for many species of marine fish and invertebrates and is an important tropical marine [7] . Variations in environmental variables are the major factors that govern marine fish communities in terms of species richness and distribution of individual species. They are important in terms of fish biology and ecology and related studies of life history and population dynamics. Understanding these relationships is tremendously important to support fisheries management, economics, social aspects and environmental protection. Seagrass meadows in particular are very sensitive ecosystems with a huge diversity of species and related assemblage patterns. In Thailand, huge seagrass meadows of 12 species extending to 18,986 hectares (ha) can be found. The biggest seagrass meadows in Thailand are located in Trang Province (3,435 ha) with 11 species, where Sikao Bay has 350 ha of seagrass meadows. However, fish diversity and assemblage in the seagrass meadows of Sikao Bay have not been studied thoroughly, and environmental factors in particular have not been considered recently. Our study examines fish diversity and assemblages, and relationships between species richness and environmental parameters. The results can be used in the future as a basis for comparison with other seagrass meadows in South East Asia and other tropical marine systems. They also support aquatic resources conservation strategies, which should be prepared in the near future to protect this sensitive seagrass meadow ecosystem.

2. MATERIALS AND METHODS

2.1. Study Area Description

The study area was located in Sikao Bay (Kham Bay, 7°30'9.21''N, 99°18'7.67''E; Boonkong Bay, 7°31'2.49''N, 99°17'40.09''E; Ban Pak Klong, 7°36'17.67''N, 99°16'32.89''E and Sai Cape, 7°38'30.19''N, 99°14'49.85''E) in the Andaman Sea, Trang Province, Southern Thailand. Sikao Bay is 126 km^2 in area with a shore length of 40 km. The most frequent seagrass species in this area are Enhalus acoroides, Cymodocea rotundata, Thalassia hemprichii, Halophila ovalis, Halophila minor and Halodule pinifolia.

2.2. Fish Sampling

Fish sampling was carried out during the night at four sites on Sikao Bay from January to December 2012. Fishing was done with three experimental gillnets of different mesh sizes (2, 3.5 and 5 cm), 180-m long and 1.5 m deep, connected into one net for fish sampling. Specimens were fixed in 10% formalin for 1 month and then changed to ethanol 30%, 50% for a month. Finally, they were preserved in 70% ethanol and re-checked and identified taxonomically to species level level [8] -[12] at the Maejo Aquatic Resources Natural Museum (MARNM), Faculty of Fisheries Technology and Aquatic Resources, Maejo University, Chiangmai, Thailand.

2.3. Environmental Parameter Sampling

The physico-chemical water quality parameters measured in situ using a YSI® 556 multi-Probe System (MPS) made in the USA were conductivity (Con), total dissolved solid (Tds), salinity (Sal), dissolved oxygen (Dis), pH (pH), water temperature (Tem) and transparency (Tra). Other parameters were sampled for laboratory analyses using the standard methods: ammonia (Amm), nitrite (Nti), nitrate (Nta), orthophosphate (Ort) and chlorophyll a (Chla) [13] . Data were collected in Sikao Bay at the four fish sampling sites on a monthly basis from January to December 2012. Meteorological parameters data were maximum air temperature (Mat), minimum air temperature (Mit), average air temperature (Aat), wind direction (Wid), wind speed (Wis) and rainfall (Rai). These data were provided by the Thai Meteorological Department [14] . The highest water level (Hwa) and lowest water level (Lwa) were provided by the Hydrographic Department, Royal Thai Navy [15] .

2.4. Data Analyses

The physico-chemical water quality parameters, meteorological parameters, fish species (Fsp) and fish numbers (Fnu) were analyzed using Pearson rank correlation tests and one-way ANOVA (Tukey honestly significant difference test, P = 0.05) conducted with the SPSS© software package. The data set is presented as the percentage of occurrence frequency (% OF), species richness and Shannon diversity index [16] , to examine the relationships between species richness and environmental parameters using Canonical Correspondence Analysis (CCA) and Principal Component Analysis (PCA), an ordination technique designed for direct analysis of relationships between multivariate ecological data [17] . All statistical analyses were performed using R-statistical software, packages "stats" [18] [19] .

3. RESULTS

3.1. Fish Diversity

A total of 10,596 fish specimens of 97 taxa in 48 families was collected (Table 1). Leiognathidae was the most abundant with eight species (15%) (Leiognathus stercorarius, L. jonesi, L. decorus, L. equulus, L. fasciatus, L. splendens, Secutor insidiator and Gazza minuta). This was followed by seven species (13%) of Carangidae (Alepes kleinii, Carangoides praeustus, Megalaspis cordyla, Scomberoides lysan, Siganus commersonnianus, Decapterus kurroides and Caranx sexfasciatus) and seven species (13%) of Tetraodontidae (Tetraodon fluviatilis, T. palembangensis, T. nigroviridis, Lagocephalus spadiceus, L. lunaris, Chelonodon patoca and Arothron reticularis) (Figure 1). The dominant species were Atherinomorus duodecimalis, Sillago sihama and Pelates quadrilineatus. The highest frequencies of occurrence were S. sihama (79.17%), Atherinomorus duodecimalis (72.92%) and Hyporhamphus limbatus (68.75%), respectively (Figure 2). Specimens were highly abundant in July and less so in January (Figure 3(a)), and species richness was high in July and lain March, (Figure 3(b)). The average Shannon-Wiener index was 2.7.

Table 1. Fish species found in the seagrass medowsin Sikao Bay, Trang Province, Thailand.

Family	Scientific name	Abb	Family	Scientific name	Abb
Dasyatidae	*Himantura imbricata*	Himb		*Secutor insidiator*	Sins
Megalopidae	*Megalops cyprinoides*	Mcyp		*Gazza minuta*	Gmin
Engraulidae	*Thryssa hamiltonii*	Tham	Lutjanidae	*Lutjanus russelli*	Lrus
	Stolephorus indicus	Sind		*Lutjanus fulviflamma*	Lful
	Thryssa setiorostris	Tset	Lobotidae	*Lobotes surinamensis*	Lsur
	Thryssa scratchleyi	Tscr	Gerreidae	*Gerres erythrourus*	Gery
Clupeidae	*Sardinella albella*	Salb		*Gerres oyena*	Goye
	Anodontostoma chacunda	Acha		*Gerres filamenttosus*	Gfil
Plotosidae	*Plotosus lineatus*	Plin	Haemulidae	*Plectorhinchus gibbosus*	Pgib
Bagridae	*Mystus gulio*	Mgul		*Pomadasys kaakan*	Pkaa
Synodontidae	*Saurida nebulosi*	Sneb	Lethrinidae	*Lethrinus lentjan*	Llen
Mugilidae	*Chelon subviridis*	Csub	Sciaenidae	*Pennahia anea*	Pane
	Ellochelon vaigiensis	Evai	Mullidae	*Upeneus tragula*	Utra
	Moolgarda cunnesius	Mcun		*Parupeneus heptacanthus*	Phep
	Valamugil perusii	Vper	Toxotidae	*Toxotes chatareus*	Tcha
Atherinidae	*Atherinomorus duodecimalis*	Aduo	Drepanidae	*Drepane punctate*	Dpun
Hemiramphidae	*Hyporhamphus limbatus*	Hlim	Teraponidae	*Pelates quadrilineatus*	Pqua
	Zenarchopterus buffonis	Zbuf		*Terapon puta*	Tput
Belonidae	*Tylosurus crocodiles crocodilus*	Tcro		*Terapon jarbua*	Tjar
	Strongylura strongylura	Sstr	Labridae	*Halichoeres bicolor*	Hbic
Syngnathidae	*Syngnathoides biaculeatus*	Sbia	Blenniidae	*Petroscirtes variabilis*	Pvar
Scorpaenidae	Vespiculatrachinoides	Vtra	Callionymidae	*Repomucenus schaapii*	Rsch
Synanceiidae	*Synanceia horrida*	Shor	Eleotridae	*Butis butis*	Bbut
Platycephalidae	*Grammoplites scaber*	Gsca	Gobiidae	*Oxyurichthys microlepis*	Omic
	Platycephalus indicus	Pind		*Acentrogobius caninus*	Acan
	Cociella punctata	Cpun		*Yongeichthys nebulosus*	Yneb
	Inegonia japonica	Ijap		*Psammogobius biocellatus*	Pbio
Ambassidae	*Ambassis nalua*	Anal	Scatophagidae	*Scatophagus argus*	Sarg
	Ambassis vachellii	Avac	Siganidae	*Siganuscanaliculatus*	Scan
	Ambassis urotaensis	Auro		*Siganus javus*	Sjav
	Ambassis interruptus	Aint		*Siganus fuscescens*	Sfus
Latidae	*Lates calcarifer*	Lcal	Sphyraenidae	*Sphyraena barracuda*	Sbar
Apogonidae	*Apogon truncates*	Atru	Scombridae	*Rastrelliger brachysoma*	Rbra
Sillaginidae	*Sillago sihama*	Ssih	Paralichthyidae	*Pseudorhombus arsius*	Pars

	Sillago aeolus	Saeo	Soleidae	*Zebrias quagga*	Zqua
Rachycentridae	*Rachycentron canadum*	Rcan	Cynoglossidae	*Cynoglossus puncticeps*	Cpuc
Carangidae	*Alepes kleinii*	Akle		*Cynoglossus abbreviatus*	Cabb
	Carangoides praeustus	Cpra	Triacanthidae	*Triacanthus nieuhofii*	Tnie
	Megalaspis cordyla	Mcor	Monacanthidae	*Monacanthus chinensis*	Mchi
	Scomberoides lysan	Slys	Ostraciidae	*Lactoria cornuta*	Lcor
	Scomberomorus commersonnianus	Scom	Tetraodontidae	*Tetraodon fluviatilis*	Tflu
	Decapterus kurroides	Dkur		*Tetraodon palembangensis*	Tpal
	Caranx sexfasciatus	Csex		*Tetraodon nigroviridis*	Tnig
Leiognathidae	*Leiognathus stercorarius*	Lste		*Lagocephalus spadiceus*	Lspa
	Leiognathus jonesi	Ljon		*Lagocephalus lunaris*	Llun
	Leiognathus decorus	Ldec		*Chelonodon patoca*	Cpat
	Leiognathus equulus	Lequ		*Arothron reticularis*	Aret
	Leiognathus fasciatus	Lfas	Diodontidae	*Diodon liturosus*	Dlit
	Leiognathus splendens	Lspl			

Abb. = Abbreviation.

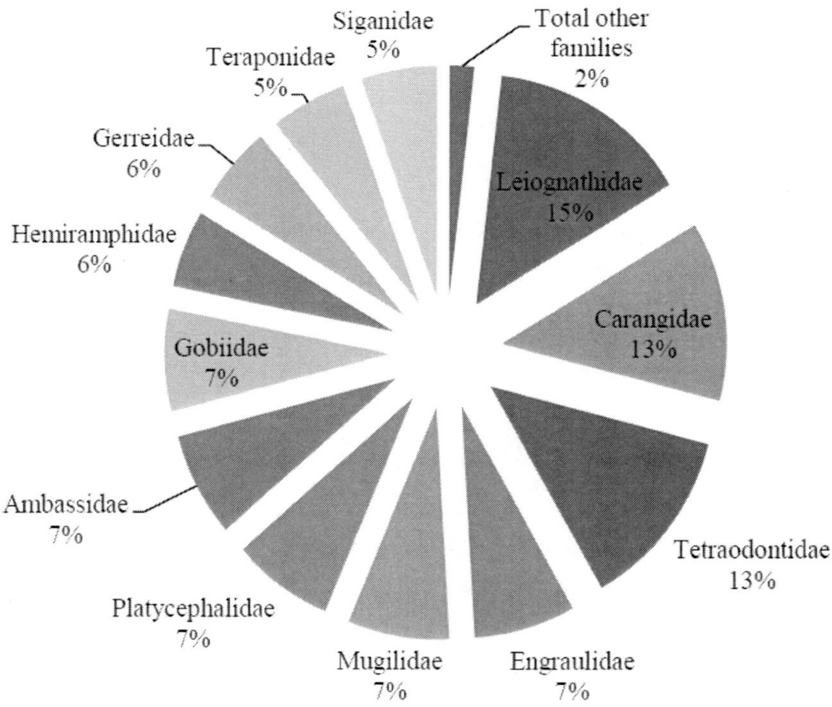

Figure 1. Percentage dominat families of fishes in seagrass areas at Sikao Bay, Trang Provice, Thailand.

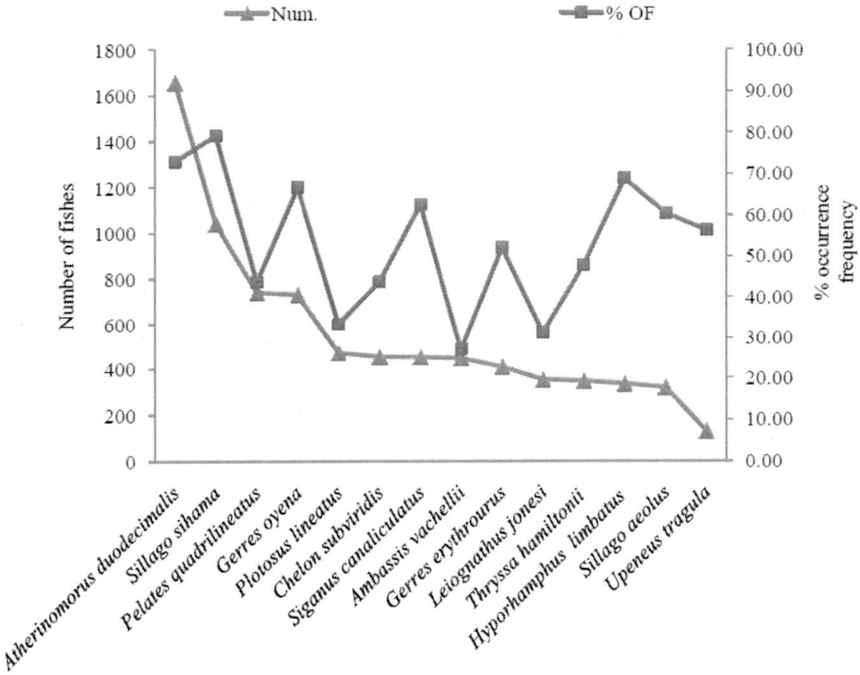

Figure 2. Number of dominant species and percentage of occurrence frequence of fishes in seagrass areas at Sikao Bay, Trang Province, Thailand.

3.2. Environmental Parameters

The 20 environmental parameters, fish species and fish numbers were analyzed by Pearson rank correlation tests. The results showed significance of conductivity, ammonia, fish numbers, fish species and highest water level ($P < 0.05$). The results analyzed by one-way ANOVA, the environmental parameters measured in each month did not show significant difference ($P > 0.05$) of pH, orthophosphate (Ort), wind speed (Wis) and rainfall (Rai). Four seagrass meadows and environmental parameters showed no significant relationships ($P > 0.05$). The environmental parameters measured in dry seasons and rainy seasons showed a significant difference ($P < 0.05$) of water temperature (Tem), wind direction (Wid), lowest water level (Lwa) and highest water level (Hwa).

3.3. Prediction of Species Richness and Environmental Parameters

Ninety-seven fish species and 20 environmental variables from four sampling sites were log-transformed and fed to a CART model as a response variable to be predicted using fish species and environmental variables. The major factor in physico-chemical and meteorological parameters was Chlorophyll a. The CART model using the tree "pruning" process and optimal tree selection favored three parameters including Chlorophyll a (Chl), lowest water level (Lwa) and total dissolved solid (Tds). Chlorophyll a was used in both the first and the second splits. This model showed that if chlorophyll a levels were higher than 0.5 μg/l, species diversity was high with about 26 species (Figure 4).

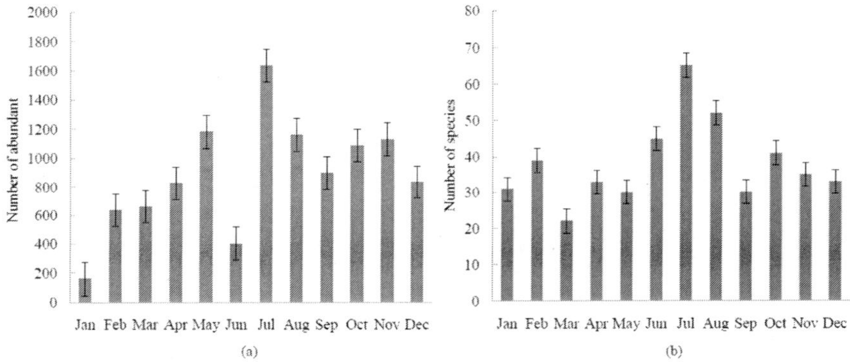

Figure 3. The abundant species (a) and species richness (b) of fishes in seagrass meadowns at Sikao Bay, Trang Province, Thailand.

3.4. Relationships of Fish Assemblage Structure and Environmental Parameters

Ninety-seven fish species and 20 environmental variables were analyzed by CCA, including common fish species such as S. sihama (Ssih), A. duodecimalis (Aduo), H. limbatus (Hlim), Gerres oyena (Goye), P. quadrilineatus (Pqua), S. aeolus (Saeo), Upeneus tragula (Utra) and G. erythrourus (Gery) which can be classified into four groups (Figure 5(a), Figure 5(b)).

The first group (I) represented species including Rastrelliger brachysoma (Rbra), Cynoglossus puncticeps (Cpuc) and Lutjanus russelli (Lrus), related to rainfall (Rai) and nitrite (Nti). In group II were found most dominant species such as S. sihama (Ssih) and Siganus janus (Sjan), and Triacanthus nieuhofi (Tnie), related to highest water levels (Hwa) and wind direction (Wid). Group III represented species such as Sig. canaliculatus (Scan), Cynoglossus abbreviatus (Cabb) and Monacanthus chinensis (Mchi), related to average air temperature (Aat), maximum air temperature (Mat) and conductivity (Con). Finally, Group IV included species such as Thryssa scratchleyi (Tscr), Cociella punctata (Cpun) and Petroscirtes variabilis (Pvar), related to salinity (Sal) and total dissolved solid (Tds).

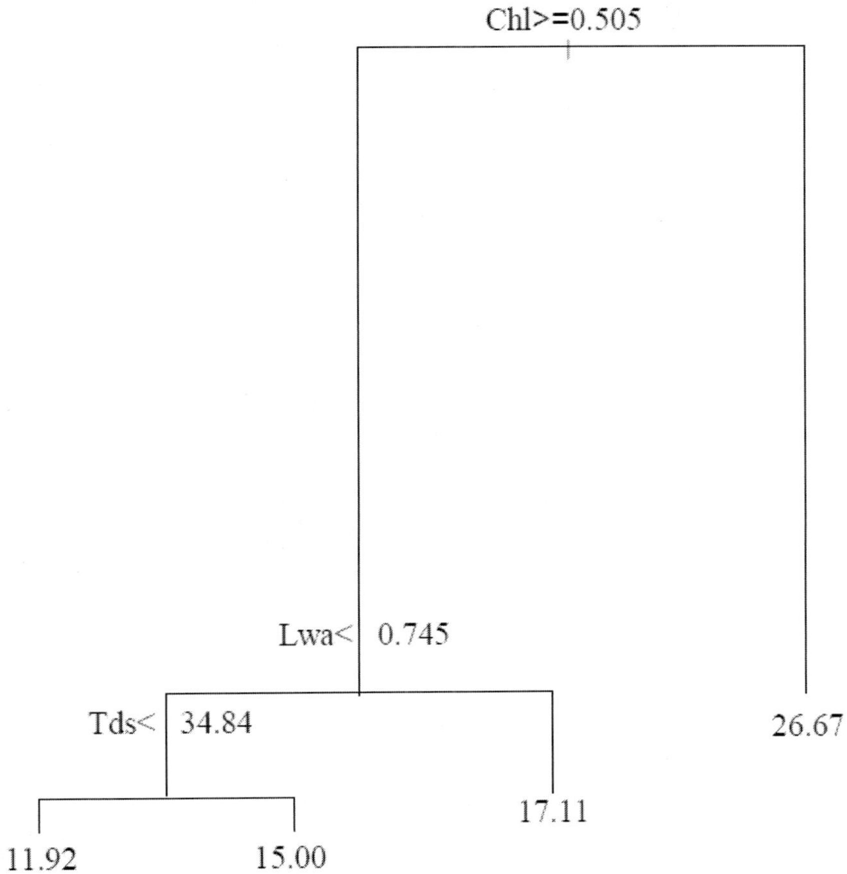

Figure 4. CART model to predict species richness in seagrass meadowns at Sikao Bay, Trang Province, Thailand.

The results are shown in a cluster dendrogram, summarizing the similarity of species richness with environmental parameters (plotted by CCA), including the cluster composition of species at the sampling site through all the year (Figure 5(c)), which can be classified into two main groups. The first group (A) represents the similarity of species richness related to the environmental variables in the dry season (January, February, March and April), the beginning of the rainy season (May and June) and in the central part and end of the rainy season (September, October, November and

December). The dominant fish species were A. duodecimalis (Aduo), G. oyena (Goye), Sillago aeolus (Saeo)and Siganus canaliculatus (Scan), (January, February, March and April); H. limbatus (Hlim), Sillago sihama (Ssih), and G. oyena (Goye) (May and June); and S. sihama (Ssih), H. limbatus (Hlim), A. duodecimalis (Aduo) and Siganus canaliculatus (Scan), (September, October, November and December). The second group (B) represents similarity of species richness related to the environmental variables in the rainy season (July and August) with dominant species as Sillago sihama (Ssih), S. aeolus (Saeo), A. duodecimalis (Aduo), G. erythrourus (Gery), Pennahia anea (Pane), Drepane punctate (Dpun), Leiognathus jonesi (Ljon), Platycephalus indicus (Pind), Ambassis nalua (Anal), Thryssa hamiltonii (Tham) and Chelon subviridis (Csub).

Further results then appear in a cluster dendrogram, summarizing the similarity of species in terms of environmental parameters (plotted by CCA) and showing the composition at the sampling site throughout the year (Figure 5(d)), which can also be classified into two main groups. The first group (A) represents the similarity in environmental variables at the end of the dry season (March and April) and at the beginning and central part of the rainy season (May, June, July, August and September). The second group (B) represents the similarity in environmental variables at the end of the rainy season (October, November and December) and at the beginning of the dry season (January and February).

4. DISCUSSION

In this study, we found fish diversity in seagrass meadows in Sikao Bay, Trang Province, Thailand, was 97 taxa in 48 families relative with seagrass meadows, which is similar to the findings, [20] who found 81 fish species and to [21] , who found 87 species of juveniles in their study of fish assemblages in Caribbean seagrass beds. [22] reported that 42 fish species, 36 species and 27 species were collected from seagrass. [23] , reported 37 fish species belonging to 22 families when studying differences in fish assemblage structures between fragmented and continuous seagrass beds in Trang, Southern Thailand. [24] , reported 36species of fish from 24 families from spatial patterns in fish herbivory in a temperate Australian seagrass meadow; and [25] noted 35 families. [26] found 65 taxa in 35 families of fish in seagrass beds at Kham Bay, Trang Province, Thailand; [27] identified 70 taxa from diverse species of fish in seagrass beds at Sai Cape, Trang Province, Thailand; and [28] were reported 62 taxa in 35 families of fish in seagrass beds at Ban Pak Klong, Trang Province, Thailand; thier fish diversity were less than in this study. [29] used poison as a sampling technique and found 189 species in 46 families, mainly from

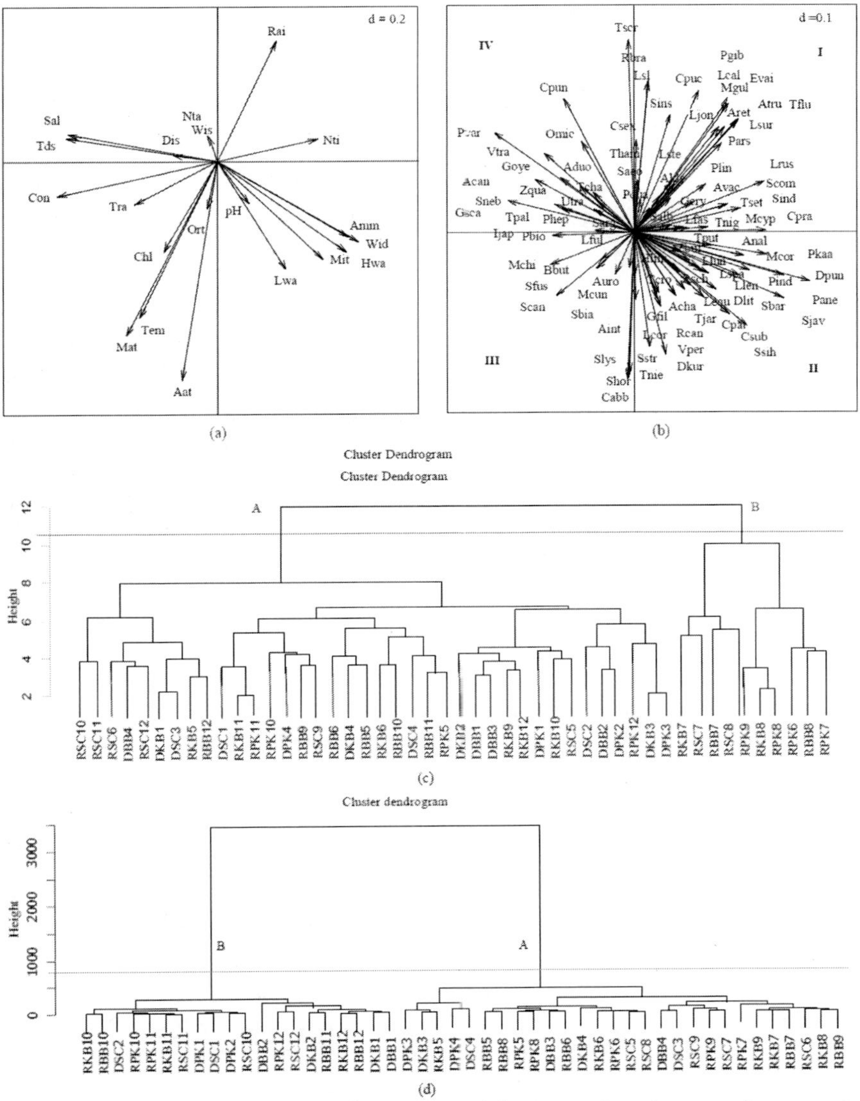

Figure 5. CCA ordination with species richness related to environmental factors in seagrass meadowns at Sikao Bay, Trang Province, Thailand. (a) Environmental variables loading to CCA axex; (b) PCA analysis of fish assemblage; (c) dendrogram of fish assemblage and; (d) dendrogram of environmental parameters and fish assemblage (RKB = Rain season of Kham Bay, RBB = Rain seasion of Boonkong Bay, RPK = Rain season of Ban Pak Klong, RSC = Rain season of Sai Cape), (DKB = Dry season of Kham Bay, DBB = Dry seasion of Boonkong Bay, DPK = Dry season of Ban Pak Klong, DSC = Dry season of Sai Cape) and (1-12 = Jan. - Dec.).

Thalassia. [30] found 249 fish species in 62 families from seine net catches; their fish diversity was greater than in this study.

In another study [31] found similar results, as the most abundant fish species were from the families Labridae, Siganidae, Atherinidae, Pomacentridae and Nemipteridae, with variations between the study sites. [23] , reported fish species such as Clupeidae, Plotosidae, Leiognathidae, Sillaginidae, Lutjanidae, Lethrinidae and Gerreidae in continuous and fragmented beds at the study sites. [32] found that permanent residents were defined by the presence of all life history stages within the habitat. Some common species (Cheilio inermis, Halichoeres argus, Halichoeres chloropterus, Pentapodus trivittatus, Apogon margaritiphorus) were found regularly as both adults and juveniles in the seagrass beds. Other species of reef-associated families that might use adjacent seagrass beds as nurseries were found exclusively as juveniles, such as Chaetodontidae, Haemulidae and Ephippidae.

Another finding was that environmental parameters are important factors for fish distribution patterns. This was proved by CCA, which showed the relation with species richness and indicated patterns in seagrass meadows. The major factors important for fish assemblage were Chl a, lowest water level (Lwa) and total dissolved solid (Tds), in relation with season change, fish abundance, fish species and fish species richness. These results are similar with [33] where spatial and seasonal variations of fish assemblages in mangrove creek systems in Zanzibar (Tanzania) were investigated.

5. CONCLUSIONS

Seagrass meadows provide a permanent and temporary habitat for commercial fishes such as Alepes kleinii, Carangoides praeustus, Megalaspis cordyla, Scomberoides lysan, S. commersonnianus, Decapterus kurroides, Caranx sexfasciatus, Sillago sihama, S. aeolus, Gerres erythrourus, Sardinella albella, Thryssa hamiltonii, Plotosus lineatus, G. oyena and Atherinomorus duodecimalis. The dominant species in our study are A. duodecimalis, Sillago sihama and Pelates quadrilineatus. The highest frequencies of occurrence are S. sihama and A. duodecimalis. Fish abundance differs because of varying environmental conditions. Observations of physico- chemical water qualities parameters are based on coastal water quality standards.

Overall, 97 fish species and 20 environmental variables are analyzed by CCA, classifying the relationships of fish assemblage structure and environmental parameters into four groups. Two cluster dendrograms summarize the similarity of species richness with environmental parameters, and of environmental parameters with composition in the sampling site throughout the year, respectively. Our study has clearly improved knowledge of fish diversity and the relationship of fish assemblage with environmental parameters in seagrass meadows, and this is important for understanding the functioning and integrity of these ecosystems. Our results can be passed on to local fishermen and responsible stakeholders to encourage them to begin a Thai seagrass meadows conservation program for ecological sustainability in an area with growing tourism.

ACKNOWLEDGEMENTS

We are grateful to Rajamangala University of Technology Srivijaya, Faculty of Science and Fisheries Technology, for supporting this study, and to Shell Centennial Education Fund, Shell Companies in Thailand for research funding support. Further thanks to everyone at the Aquatic Resource Research Center (ARC Team) and the Faculty of Fisheries Technology and Aquatic Resources at Maejo University.

REFERENCES

1. Heck, K.L., Hays, G. and Orth, R.J. (2003) Critical Evaluation of the Nursery Role Hypothesis for Seagrass Meadows. Marine Ecology Progress Series, 253, 123-136. http://dx.doi.org/10.3354/meps253123
2. Jackson, E.L., Rowden, A.A., Attrill, M.J., Bossey, S.J. and Jones, M.B. (2001) The Importance of Seagrass Beds as a Habitat for Fishery Species. Oceanography and Marine Biology, 39, 269-303.
3. Aaron, A.J., Dahlgren, C.P., Kellison, G.T., Kendall, M.S., Layman, C.A., Ley, J.A., Nagelkerken, I. and Serafy, J.E. (2006) Nursery Functions of Tropical Back-Reef Systems. Marine Ecology Progress Series, 318, 287-301.http://dx.doi.org /10.3354/meps318287
4. de la Morinière, E.C., Pollux, B.J.A., Nagelkerken, I. and Van der Velde, G. (2002) Post-settlement life cycle migration patterns and habitat preference of coral reef fish that use seagrass and mangrove habitats as nurseries. Estuarine Coastal and Shelf Science, 55:309-321. http://dx.doi.org/10.1006/ecss.2001.0907
5. Duarte, C.M. (2002) The Future of Seagrass Meadows. Environmental Conservation, 29, 192-206. http://dx.doi.org/10.1017/S0376892902000127
6. Fortes, M.D. (1990) Seagrasses: A Resource Unknown in the ASEAN Region. ICLARM Education Series, International Centre for Living Aquatic Resource Management, Manila, 46 p.
7. Parrish, J.D. (1998) Fish Communities of Interacting Shallow-Water Habitats in Tropical Oceanic Regions. Marine Ecology Progress Series, 58, 143-160. http://dx.doi.org/10.3354/meps058143
8. Carpenter, K.E. and Niem, V.H. (1999a) FAO Species Identification Guide for Fishery Purposes. The Living Marine Resources of the Western Central Pacific Volume 3:

Batoid Fishes, Chimaeras and Bony Fishes Part 1 (Elopidae to Linophrynidae). FAO, Rome, 1397-2068.

9. Carpenter, K.E. and Niem, V.H. (1999b) FAO Species Identification Guide for Fishery Purposes. The Living Marine Resources of the Western Central Pacific Volume 4: Bony Fishes Part 2 (Mugilidae to Carangidae). FAO, Rome, 2069-2790.
10. Carpenter, K.E. and Niem, V.H. (2001) FAO Species Identification Guide for Fishery Purposes. The Living Marine Resources of the Western Central Pacific Volume 5: Bony Fishes Part 3 (Menidae to Pomacentridae). FAO, Rome, 2791-3380.
11. Carpenter, K.E. and Niem, V.H., Eds. (2001) FAO Species Identification Guide for Fishery Purposes. The Living Marine Resources of the Western Central Pacific Volume 6: Bony Fishes Part 4 (Labridae to Latimeriidae), Estuarine Crocodiles, Sea Turtles, Sea Snakes and Marine Mammals. FAO, Rome, 3381-4218.
12. Nelson, J.S. (2006) Fishes of the World. 4th Edition, John Wiley & Sons, Hoboken, 601 p.
13. American Public Health Association (APHA), Eton, A.D., Clesceri, L.S., Rice, E.W. and Greenberg, A.E., Eds. (2005) Standard Methods for the Examination of Water and Waste Water. 21st Edition, American Water Work Association, Water Pollution Control Federation, Washington DC, 1368 p.
14. Information Services, the Climate Bureau of Meteorology (2013) Climate. Thai Meteorological Department, Thailand. http://www.tmd.go.th/services/services.php
15. Hydrographic Department (2012) Highest and Lowest Water Level of 2012. Hydrographic Department, Royal Thai Navy, Thailand. http://www.hydro.navy.mi.th/services2014.htm
16. Shannon, C.E. and Weaver, W. (1949) The Mathematical Theory of Communication. University of Illinois, Urbana.
17. TerBraak, C.J.F. (1986) Canonical Correspondence Analysis: A New Eigenvector Technique for Multivariate Direct Gradient Analysis. Ecology, 67, 1167-1179. http://dx.doi.org/10.2307/1938672
18. Lek, S. and Guegan, J.F. (2000) Artificial Neural Networks: Application to Ecology and Evolution. Springer Verlag, Berlin.
19. R Development Core Team (2013) R: A Language and Environment for Statistical Computing. Reference Index Version 2.10, R Foundation for Statistical Computing, Vienna. http://www.r-project.org
20. Unsworth, R.K.F., Elizabeth, W., David, J.S. and James, J.B. (2007) Diel trophic Structuring of Seagrass Bed Fish Assemblages in the Wakatobi Marine National Park, Indonesia. Estuarine, Coastal and Shelf Science, 72, 81-88. http://dx.doi.org/10.1016/j.ecss.2006.10.006
21. Dorothie, K., Yolande, B.N., Max, L., David, M. and Claude, B. (2010) Juvenile Fish Assemblages in Caribbean Seagrass Beds: Does Nearby Habitat Matter? Journal of Coastal Research, 26, 1133-1141.
22. Elbart, J.E., Pauline, M.R. and Rod, M.C. (2007) Fish Assemblages in Seagrass Beds Are Influenced by the Proximity of Mangrove Forests. Marine Biology, 150, 993-1002.
23. Horinouchi, M., Tongnunui, P., Nanjyo, K., Nakamura, Y., Sano, M. and Ogawa, H. (2009) Differences in Fish Assemblage Structures between Fragmented and Continuous Seagrass Beds in Trang, Southern Thailand. Fisheries, 75, 1409-1416. http://dx.doi.org/10.1007/s12562-009-0166-1
24. White, K.S., Westera, M.B. and Kendrick, G.A. (2011) Spatial Patterns in Fish Herbivory in a Temperate Australian Seagrass Meadows. Estuarine, Coastal and Shelf Science, 93, 366-374. http://dx.doi.org/10.1016/j.ecss.2011.05.006
25. Khalaf, M.A., Al-Rousan, S. and Al-Horani, F.A. (2012) Fish Assemblages in Seagrass Habitat along the Jordanian Coast of the Gulf of Aqaba. Natural Science, 4, 517-525. http://dx.doi.org/10.4236/ns.2012.48069
26. Phinrub, W., Promya, J., Montien-Art, B. and Suvarnaraksha, A. (2013) Fish Diversity and Water Quality in Seagrass Beds at Kham Bay, Trang Province, Thailand.

Proceedings of the International Conference on Frontiers of Environment, Energy and Bioscience (ICFEEB 2013), Beijing, 24-25 October 2013, 399-404.

27. Phinrub, W., Promya, J., Montien-Art, B. and Suvarnaraksha, A. (2013) Fish Diversity in Seagrass Beds at Sai Cape, Trang Province, Thailand. Proceedings of the Sixth International Fisheries Conference with the Theme "Climate Change: Impact on Aquatic Resources and Fisheries, 4 December 2013, Chiang Mai, 7.
28. Phinrub, W., Promya, J., Montien-Art, B. and Suvarnaraksha, A. (2014) Fish Diversity in Seagrass Beds at Ban Pak Klong, Trang Province, Thailand. Proceedings of the Asian Fish Biodiversity Conference 2014, 12-13 February 2014, Penang, 59.
29. Vivien, M.L. (1974) Ichtyofaune des herbiers de phanerogames marines du Grand Recif de Tuear (Madagascar). I: Les peuplements et leur distribution ecologique. Tethys, 5, 425-436.
30. Fiona, R.G. and Mark, W.W. (2002) Diversity of Fishes in Seagrass Beds in the Quirimba Archipelago, Northern Mozambique. Marine & Freshwater Research, 53, 115-121. http://dx.doi.org/10.1071/MF01125
31. Nakamura, Y. and Sano, M. (2004) Comparison between Community Structures of Fishes in Enhalus acoroides and Thalassia hemprichii Dominated Seagrass Beds on Fringing Coral Reefs in the Ryukyu Islands, Japan. Ichthyological Research, 51, 38-45. http://dx.doi.org/10.1007/s10228-003-0191-5
32. Kuriandewa, T.E., Kiswara, W., Hutomo, M. and Soemodihardjo, S. (2003) The Seagrasses of Indonesia. In: Green, E.P. and Short, F.T., Eds., World Atlas of Seagrasses, University of California Press, Barkeley, 172-182.
33. Mwandya, A.W., Martin, G., Mathias, H.A., Marcus, C.O., Yunus, D.M. and Ian, B. (2010) Spatial and Seasonal Variations of Fish Assemblages in Mangrove Creek Systems in Zanzibar (Tanzania). Estuarine, Coastal and Shelf Science, 89, 277-286. http://dx.doi.org/10.1016/j.ecss.2010.08.002

CHAPTER 10

Stable Isotope Enrichment in Muscle, Liver, and Whole Fish Tissues of Brown-Marbled Groupers (*Epinephelus Fuscoguttatus*)

Gang Chen[1,2], Hui Zhou[1], Duoliang Ji[1] and Binhe GuEmail author[1*]

[1]*College of Fishery, Guangdong Ocean University, Zhanjiang 524025, China.*

[2]*Guangdong South China Sea Key Laboratory of Aquaculture for Aquatic Economic Animals, Zhanjiang524025, China.*

ABSTRACT

Introduction: The purpose of this study was to assess enrichments in stable carbon and nitrogen isotopes ($\delta^{13}C$ and $\delta^{15}N$) in brown-marbled groupers (Epinephelus fuscoguttatus), a marine fish that has been widely used in aquaculture. Stable isotope analysis has been used to evaluate dietary sources and the trophic position of fish. There is the need to better understand the pattern of isotope enrichment between consumers and diets under laboratory conditions.

Methods: We studied the changes in stable carbon and nitrogen isotopes of juvenile brown-marbled groupers during a feeding experiment in 2009. Fish were grown in aquaria and fed a sole source of protein for 56 days and analyzed for $\delta^{13}C$ and $\delta^{15}N$ ratios in whole fish, muscle, and liver tissues.

Results: At the end of the 56-day feeding experiment, fish grew to an average of 55.6 g from an average of 22.5 g. Compared to the dietary isotope compositions, whole fish and muscle tissues of the juvenile groupers were enriched in $\delta^{13}C$ by 1.6 and 0.5‰, while the liver was depleted by 1.3‰. The $\delta^{15}N$ enrichments were 1.6‰ for whole fish, 1.3‰ for muscle, and 1.0‰ for liver. Except for liver, $\delta^{15}N$ isotope values increased significantly with time.

Conclusions: The small change in $\delta^{13}C$ between the diet and fish suggests that little isotope alteration is occurring during the assimilation of dietary carbon. This provides a basis for estimates of the importance of different sources of dietary components when contrasted with the isotope values from a formulated diet with known isotope values of the different components. The smaller than expected $\delta^{15}N$ enrichment in all tissue suggests that isotope values from a wild fish sample may not always reach isotope equilibrium with the current diet, and an inference about recent dietary sources and an estimate of the consumer's trophic position should be evaluated with caution.

Keywords: Stable isotopes Diet shift Fish Isotope enrichment

INTRODUCTION

Carbon and nitrogen stable isotopes ($\delta^{13}C$ and $\delta^{15}N$) are increasingly used to evaluate the relative contributions of different food sources and the trophic position of fish (Anderson et al. 1987; Lochman and Phillips 1996; Gu et al. 1996a, b; Gamboa-Delgado et al. 2008). Because isotope compositions reflect the organic compounds that have been incorporated into the bodies of consumers, the measurements of $\delta^{13}C$ and $\delta^{15}N$ provide insightful information on the dietary component assimilated by the consumers. At the same time, conventional studies such as feeding observations or gut content analyses reveal only the materials ingested and not necessarily those assimilated by the consumers. These approaches can also be labor intensive and time-consuming while providing little information on the sources of energy and nutrients for animal growth (Bitterlich and Gnaiger 1984; Gu et al. 1996a, b).

The use of stable isotope analysis in trophic ecology is based on the premise that the $\delta^{13}C$ and $\delta^{15}N$ values of consumers reflect those of the assimilated diet and that the alteration of isotope compositions during the assimilation of food by consumers can be followed in a predicable manner (DeNiro and Epstein 1978, 1981; Minagawa and Wada 1984; Post 2002). Isotope enrichment in consumers takes place during the assimilation of carbon and nitrogen from the diet. A review of the literature indicates that the average isotope enrichment during each trophic transfer is 0.5‰ for $\delta^{13}C$ and 3.4‰ for $\delta^{15}N$, but with large variations in both stable isotopes (Post 2002). These data are largely derived from field studies with many unknown and uncontrolled factors such as food sources (Gu et al. 1996a, b; Post 2002) and differences in growth rate, age, and food quality, which may affect the magnitude of isotope fractionation. Muscle tissue of fish has been widely used to represent the whole fish isotope characteristic. However, previous and current studies also indicate wide variations in isotope composition among consumer tissues (DeNiro and Epstein 1978, 1981; Tieszen et al. 1983; Pinnegar and Polunin 1999; Miller 2006). More studies are needed to validate the use of a certain tissue type during a food web study to establish the precise trophic relationship between consumers and their diets.

This study was designed to analyze the magnitude of isotope enrichment in fish and thus to provide the basis for a better understanding of the dietary sources and trophic position of fish. The experiment utilized brown-marbled groupers, Epinephelus fuscoguttatus, a carnivorous fish widely distributed in the subtropical and tropical marine regions and grown as a high quality human food in Southeast Asia.

Methods

The experiment was conducted at the aquaculture facility of Long Hai Tian Science Garden, Dong Hai Island, Guangdong, China, between July and October 2009. The experimental units were plastic aquaria with the dimensions of 70 × 50 × 70 cm (length × width × height). Each aquarium was filled with continuously aerated natural seawater prefiltered using a sand medium. Water quality was monitored following the standard methods (APHA 1998). The experiment was conducted indoors using a natural dark:light cycle. The fish diet was of trash fish (Engraulis ringens) powder imported from Peru, which was formed into pellets. The composition of the formulated pellets is presented in Table 1. The average stable isotope values of the diet were 16.4‰ for $\delta^{15}N$ and −20.4‰ for $\delta^{13}C$ (Table 1).

Table 1 Chemical composition and stable isotope values of the diet (trash fish) used in the current experiment

Component	Values
Crude protein (% dry mass)	50.17
Crude fat (% dry mass)	10.32
Crude fiber (% dry mass)	5.6
Gross energy (kJ/g)	19.25
$\delta^{13}C$ (‰)	−20.4
$\delta^{15}N$ (‰)	16.4

Newly hatched larval fish were obtained from the Guangdong Ocean University hatchery and raised in a large cement tank with a commercial feed. After reaching an average weight of 22.5 ± 3.2 g, the fish were transferred to six aquaria stocked with 12 fish per aquarium. Fish were fed with the formulated pellets as the only protein source twice daily at 0900 and 1300 local time. Samples for stable isotope analysis were taken at day 0, 3, 7, 14, 28, 42, and 56. Prior to sampling, all fish were starved for 24 h to expunge the food content of their stomachs and intestines. All fish from a randomly selected aquarium were harvested during each sampling event. Three individuals were used for whole fish, muscle, and liver ($n = 9$ fish), respectively. The remaining fish served as backup alternates. For whole fish, the entire fish was cut into several pieces. For muscle and liver tissue, the dorsal muscle tissue and liver were removed from each fish and washed with deionized water to remove blood. Fish feces were collected by siphoning from the bottom of the aquarium. All samples were dried at 70°C and ground to a fine power.

For stable isotope analysis, about 1 mg of the sample was loaded into a tin capsule and shipped to the Stable Isotope Facility of the University of California, Davis for ^{13}C and ^{15}N analysis using a PDZ Europa ANCA-GSL elemental analyzer interfaced to a PDZ Europa 20–20 isotope ratio mass spectrometer (Sercon, Cheshire, UK). Samples were combusted at 1,000°C in a reactor packed with chromium oxide and silvered cobaltous/cobaltic oxide. During analysis, samples were interspersed with several replicates of at least two different laboratory standards. The long term standard deviation is 0.2‰ for 13C and 0.3‰ for 15N. The heavy to light isotope ratios ($^{13}C/^{12}C$ and $^{15}N/^{14}N$) are reported as the conventional delta notation (δ) defined as $\delta X = [(R_{sample}/R_{standard}) - 1] \times 1{,}000$, where X is ^{13}C or ^{15}N and R is the ratio of $^{13}C/^{12}C$ or $^{15}N/^{14}N$. Isotopic ratios are expressed relative to VPDB (Vienna Pee Dee belemnite) for $\delta^{13}C$ and to atmospheric nitrogen for $\delta^{15}N$.

RESULTS

During the experimental period, water temperature ranged from 24 to 28°C, salinity from 29 to 31‰, pH from 7.9 to 8.2, dissolved oxygen was > 6 mg/L, ammonium < 0.02 mg/L, and nitrate < 2.0 mg/L. At the end of the experiment, fish grew to an average of 55.6 g from an average of 22.5 g, more than doubling their initial weight (Table 2). The average growth rate (k) of fish was calculated to be 0.023.

Table 2 Body length, total length and wet weight of brown-marbled groupers before and during the experimental period

Time	Body length (cm)		Total length (cm)		Wet weight (g)		Sample number
Day	Mean	SD	Mean	SD	Mean	SD	
Before	9.0	0.6	10.7	0.7	22.5	3.8	10
3	9.1	0.5	10.7	0.4	24.5	3.8	10
7	9.4	0.5	11.6	0.9	31.1	5.1	10
14	10.7	0.4	12.8	0.5	35.4	5.1	10
28	11.6	0.6	13.4	0.6	42.7	5.9	10
42	12.7	0.6	14.4	0.8	45.6	7.5	10
56	10.5	1.2	13.6	1.5	55.6	14.1	10

Changes in both $\delta^{13}C$ and $\delta^{15}N$ values of whole fish, muscle, and liver tissues during the experimental period are presented in Figure 1. The initial whole fish δ13C value (−21.1‰) was below the dietary value (−20.4‰) and increased linearly to −18.8‰ by day 56 (Figure 1A). In contrast, the initial muscle $\delta^{13}C$ value (−18.6‰) was above the dietary value and increased to −17.5‰ by day 56 (Figure 1B). The initial liver $\delta^{13}C$ value (−23.2‰) was well below the dietary value and reached a peak value (−21.3‰) at day 28 and then slightly decreased and increased at day 42 and day 56, respectively (Figure 1C). At the end of the experiment, whole fish and muscle tissue of the juvenile groupers were enriched in $\delta^{13}C$ by 1.6 and 0.5‰ while the liver was depleted by 1.3‰. All regression models showed significant increases in $\delta^{13}C$ values during the experimental period.

The initial $\delta^{15}N$ values of whole fish, muscle, and liver tissues were all below the dietary value and increased linearly as fish grew (Figure 1). The whole fish $\delta^{15}N$ value increased from 15.7‰ at day 0 to 18.0‰ at day 56 (Figure 1D). The muscle $\delta^{15}N$ value increased from 16.1‰ at day 0, reached its peak value (18.1‰) at day 42, and then decreased slightly to 17.8‰ at day 56 (Figure 1E). The initial liver $\delta^{15}N$ value (14.5‰) was well below the dietary value and remained little changed between days 3 and 27 before rapidly increasing above the dietary value during the final two sampling dates (Figure 1F). The $\delta^{15}N$ enrichments were 1.6‰ for whole fish, 1.3‰ for muscle, and 1.0‰ for liver. Except for liver, regression models for whole fish and muscle tissue showed significant increases in $\delta^{15}N$ values during the experimental period.

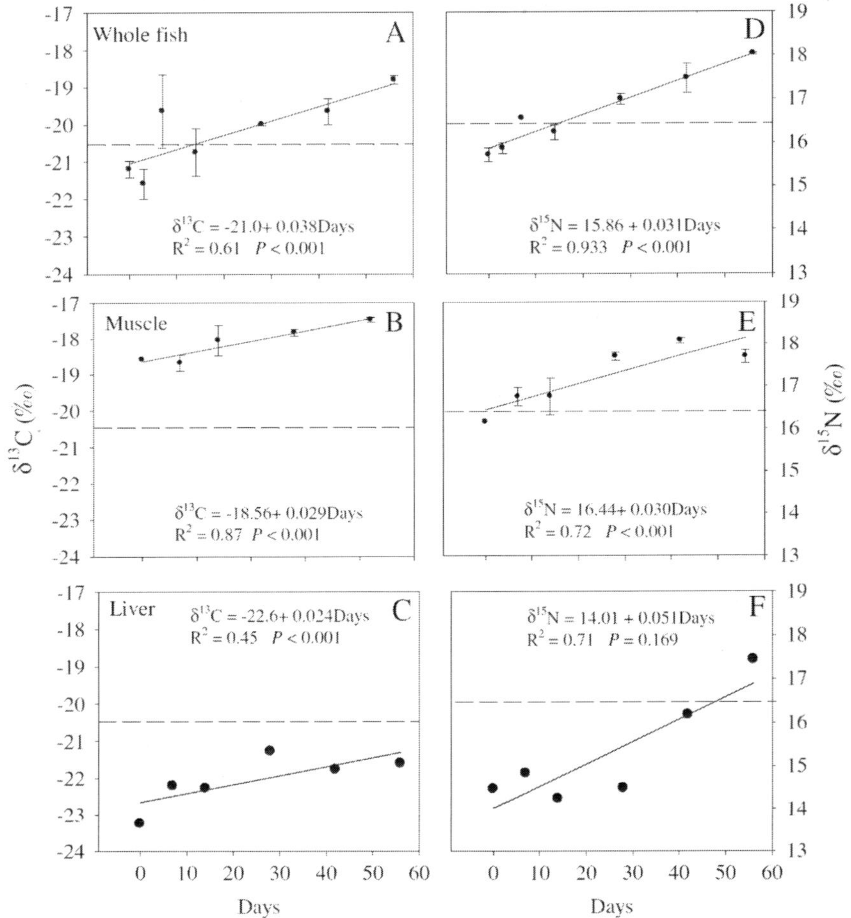

Figure 1 Changes in δ 13 C and δ 15 N for whole fish, muscle, and liver tissues of juvenile brown-marbled groupers during the 56-day experimental period. The dashed lines represent the dietary values of $\delta^{13}C$ (−20. 4‰) or $\delta^{15}N$ (16.4‰).

DISCUSSION AND CONCLUSIONS

Previous studies have shown various magnitudes of isotope enrichment in whole animals and individual tissues such as muscle, liver, kidney, blood, and hair (DeNiro and Epstein 1978, 1981; Tieszen et al. 1983; Pinnegar and Polunin 1999; Miller 2006). Isotope enrichments were the result of preferential retention of the heavy isotopes ^{13}C and ^{15}N during assimilation of dietary carbon and nitrogen (DeNiro and Epstein 1978, 1981). This was also confirmed in this experiment by noting that the feces were isotopically depleted (Table 3), leading to ^{13}C and ^{15}N enrichment in the tissues of juvenile groupers. Our results also show that, compared to the dietary isotope value, liver was depleted in ^{13}C. This is likely attributed to lipid storage in the liver that is depleted in ^{13}C.

Table 3 Average (standard deviation) of stable carbon and nitrogen isotope ratios (δ^{13}C and δ^{15}N) and percent carbon (%C) and nitrogen (%N) in the feces of brown-marbled grouper

Day	δ^{13}C (‰)	δ^{15}N (‰)	%N	%C	Sample number
0	−21.9 (0.2)	10.3 (0.0)	2.2 (0.2)	23.1 (2.0)	4
7	−22.3 (0.6)	15.5 (0.5)	1.9 (0.4)	19.9 (3.1)	5
14	−21.5 (0.3)	14.8 (0.8)	1.9 (0.3)	17.7 (3.3)	5
28	−21.2 (0.6)	14.8 (0.9)	1.8 (0.3)	18.0 (4.0)	4
42	−21.6 (0.3)	14.3 (0.2)	1.8 (0.1)	18.0 (0.7)	2
56	−20.1	18	3.3	20.6	1

Various factors have been proposed to account for the differences in δ^{13}C and δ^{15}N values between animal tissues and their diet. Earlier studies showed that different individuals of a species differ in isotope fractionation when fed the same diet, but that isotope fractionation differs more between species than among individuals of the same species (DeNiro and Epstein 1978, 1981). Our results also showed differing isotope enrichments or depletions compared with the dietary δ^{13}C. However, in general, the δ^{13}C enrichment of the juvenile groupers was less than 2‰. The muscle tissue of the juvenile groupers was enriched by 0.5‰, nearly identical to the reported average enrichment (Post 2002). These results suggest that muscle tissue is the most consistent proxy for carbon isotope fractionation when compared to the findings of many previous studies on fish (Hesslein et al. 1993; Pinnegar and Polunin 1999).

Our results also show that the δ^{15}N enrichment of muscle and liver tissues or whole fish is only 1.0–1.6‰, which is far below the reported average value (3.4‰). Several mechanisms may explain this discrepancy. First, different consumers may have their intrinsic isotope fractionation during assimilation of protein. For example, (Hesslein et al. 1993) found a maximum value of 3.0 of $\delta^{15}N$ enrichment for broad white fish (Coregonus nasus). Pinnegar and Polunin (1999) reported $\delta^{15}N$ enrichment in rainbow trout (Oncorhynchus mykiss) of 2.6‰. Second, consumers might assimilate a specific dietary component isotopically different from the whole diet, especially when they have access to a variety of diets in nature or are fed a formulated diet that is made from several sources of protein (Robbins et al. 2005a, b). This is not the case in our study because only a single source of protein was used. Third, dietary nutritional quality (Robbins et al. 2005a, b) and growth conditions (Hobson et al. 1993; Gaye-Siessegger et al. 2004) may also affect the isotope compositions of the consumers. These cannot be applied to our study because the formulated diet was high in nutritional value (Table 1), and good water quality was maintained during the experimental period. Lastly, it is also likely that the consumers had not reached isotopic equilibrium with the new diet during this study. This is evident from the linear relationship between time and the isotope values of the liver tissue and whole fish tissues (Figure 1), which never attained asymptotic values for either $\delta^{13}C$ or $\delta^{15}N$. It appears that the muscle $\delta^{15}N$ likely reached isotope equilibrium, but the experimental period was not long enough to provide sufficient data to confirm the asymptotic isotope stage. This finding is important when the measured values are used to determine the trophic position of fish that may switch diets due to migration or growth or that are affected by rapid changes in $\delta^{13}C$ and $\delta^{15}N$ of the dietary organisms over a seasonal cycle (Gu 2009; Gu et al. 2011).

Results from this experiment have important implications for the studies of animal dietary composition and trophic positions. The small change in $\delta^{13}C$ between the diet and fish suggests that little isotope alteration is occurring during the assimilation of dietary carbon. This provides the basis for the estimates of the importance of different sources of dietary components when contrasted with the isotope values from a formulated diet with known isotope values of the different components (Lochman and Phillips 1996; Gamboa-Delgado et al. 2008). The data from this study also suggest that isotope values from a wild fish sample may not reach isotope equilibrium with the current diet and hence an inference about recent dietary sources or trophic position should be evaluated with caution because the isotope values are likely the results of previous and current dietary uptake. This is especially true in aquatic ecosystems where environmental changes or fish migration often result in a shift in diet or changes in the isotope composition of dietary organisms such as plankton (see review by Gu 2009; Gu et al. 2011).

ACKNOWLEDGMENTS

We appreciate Dr. Xiaohui Dong for assistance with the design and formulation of the formulated pellets. This study was supported by grants A200608C02, A200908D03, 2006B20201059, and 2009B020308005 from several funding agencies of the Guangdong Province government, China. Dr. Thomas Dreschel provided language improvements on the final draft of this manuscript.

REFERENCES

1. Anderson R, Parker PL, Lawrence A: A 13C/12C tracer study of the utilization of presented feed by a commercially important shrimp Penaeus vannamei in a pond growout system. J World Aquacult Soc 1987, 18: 148–155. 10.1111/j.1749-7345.1987.tb00433.x
2. APHA: Standard methods for the examination of water and wastewater. 20th edition. American Public Health Association, American Water Works Association and Water Pollution Control Federation, Washington DC; 1998.
3. Bitterlich G, Gnaiger E: Phytoplanktivorous or omnivorous fish? Digestibility of zooplankton by silver carp, Hypophthalmichthys molitrix (Val.). Aquaculture 1984, 40: 261–263. 10.1016/0044-8486(84)90194-7
4. DeNiro MJ, Epstein S: Influence of diet on the distribution of carbon isotopes in animals. Geochim Cosmochim Acta 1978, 42: 495–506. 10.1016/0016-7037(78)90199-0
5. DeNiro MJ, Epstein S: Influence of diet on the distribution of nitrogen isotopes in animals. Geochim Cosmochim Acta 1981, 45: 341–351. 10.1016/0016-7037(81)90244-1
6. Gamboa-Delgado J, Cañvate JP, Zerolo R, Le Vay L: Natural carbon stable isotope ratios as indicators of the relative contribution of live and inert diets to growth in larval Senegalese sole (Solea senegalensis). Aquaculture 2008, 280: 190–197. 10.1016/j.aquaculture.2008.04.036
7. Gaye-Siessegger J, Focken U, Muetzel S, Abel H, Becker K: Feeding level and individual metabolic rate affect δ13C and δ15N values in carp: implications for food web studies. Oecologia 2004, 138: 175–183. 10.1007/s00442-003-1429-7
8. Gu B: Variations and controls of nitrogen stable isotopes in particulate organic matter of lakes. Oecologia 2009, 160: 421–431. 10.1007/s00442-009-1323-z
9. Gu B, Schelske CL, Hoyer MV: Stable isotopes of carbon and nitrogen as indicators of diet and trophic structure of the fish community in a shallow hypereutrophic lake. J Fish Biol 1996, 49: 1233–1243. 10.1111/j.1095-8649.1996.tb01792.x
10. Gu B, Schell DM, Huang X, Yie F: Stable carbon and nitrogen isotope evidence for dietary overlap between two planktivorous fish in aquaculture ponds. Can J Fish Aquat Sci 1996, 53: 2814–2818. 10.1139/f96-242
11. Gu B, Schelske CL, Waters MN: Patterns and controls of seasonal variability of carbon stable isotopes of particulate organic matter in lakes. Oecologia 2011, 165: 1083–1094. 10.1007/s00442-010-1888-6
12. Hesslein RH, Hallard KA, Ramlal P: Replacement of sulfur, carbon, and nitrogen in tissue of growing broad whitefish (Coregonus nasus) in response to a change in diet traced by 34S, 13C, and 15N. Can J Fish Aquat Sci 1993, 50: 2071–2076. 10.1139/f93-230

13. Hobson KA, Alisauskas RT, Clark RG: Stable-nitrogen isotope enrichment in avian tissues due to fasting and nutritional stress: implications for isotopic analyses of diet. Condor 1993, 95: 388–394. 10.2307/1369361
14. Lochman R, Phillips H: Stable isotopic evaluation of the relative assimilation of natural and artificial foods by golden shiners Notemigonus crysoleucas in ponds. J World Aquacult Soc 1996, 27: 168–177.
15. Miller TW: Tissue-specific response of 15N in adult Pacific herring (Clupea pallasi) following an isotopic shift in diet. Env Biol Fish 2006, 76: 177–189. 10.1007/s10641-006-9020-9
16. Minagawa M, Wada E: Stepwise enrichment of 15N along food chains: further evidence and the relation between δ15N and animal age. Geochim Cosmochim Acta 1984, 48: 1135–1140. 10.1016/0016-7037(84)90204-7
17. Pinnegar JK, Polunin NVC: Differential fractionation of 13C and 15N among fish tissues: implications for the study of trophic interactions. Funct Ecol 1999, 13: 225–231. 10.1046/j.1365-2435.1999.00301.x
18. Post DM: Using stable isotopes to estimate trophic position: models, methods, and assumptions. Ecology 2002, 83: 703–718. 10.1890/0012-9658(2002)083 [0703 :USITET]2.0.CO;2
19. Robbins CT, Felicetti LA, Sponheimer M: The effect of dietary protein quality on nitrogen isotope discrimination in mammals and birds. Oecologia 2005, 144: 534–540. 10.1007/s00442-005-0021-8
20. Robbins CT, Felicetti LA, Florin TF: The impact of protein quality on stable nitrogen isotope ratio discrimination and assimilated diet estimation. Oecologia 2005, 162: 571–579.
21. Tieszen LL, Boutton TW, Tesdahl KG, Slade NA: Fractionation and turnover of stable carbon isotopes in animal tissues: implications for 13C analysis of diet. Oecologia 1983, 57: 32–37. 10.1007/BF00379558

CHAPTER 11

Ecological Devastation in Lake Victoria: Part B: Plankton and Fish Communities

Moshe Gophen

Migal-Scientific Research Institute, Kiryat Shemone, Israel

ABSTRACT

A few specimens of the exotic fish, Nile Perch, (Lates niloticus, Linnaeus 1758) were transferred into Lake Victoria in early 1950's. In early 1980's this fish occupied the Lake Victoria ecosystem eliminating the local endemic Haplochromines (app. 400 species). As a result of this change, together with intensification of pollution constrains from the catchment and dust deposition the ecology of Victoria's ecosystem was modified: cyanobacteria replaced diatoms, anoxia enhanced, secchi depth became shallower, euphotic zone became thinner, fishery enhanced fully comprised of Nile Perch and Rastrineobola. The ecological significances are discussed and future propositions are presented.

Keywords: Lake Victoria, Nile Perch, Ecological Changes, Fish, Plankton

1. INTRODUCTION

Lake Victoria, the second (excl. Caspian Sea) largest lake in the world by surface area (68,800 km^2) and 7th by Volume (2800 km^3) has undergone fundamental changes of its limnological structure. The lake is located at an altitude of 1134 m above Mean Sea Level and stretched between latitudes of

0°20'N - 3°0'S and 31°39'W - 34°53'E. Limnological changes include not only internal modifications enhanced by the Nile Perch invasion [1] but also by external anthropogenic activities in the catchment resulted in input of 20 kg /km²/year of total phosphorus and 400 kg/km²/year, in addition to atmospheric contaminated dust deposition [2] [3] and as a consequence Lake Victoria clearly exhibit symptoms of eutrophication (Table 1). The lake and the drainage Basin territories are shared between three countries (Figure 1): Kenya, Tanzania and Uganda. The ratio between Cat-

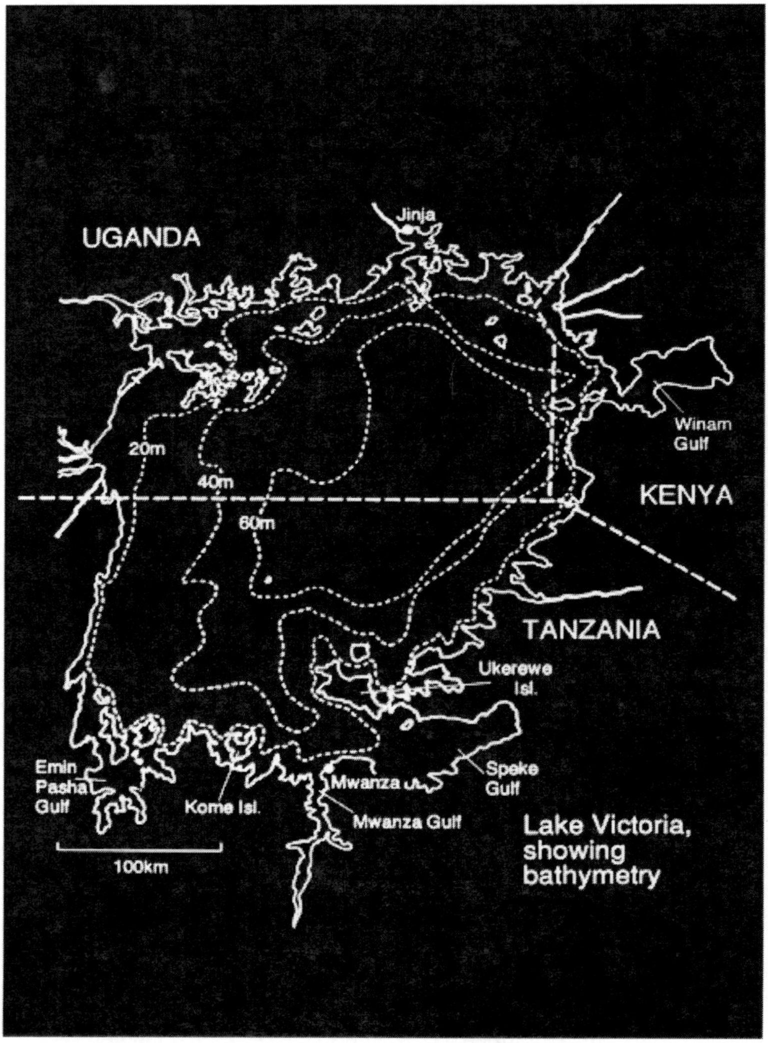

Figure 1. Bathimetrical-Geographical Map of Lake Victoria.

Table 1. Annual ranges of nutrients (mg/l, µg/l), Primary Production (PP) (mg DO/m³) and Chlorophyll Concentration (µg/l) in Lake Victoria during 1966-1990 [4] - [9].

Parameter	1966	1977	1979	1984	1989	1990
PO_4 (µg/l)	7 - 120	1 - 122		2 - 75	4.0 - 37	1 - 19
NO_3 (µg/l)	10 - 112	0.5 - 122	0.16 - 0.18	21 - 237	0.1 - 513	1 - 30
SO_4 (mg/l)	0.4 - 4.0			0.1 - 5.0		
SiO_2 (mg/l)	4 - 8	0.2 - 3	2 - 7.9	0.1 - 7.6		0.06 - 0.72
PP (mg DO/m³)	100 - 130		400 - 600		180 - 600	100 - 1400
Chlorophyll (µg/l)	0.5 - 22.3	2.1 - 8.5		1.8 - 23.5	8 - 120	35.8 - 115.2

chment area and lake Volume is 11.7 m³/m² i.e. each m³ of the lake accept nutrients from 11.7 m² of the catchment. The D number for Lake Victoria is 3.7. Those two indicators exhibit high potential of nutrient supply from the catchment (Table 1). The major inflow rivers are Kagera and Nzoia (Tanzanian part of the lake), the only outflow is through the White Nile river, outlet located North, nearby Jinga, Uganda.

Phytoplankton research in Lake Victoria started in 1898 when expeditions were carried out to collect net hauling samples, therefore, information until the 1960's is limited to algal taxonomy. Talling [4] [5] was the first who studied long term cycle of community composition, quantitative abundance, population dynamics and vertical distribution and later on Kenyan and other scientists continued [6] [7] [9]. The Zooplankton of Lake Victoria was studied by Rzoska and others [10] - [14]. The research of fish biology and fisheries management in lake Victoria was enhanced by local and external scientists, resulted by the Nile Perch Invasion [15] - [18]. The objective of our study was to initiate an international effort to collect data about the ecological changes in Lake Victoria caused by the Nile Perch invasion and to establish a Data Base for future investigations. Our studies and consequently the Lake Victoria Data Base (LVDB) information were previously summarized in technical reports which are publicize here as a scientific paper.

2. MATERIAL AND METHODS

2.1. Sampling Stations (1989-1992)

Eighteen sampling stations (Figure 2) were monthly monitored for physico-chemical parameters (DO, pH and Temperature), and plankton: 8 stations in Wynam Gulf (numbered: 1,2,3,17, 30, 31, 36, 37), 4 Stations at the transition zone between Wynam (Kavirondo) Gulf and the open waters (offshore) (numbered: 5, 6, 7, 8,), 1 (numbered: 34) station in the Rusinga Channel, and 5 stations in offshore zone (numbered: 32, 33, 55, 100, 103).

2.2. Zooplankton Sampling

Samples were collected from discrete depths and 3 liters were sieved through 100 m mesh size net on board, and all organisms were flushed into bottles (final volume of 30-50 ml) and 5 ml formalin 10% were added. Counts were carried out in the Kinneret Limnological Laboratory (Israel) under Wild M5 stereoscope.

2.3. Phytoplankton Sampling

Original sampled waters (125 ml) were placed in brown bottles and a few drops of Lugol were added. Counts were carried out in Kinneret Limnological Laboratory and IOLR-Haifa Institute by Utermol's sedimentation method in 10 ml chambers, under Inverted Microscope (Zeise M20) and counts were done.

2.4. Fisheries

Information on fishery in Lake Victoria (1968-1991) was supported by Kenya Marine and Fisheries Research Institute. Statistical evaluation of the data was evaluated by Simple Linear Regressions between annual landings (dependent variable) and years (independent variable), r^2 (correlation coefficient) which indicates the strength of the association between the two variables and p (probability) values are given.

2.5. Fish Gut Content, Body Condition & Index of Satiation (IOS)

The Tilapias, T. niloticus, T. variailis, T. zillii, T. niloticus specimens were collected from fishermen yields, total length and weight were measured, then the viscera (Stomach & intestine) was removed individually and weighed, placed in plastic bottles and preserved by 5 ml 10% formalin. Samples were taken to Kinneret Limno logical Laboratory for second weight measurement, gut content was squeezed and food items analysis under both Wild M5 Binocular and Light Microscope (M20) was carried out. Index of Satiation (IOS) was calculated as follows:

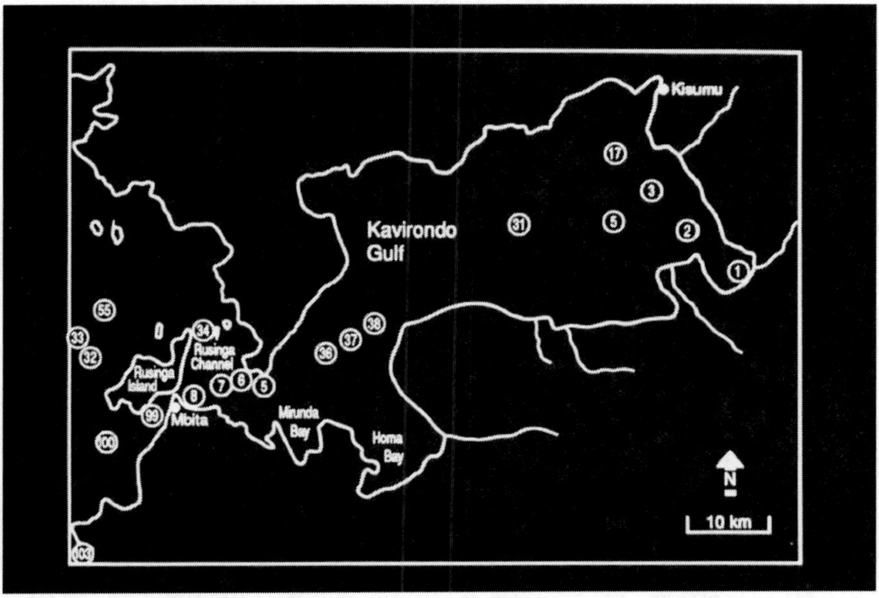

Figure 2. The Kenyan Part of Lake Victoria Kavirondo, (Wynam) Gulf Sampling Stations are numbered.

Viscera (stomach-intestine) Weight (g)—V.

Body Weight (g)—B.

$(V/B) \times 100 = IOS$.

2.6. Zooplankton Removal by Fish: Feeding Experiments

Living specimens of Tilapias, (T. leucostictis, T. variabilis, T. zillii, T. niloticus T. esculentus) were collected alive from beach seine catches and immediately transferred into Aquaria containing 60 - 70 filtered (63 μ mesh size) lake water in duplicates (two per species) and fishless aquaria for the determination of zooplankton natural mortality. Freshly collected (100 μ sieving) live zooplankton were added in similar aliquots to the aquaria. Water samples were taken from the tanks at initial (zero) time and 3 times with intervals of 30 and 40 minutes and last sample was taken after 1 hour. The samples were taken by plastic pipe (2.5 cm diameter) closed by rubber cup placed randomly on the bottom of the aquaria. Samples were filtered through 63 μ mesh size net and all zooplankton organisms were counted under M5 Wild Binocular. Zooplankton mortality was subtracted from counted organisms and % removal was evaluated. Seven trials were carried out.

Fishes body weight and number of trials were as follows:

T. zillii, 2 fishes, 1 trial TL: 20, 12 cm .

T. niloticus, 4 fishes, 2 trials, TL: 13, 12.6, 14, 12.3 cm .

T. variabilis, 21 fishes, 4 trials, TL: 8.1 - 22.7 cm .

T. leucostictus, 4 fishes, 1 trial, TL: 8.1 - 14.0 cm .

The Aquaria were slightly aerated.

3. RESULTS

3.1. Nutrients

Samples were collected in stations numbered (Figure 2): 1, 6, 17, 31, 32, 33, 34, 36, 37, 38, 100, 103 at depths of (m): 0, 1, 2, 3, 5, 9, 11, 15, 20, 23, 25, 30, 35, 39, 40, and 42 with respect to the actual depth. The following parameters were measured: Total Nitrogen (TN), Ammonium (NH_4), Nitrate (NO_3), Total Phosphorus (TP) and Ortho-Phospate (OP) (Methods are given in [3]). The mass (W/W) TN/TP ratio was calculated. Results are given in Table 2.

Results given in Table 2 indicates the followings: percentage of OP from TP is similar throughout the centire water column: 35% and 40% in the upper and lower part respectively; Similarity of % of Inorganic Nitrogen (NH_4+NO_3; NO_2 is negligible) from TN was also indicated within the entire water column; Particulate Organic Nitrogen comprised >90% of TN; Prominent increase of TP in lower part of the water column, therefore the TN/TP W/W mass ratio decline in deep waters. These chemical conditions of low TN/TP ratio are favored by Cyanobacteria.

3.2. Phytoplankton

Cyanobacteria were prominently dominant in all samples. The following genera and partly species were defined:

3.2.1. Cyanobacteria

Merismopedia, Lyngbia circumcreta, Aphanocapsa delicatissima, Microcystis, Anabaena flos aqua, Chroococcus, Cylindrospermopsis.

Table 2. Averaged (SD) concentrations (ppm) of chemical parameters measured in all stations and combined into 2 water layers: <10 m and >10 m depths [3].

Depth Layer	TN (ppm)	NH_4 (ppm)	NO_3 (ppm)	TP (ppm)	OP (ppm)	TN/TP
<10 m	1.136 (0.651)	0.073 (0.073)	0.024 (0.021)	0.075 (0.024)	0.026 (0.011)	15.6 (7.4)
>10 m	1.185 (1.363)	0.104 (0.096)	0.026 (0.020)	0.134 (0.198)	0.054 (0.084)	10.6 (7.3)

3.2.2. Bacillariophyta (Diatoms) Chlorophyta, Pyrrhophyta

Nvicula rynchocephala, Synedra cunningtonii, Melosira (Aulacoseira) granulate var. angustissima, Melosira (Aulacoseira) nyassensis var. Victoria, Nitzschia acicularis, Chlorophyta: Scenedesmus, Chlamydomonas, Scenedesmus quadricauda, Pediastrum, Cosmarium, Oocystis solitario, Pyrrhophyta: Cryptomonas, Glenodinium pulvioulus.

3.2.3. Algal Density

Densities of Cyanobacteria varied within the range of maximum of 4000/ml and minima between 200 - 1000/ml.

Dominant algae during 1966-1979 were Diatoms and later-Cyanobacteria.

Results of algal densities measured during 1989- 1990, in Wynam Gulf (Kavirondo) at 0, 1, 5, 7, and 10 m depths;, Rusinga Channel at 2, 3, 5, and 20 m depths; and in the deep offshore open waters stations at 0, 1, 1.5, 5, 16, 1nd 20 mdepths; were (No. per ml).

3.2.4. Wynam Gulf

Aphanocapsa—1200 - 3500; Lyngbia—300 - 3900; Merismopedia—200 - 1000; Microcystis—400 - 1200; Anabaena—300 - 1000; Chroococcus—200 - 900; Coenobial Cyanophytes—>3000; Cylindrospermopsis—500; Melosira (Aulacoseira)—600 - 1400; Synedra—200 - 3300; Navicula—300; Rhodomonas—300; Chlamidomonas—300; Scenedesmus—200 - 400; Cryptomonas—300 - 400.

3.2.5. Rusinga Channel

Aphanocapsa—1600; Lyngbia—900 - 17,000; Microcystis—600; Chroococcus—400; Cylindrospermopsis— 600 - 4100; Synedra—2400 - 4900; Rhodomonas—500; Cryptomonas—400.

3.2.6. Offshore

Lyngbia—2100 - >10,000; Cylindrospermopsis—2700 - >4000; Synedra—2000 - >3000. Algal species diversity is much higher in the shallow water of Wynam Gulf and the dominance of Cyanobacteria is prominent.

3.3. Zooplankton

The Following species were identified in samples collected during 1989-1990:

3.3.1. Copepoda

Thermocyclops neglectus, Thermocyclops hyalinus, Mesocyclops leuckarti, Microcyclops minutus.

3.3.2. Calanoida

Thropodiaptomus neumanii.

3.3.3. Cladocera

Diaphanosoma excisum, Ceriodaphnia rigaudii, Daphnia longispina, Moina macrura, Bosmina longirostris, Chydorus sphaericus.

3.3.4. Rotifera

Brachionus caudatus, Brachionus calyciflorus, Keratella tropica, Polyarthra remata, Filinia longispina, Asplanchna brightwelli, Collotheca sp., Trichocerac sp., Ascomorpha sp. Samples for zooplankton density counts (excluding Caridina nilotica) were collected in all stations during 1989-1990; The dominant (most common) group was Cyclopoida presented in all stations and depths from surface down to bottom layers (40 meters in offshore stations). Densities of cyclopoids and calanoids varied between 3 - 88 and 0 - 56 per litre respectively. In all zooplankton samples cladocerans were found only in 27 and rotifers in 26 of them. Densities of cladocerans and rotifers varied between (Maximum-Minimum) 1 - 28 and 1 - 56 per litre for cladocerans and rotifers respectively.

3.4. Zooplankton Consumption

Feeding Experiments

Consumption of zooplankters by Tilapias as concluded from feeding experiments is given in Table 3. Results presented in Table 3 indicates that the major source of zooplankton prey by 4 Tilapia species from Lake Victoria as concluded from 7 trials of feeding experiments are small organisms of young stages of cyclopoid and rotifers.

3.5. Gut content Analysis

Results in Table 4 indicate that major trophic source for the Haplochromines that were captured in Mwamza Gulf are suspended particles of detritus, phytoplankton, and zooplankton probably collected by filtration.

Results in Table 5 indicate that Tilapias collected in the Kenyan open waters collect their food mostly in bottom sediments.

Body Condition and Food Composition of Sampled Tilapias

For the study of body condition and food composition of Tilapias we sampled specimens in Wynam Gulf Stations (6, 34, 36) and offshore (100) during January February and March. Number of analyzed specimens was:

Tilapia variabilis (TV)—69.

Tilapia zillii (TZ)—22.

Tilapia nilotica (TN)—33.

Tilapia esculentus (TE)—13.

Tilapia leucostictus (TL)—13.

Total and Standard lengths of all Tilapias ranged between 10 - 55 cm and 8 - 50 cm respectively. The Body Weight of all Tilapias ranged between 100 - 600 g , and the Viscera (stomach and intestine) weights range was —0.2 - 22.0 g ; Index of Satiation (IOS) was calculated (see Methods) for individual specimen (Table 6).

Table 3. Zooplankton Predation by Tilapias from Lake Victoria: zooplankters which were removed from the water in feeding experiments at a level of >50% are listed.

Experiment #	Fish Species	Zooplankters Removed
1	*T. niloticus*	Cyclopoid Nauplii small rotifers
2	*T. zillii*	Cyclopoid Nauplii&adults small & Large rotifers
3	*T. variabilis*	Cyclopoid Nauplii, small rotifers
3a	*T. variabilis*	Adult Cyclopoids
3b	*T. variabilis*	Cyclopoid Nauplii & Copepodites
4	*T. leucostictus*	Cyclopoid Nauplii, small rotifers
5	*T. niloticus*	Cyclopoid Nauplii Copepodites, small rotifers

Table 4. Trophic composition of Haplochromine fishes in Mwanza Gulf (Tanzania), Lake Victoria.

Trophic Element	% by number	Estimated % by weight
Detritivores/Planktivores	38	31
Phytoplanktivores	22	18
Zooplanktivores	30	25
Insectivores	5	15
Molluscivores	1	3
Piscivores	2	5
Caridinia Predatores	1	2
Other	1	1

3.6. Index of Satiation (IOS)

Results in Table 6 indicated similar intensity of food ingestion by the five species of Tilapias during 1 - 3 winter months.

Food Composition of three species of Tilapia (T. nilotica, T. variabilis, T. zillii) was analyzed on 124 specimens which were sampled during January-March 1989, in Wynam Gulf at Stations 6, 34, 36 and offshore zone—Stn.100. Their gut contents were removed individually and analyzed under both Dissecting Microscope (M5 Wild) and Light Microscope (Dialux 20). Food items were counted and their biomass was calculated as measured bio-volume converted to weight (specific gravity = 1.0) and the summary is given in Table 6 & Table 7.

Results in Table 7 indicate that the 57%, 68% and 55% of the food items consumed by TZ, TV and TN respectively is collected from the bottom whilst the source of the rest is probably planktonic. It should be considered that video filming in deep anoxic waters at stations 100, 32 and 34 (results are not presented) indicated the existence of high densities of the Atyid Decapod, Caridina nilotica there. Therefore, C. nilotica might be defined as partial "bottom layer" inhabitant.

4. DISCUSSION

The phytoplankton as analyzed in this study is characterized by high densities of coccoids accompanied by small filaments of Cyanophytes (Lyngbia). Diatoms were presented by Nitzchia and Melosira (Aulacoisera) and the chlorophytes by Chlorococcales and desmids. Dinophytes and Cryptophytes were rarely observed. The densities of zooplankton were mostly low and their body size small. The most abundant group was cyclopoid copepods. The results of feeding experiments were unpredicted: most of the organisms which were removed from the water by the fish were free swimming small body zooplankton and gut content analysis of specimens collected in the lake indicated deep layer and bottom inhabited organisms. It is therefore concluded that as a result of the Nile Perch invasion Tilapias were pushed by the Perch downward to find a refuge space where they can be better protected. The bottom habitat was utilized by Tilapias as a refuge and therefore bottom organisms were preyed and indicated in the Tilapias gut contents.

Data presented in Table 1, Table 2 & Table 8 and Figure 3 summarize the ecological changes induced by the Nile Perch invasion and environmental pollution: Elimination of the native fish species (Haplochromines, Tilapias) and prominent increase of the exotic Nile Perch and the pelagic small size native Rastrineobola. The food-web structure was prominently modified: Before 1980 (Figure 3): Optimal balanced food-web, strong grazing activity by planktivorous Haplochromines [17] [18] low levels of algal biomass and consequently Primary Production. no surplus of organic matter, low sedimentation and low level of anoxia. Nevertheless an increase of pollutants input was indicated: N-after 1920 and P-after 1950. Low annual fish catches mostly comprised of small size Haplochromines and overfishing avoidance.

Table 5. Gut Content analysis of Tilapias [15] fished in the Kenyan offshore waters of Lake Victoria. % of specimens contained >50% of their gut content biomass the indicated food form.

Food Component	*T. niloticus* (n = 70)	*T. variabilis* (n = 46)	*T. zillii* (n = 24)
Benthic, Deep Water Mud, Sand	58%	74%	87%

Table 6. Averages (SD) of Maxima and Minima of IOS values of Tilapias: T. variabilis (TV), T. zillii (TZ), T. nilotica (TN), T. esculentus (TE), T. leucostictus (TL), sampled during January-March 1989, in Wynam Gulf at Stations 6, 34, 36 and offshore zone—Stn. 100.

Month	Fish Species	Minima (SD)	Maxima (SD)
1	TV, TZ, TN	0.41 (0.24)	1.62 (1.18)
2	TV, TN	0.54 (0.29)	1.8 (0.92)
3	TZ, TN, TV, TE, TL	0.66 (0.32)	1.78 (1.69)
		Average (SD) 0.55 (0.3)	Average (SD) 1.73 (1.35)

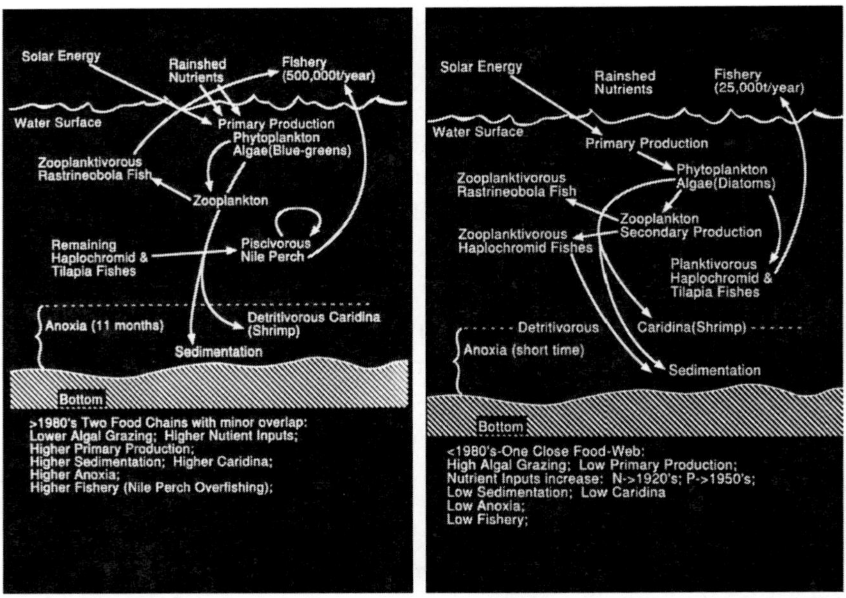

Figure 3. Food-Web structure scheme in Lake Victoria: After 1980 (left) and before 1980 (right) (by courtesy of S.L. Kaufman).

After 1980 (Figure 3): Food web structure exhibit disequilibrium. Grazing level decline due to elimination of grazer Haplochromines and dominance of the predator Nile Perch; Therefore, Increase of non-grazed algal biomass (chlorophyll elevation) followed by higher Primary Production accompanied by accumulation of surplus organic matter and sedimentation resulted in enhanced anoxia. Secchi depth and the thickness of the euphotic zone became shallower as a result of higher densities of suspended matter and decline of light penetration [19] - [21]. Decline of available Silica and SRP (Soluble Reactive Phosphorus), and the decline of TN/TP mass ratio were probably the reason for the replacement of Diatom before 1980 by

Cyanobacteria dominance after 1980 (Table 8). Another outcome of those changes was the proliferation of the detritivorous Atyid-Decapod Caridina nilotica. The total annual landings were significantly enhanced as a result of the optimal conditions for the reproduction and growth of the Nile Perch. The ecological structure modification summarized here gave scientists and managers the basis for further programming fisheries activity. Recent information does not confirm significant improvement in comparison with data presented in this paper.

Table 7. Numbers represent % of fishes contained the indicated food item as dominant (>30% by biomass).

Food Item	TZ	TV	TN
Plantonic			
Caridina	23	7	7
Phytoplankton	20	25	28
Zooplankton			20
Bottom			
Trichoptera (larvae)	21	26	15
Plant Debris	36	20	10
Mud & Sand Grains		12	13
Bottom Dwellers		10	7

Cyanobacteria replaced Diatoms, Secchi Depth became shallower, *i.e.* turbidity enhancement, the euphotic zone became thinner due to increase of turbidity and decline of light penetration, increase of algal density accompanied by elevation of primary production and organic matter content which caused increase of anoxia. Moreover, computation of total lake primary production (annual tons Carbon entire lake assimilation) indicated an increase from 1.7×10^7 during the 1960's to 10.5×10^7 during the 1990's [19] - [21]. Fish landings increased from 25,000 tons to 500,000 tons during the 1960's and 1990's respectively. Consequently % of fish catches relative to Primary Production was increased from 0.01% to 0.05% and the depth of the thermocline became shallower by 30 meters (from 60 to 30 m) [19] - [21]. The food Web pattern of energy flow was prominently modified and Lake Victoria before the 1980's is different from that established after the 1980's (Figure 3). Figures 3-5 represent clearly the changes which enhanced those modifications of the food web structure. The

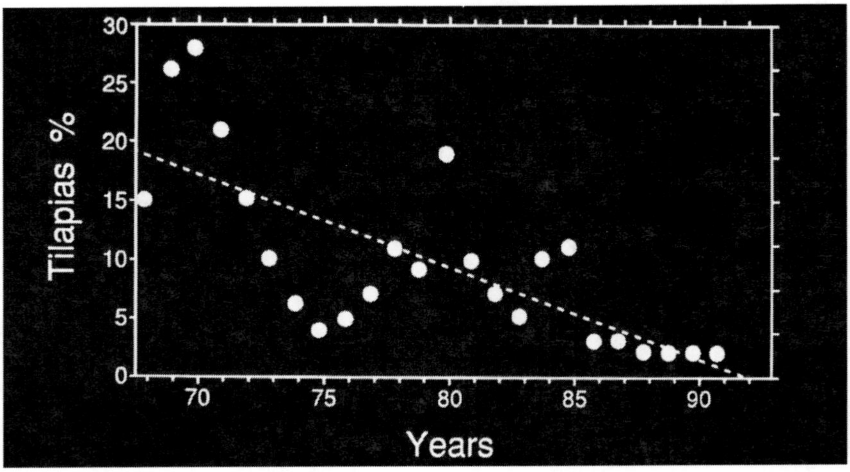

Figure 4. Percentage of Kenya Tilapias annual catches from total landings Vs years (1968-1992). R-Squared = 0.524; p = 0.0001. (Data souece: Kenya Marine and Fisheries Research Institute).

Table 8. Comparative data from Lake Victoria: 1960-Uganda-Talling [4] [5] ; Uganda-1990- [20] [21] Kenya 1990-Gophen this paper (statistical evaluation of [4] [5] , and [20] [21] are given in those articles.

Parameter	Uganda 1960	Uganda 1990	Kenya 1990
NH_4 (ppm: Epil. Hypol.)		0.028 0.056 - 0.12	0.008 - 0.013 0.021 - 0.152
NO_3 (ppm)	0.00 - 0.01	0.011 - 0.084	0.01 - 0.03
SRP (ppm)	0.091 - 0.50	0.012 - 0.043	0.004 - 0.073
TP (ppm)		0.062 - 0.124	0.07 - 0.103
TN (ppm)		0.336 - 0.448	0.44 - 1.16
TN/TP (Wt/Wt)		3.2 - 9.0	4.7 - 18.5
<4.0ppm DO Depth (m)	40	23	23
Silica (ppm)	4.2 - 8.4	0.01 - 2.5	
Chlorophyll (μg/L)	2 - 4	10 - 22	
Primary Production (mgO_2/m^3)	100 - 130	180 - 600	
Secchi Depth	2.5	1.2	1.8
Euphotic Zone	0 - 15	0 - 7.5	
Dominant algae	Diatoms	Cyanobacteria	Cyanobacteria

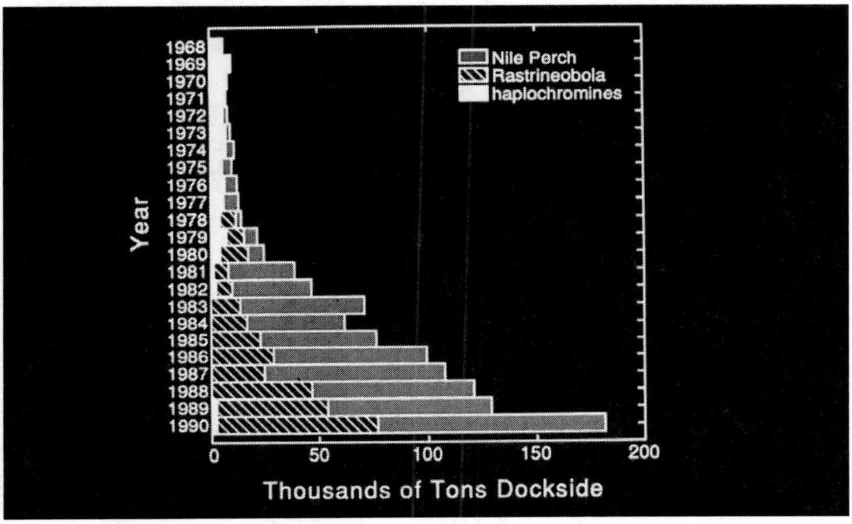

(a)

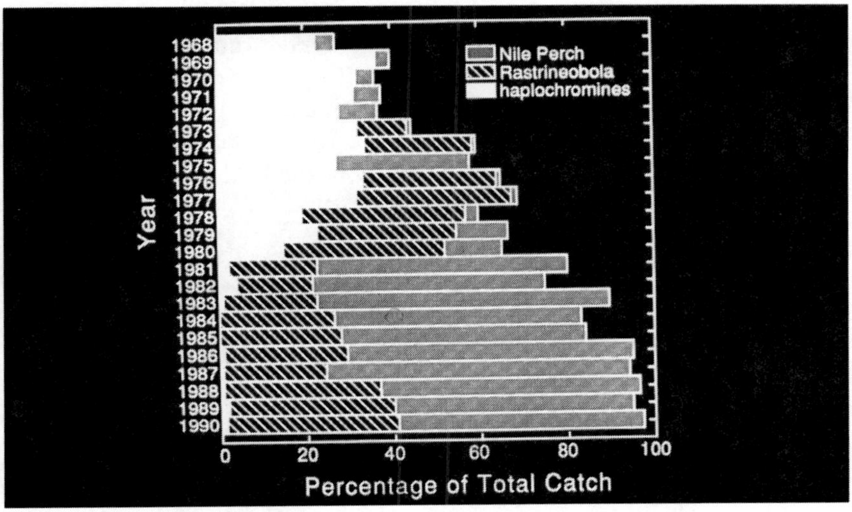

(b)

Figure 5. (a) Kenya annual (1968-1990) Landings (103 tons/year) of Nile Perch, Rastrineobola and Haplochromines. (by courtesy of S.L.Kaufman) Data: Kenya Marine and Fisheries Research Institute; (b) Annual (1968-1990) commercial catch compositions (%) in Lake Victoria, Kenya. (by courtesy of S.L.Kaufman). Data: Kenya Marine and Fisheries Research Institute.

fish faunal composition was modified from the dominant native Haplochromines which were replaced by the exotic Nile Perch and native Rastrineobobla [22]. Lake Victoria is presently exhibit traditional symptoms of eutrophication: decline of water transparency, Cyanobacteria are dominant, elevation of P and decline of N levels, increase of algal biomass (chlorophyll concentration), shallower Secchi depth and thinner euphotic zone and anoxia enhancement. Future management design of fishery in lake Victoria must therefore be done very carefully. Distinct objectives are proposed: 1) reduction of pollutant inputs resulted by anthropogenic activity; 2) increase level of phytoplankton grazing. The second target is difficult due to the intensive predation pressure produced by the Nile Perch. Relevant fish species might be planktivorous large body Tilapias. Tilapias of native and exotic origin are known to be able to survive under the new habitat if may introduce to the lake as large un-preyed individuals. The concept of the Nile Perch predation impact and its ecological implications is also confirmed by the elimination of the Haplochromines's planktivory (Table 4). Nevertheless, the small body size of the Haplochromine fishes make them vulnerable for predation and therefore not suitable for management if aimed at enhancement of planktivory. Large body size of the planktivorous Tilapias are less vulnerable to Nile Perch predation. Index Of Satiation calculated for Tilapias species collected from the natural lake population indicated high values (Table 6) in comparison with the range of 0.5 - 4.0 that was obtained from farm cultured fish given an optimally food quantity [23]. Consequently I assume that the planktivorous capabilities of Tilapias might be beneficial for ecological recovery of Lake Victoria.

5. SUMMARY

The Lake Victoria ecosystem was unique included above 400 endemic species of Haplochromine fishes. The food web structure was naturally balanced during that time with short periods of anoxia in deep waters and dominance of diatomides algal species. Nile Perch (Lates niloticus) was introduced and during the 1980's became the dominant fish. The Haplochromine species were deleted and the whole ecosystem was modified. Algal assemblages were changed to Cyanobacteria, anoxia became more frequent and in shallower waters. Primary production was enhanced and excess of organic matter was settled to the bottom. A study of Tilapias feeding habits, zooplankton and Phytoplankton distribution was carried out and results indicates an ecological devastation trait of the ecosystem. Ideas for future management are presented.

ACKNOWLEDGEMENTS

I wish to give warm thanks to the Kenyan and American collaborators:

Kenya: E. Okemwa, J. Ogari, S. Agembe, E. Odada, W. Owieko, J. Manyala, J. Ochuko.

USA: L. S. Kaufman, and W. Cooper.

Since this study was completed, three of the principal investigators and major leaders of the research team were deceased: P. B. O. Ochumba, U. Pollingher and P. Kilham. Without their contribution, the study could not be accomplished. I dedicate this paper to their unforgettable remembered collaborative personality and scientific skillful.

REFERENCES

1. Barel, C.D.N., Dorit, R., Greenwood, P.H., Fryer, G., Hughes, N., Jackson, P.B.N., Kawanabe, H. and Lowe-McConell, R.H. (1985) Destruction of Fisheries in Africa's Lakes. Nature, 325, 19-20. http://dx.doi.org/10.1038/315019a0
2. Ochumba, P.B.O., Gophen, M. and Pollingher, U. (1994) Ecological Changes in Lake Victoria after the Introduction of Nile Perch (Lates niloticus): The Catchment, Water Quality and Fisheries Management. In: Cowx, I.G., Ed., Rehabilitation of Freshwater Fisheries, Fishing News Books, Blackwell Scientific Publications Ltd., 338-347.
3. Gophen, M., Ochumba, P.B.O. and Kaufman, L.S. (1995) Some Aspects of Perturbations in the Structure and Biodiversity of the Ecosystem of Lake Victoria. Aquatic Living Resources, 8, 27-41. http://dx.doi.org/10.1051/alr:1995003
4. Talling, J.F. (1965) The Pjtosymthetic Cativity of Phtoplankton in East African Lakes. Internationale Revue der gesamten Hydrobiologie und Hydrographie, 50, 1-32. http://dx.doi.org/10.1002/iroh.19650500102
5. Talling, J.F. (1966) The Annual Cycle of Stratification and Phytoplankton Growth in Lake Victoria (East Africa). Internationale Revue der gesamten Hydrobiologie und Hydrographie, 51, 545-621.
6. Akiyama, T., Kajumulo, A. and Olsen, S. (1977) Seasonal Variation of Plankton and Physico-Chemical Condition in Mwanza Gulf, Lake Victoria. Bulletin Freshwater Fisheries Research Laboratory, 27, 49-61.
7. Melack, J.M. (1979) Photosynthetic Rates in Four Tropical African Freshwatrs. Freshwater Biology, 9, 555-571. http://dx.doi.org/10.1111/j.1365-2427.1979.tb01539.x
8. LBDA (Lake Basin Development Authority) (1984) Technical Report on Current and Future Implications of Development to the Aquatic Environment of Lake Victoria. Wynam Gulf Baseline Study Report, LBDA, Kisumu, 55 p.
9. Ochumba, P.B.O. and Kibara, D. (1989) Observation on Blue-Green Algal Blooms in the Open Waters of Lak Victoria, Kenya. African Journal of Ecology, 27, 23-34. http://dx.doi.org/10.1111/j.1365-2028.1989.tb00925.x
10. Rzoska, J. (1956) The Planktonic Crustacea of Lake Victoria. Proceedings of the Linnean Society of London, 168, 116-125. http://dx.doi.org/10.1111/j.1095-8312.1957.tb00774_116.x
11. Rzoska, J. (1957) Notes on the Crustaceans of Lake Victoria. Proceedings of the Linnean Society of London, 168, 116-125.

12. Rzoska, J. (1978) Zooplankton of the Nile System. In: Rzoska, J., Ed., The Nile, Biology of an Ancient River, Dr Junk Publishers, The Hgue, 333-343.
13. Mavuti, K.M. and Litterick, M.R. (1991) Composition, Distribution and Ecological Role of Zooplankton Community in Lake Victoria, Kenia Waters. Verhandlungen des Internationalen Verein Limnologie, 24, 1117-1122.
14. Green, J. (1971) Association of Cladocera in the Zooplankton of the Lake Sources of the White Nile. Journal of Zoology, 165, 373-414. http://dx.doi.org/10.1111/j.1469-7998.1971.tb02195.x
15. Fryer, G. and Iles, T.D. (1972) The Cichlid Fishes of the Great Lakes of Africa: Their Biology and Evolution. Oliver and Boyd, Edinburgh, 641 p.
16. Kaufman, L.S. (1991) The Anatomy of Vertebrate Mass Extinction as Revealed by a Faunal Survey of Lake Victoria. Annual Report, New England Aquarium, Boston, 51 p.
17. Greenwood, P.H. (1974) The Haplochromine Fishes of Lake Victoria, East Africa: The Biology and Evolution of a Species Flock. Bulletin of the British Museum of Natural History Zoology, 6, 1-134.
18. Greenwood, P.H. (1981) The Haplochromine Fishes of East African Lakes: Taxonomy, Biology and Evolution. Kraus International Publications, Munich, 839 p.
19. Silsbe, G.M., Hecky, R.E., Guildford, S.J. and Mugidde, R. (2006) Variability of Chlorophyll A and Photosynthetic Parameters in a Nutrient-Saturated Tropical Great Lake. Limnology and Oceanography, 51, 2052-2063 http://dx.doi.org/10.4319/lo.2006.51.5.2052
20. Kling, H.J., Mugidde, R. and Hecky, R.E. (2001) Recent Changes in the Phytoplankton Community of Lake Victoria in Response to Eutrophication. In: Munawar, M. and Hecky, R.E., Eds., The Great Lakes of the World (GLOW): Food Web, Health and Integrity, Backhuys Publishers, Leiden, 47-65.
21. Hecky, R.E., Mugidde, R., Ramlal, P.S., Talbot, M.R. and Kling, G.W. (2010) Multiple Stressors Cause Rapid Ecosystem Change in Lake Victoria. Freshwater Biology, 55, 19-42. http://dx.doi.org/10.1111/j.1365-2427.2009.02374.x
22. Kaufman, L.S. (1992) Catastrophic Change in Species-Rich Freshwater Ecosystems: The Lesson of Lake Victoria. BioScience, 42, 846-858. http://dx.doi.org/10.2307/1312084
23. Shi, Z.F., Liu, Z.Y., Li, J.L. and Mei, Z.P. (1997) Effects of Dissolved Oxygen on the Feeding Intensity of Silver Carp (H. molitrix), and Bighead Carp (A. nobilis). In: Mathias, J.A., Charles, A.T. and Baotong, H., Eds., Proceedings of a Workshop on Integrated Fish Farming, CRC Press, LLC, Boca Raton, 205-212.

CHAPTER 13

Initial Response of Fish Trophic Niche To Hydrological Alteration in The Upstream of Three Gorges Dam

Jianzhu Wang[1],Lei Li[1],Jun Xu[2] andBinhe Gu[1,3*]

[1]*Engineering Research Center of Eco-environment in Three Gorges Reservoir Region, Ministry of Education, China Three Gorges University, Yichang 443002, China*

[3]*Soil and Water Science Department, University of Florida, Gainesville, FL, USA*

ABSTRACT

Introduction

Damming is one of the dramatic impacts to river food webs due to habitat fragmentation and changes in the hydrological regimes. The world's largest dam, the Three Gorges Dam (TGD), has been constructed in the middle of Yangtze River since 2003. The objective of this study was to understand the effects of altered hydrological regime on the trophic structure of fish food web along the hydrological gradient in the TGD.

Methods

Fish samples were collected from an upstream, a midstream and a downstream site of the Yangtze River upstream of the TGD in May and September of 2004. Muscle tissue of each fish species was determined for stable isotopic compositions ($\delta^{13}C$ and $\delta^{15}N$) which were used to calculate isotope-based Bayesian community-wide trophic metrics.

Results

Stable isotope trophic niche analysis reveals reduced utilization of basal resource and trophic niche space at the midstream and downstream sites. By contrast, community trophic diversity and species redundancy were higher at the downstream than the upstream and midstream, likely as the result of reduced the spectrum of basal resources. These findings suggested that the negative impacts on the overall trophic niche space from reduced flow, increased water depth and stagnancy occurred only one year after the completion of the TGD.

Conclusions

Altered hydrological regimes posed multiple impacts to the aquatic food web. Reduced flow and increased water depth within TGD has posed initial effects to the fish community trophic structure. Further collections ana analysis of ecological data to compare the fish trophic structure revealed in this study are needed to assess the long-term impacts of TGD on fish community structure and resources utilization.

Keywords: Three Gorges Dam (TGD) Fish $\delta^{13}C$ $\delta^{15}N$ Trophic niche Yangtze River

INTRODUCTION

River damming is a dramatic anthropogenic factor affecting freshwater environments (Dynesius & Nilsson 1994; Xu et al. 2013). Damming leads to habitat fragmentation, changes in water depth, alter the rates, seasonal timing, and durations of flow. These changes may in turn affect the physical, chemical, and biological features of the river systems (Sternberg 2006; Dudgeon 2011). Numerous studies on the negative impacts of damming to the river ecosystem have been reported around the world although damming may provide other benefits to the human societies (Finer and Jenkins 2012; McClain 2013).

The construction of the Three Gorges Dam (TGD) at the middle of Yangtze River in south-central China would also have a series of environmental consequences (Wu et al. 2003; Fu et al. 2003; Xu et al. 2013). Since TGD started storing water in late 2003, the water depth increased by 66 m. The formerly narrow channel with torrential water flow has been converted to extensive stagnant water similar to the limnetic zone of a large lake system as increases in the water depth, alternations of hydrology, and other physical, chemical, and biological characteristics of Three Gorges Reservoir (TGR) and the downstream (Liu and Cao 1992; Salazar 2000; Wu et al. 2003; Xu et al. 2013).

The Yangtze River has high diversity for freshwater fish species (Xie 2003). The changes in the ecological environment have been reported to affect the fish community of TGR and the downstream waters (Fu et al. 2003; Xie 2003; Chen et al. 2009). Gao et al. (2010) reported the impacts of TGD to the fish community structure such as shift of fish species from lotic to lentic dominance. Chen et al. (2011) reported changes in fish assemblages following the completion of TGD. Other studies focused on the reproduction of major commercially important fish such as the major Chinese carp (Zhang et al. 2012; Hu et al. 2014) or endangered species (Chen et al. 2011; Zhang et al. 2014) while some studies focused on heavy metal accumulation (Cai et al. 2012; Yi and Zhang 2012) or conservation biology (Zhang et al. 2014; Ye et al. 2014). There is a severe lack of research work on trophic ecology of fish related to the TGD construction. Zhang et al. (2007) reported the suspended organic matter as the major sources of a planktivorous fish in Yangtze River. Wang et al. (2014) analyzed carbon sources for select invertebrates and fish at three river sites upstream of the TGD.

Stable isotope analysis has become a powerful tool for the study of aquatic food web (Kling et al. 1992; Vander Zanden 2001). Early stable isotope studies on fish community were focused on dietary sources, feeding relationships, and trophic levels (Cabana and Rasmussen 1996; Gu et al. 1996; Pinnegar and Polunin 2000). Recently, stable isotope data from aquatic consumers have been used to conduct trophic niche analysis (Bearhop et al. 2004; Newsome et al. 2012). Layman et al. (2007) proposed six community-wide trophic metrics based on fish assemblages. The six niche metrics are measures of the total extent of spacing and trophic redundancy within a $\delta^{13}C$ and $\delta^{15}N$ bi-plot. This approach is considered as a more in-depth utility of stable isotope data for the analysis of consumer's trophic niche. Jackson et al. (2011) proposed a Bayesian framework to standardize the calculation of these trophic niche metrics, which has been applied to several studies of terrestrial (Perkins et al. 2014) and aquatic trophic niches (Jackson et al. 2011; Abrantes et al. 2014).

We analyzed the stable carbon ($\delta^{13}C$) and nitrogen ($\delta^{15}N$) isotopic compositions of fish and invertebrates in three sites with different hydrological regimes located in the upstream of the TGD. Since damming of rivers is the most dramatic anthropogenic factor affecting freshwater environments (Dynesius and Nilsson 1994; Dudgeon, 2000; 2011), the construction of the TGD would also have severe environmental consequences (Park et al. 2003: Dudgeon 2011). The main objective of this study was to compare the stable isotope niches and the fish trophic positions of the fish collected in the three study sites of Yangtze River. Information from this study may assess the initial changes of hydrology and the resultant ecological changes which may affect fish community structure as the result of the TGD construction. Findings may provide important insights of the changes in fish trophic structure and will be useful for wildlife conservation and fisheries management of the TGD and dammed rivers in the world.

METHODS

Description of study sites

The Yangtze River is the largest river in China and the third longest river in the world, with a length of 6300 km and a drainage basin of 1,808,500 km^2 (Fig. 1). Tremendous periodic flood events occur in the summer with a dry period during the winter and spring. The Three Gorges Dam (TGD), which is the world's largest dam, has been constructed in the middle Yangtze River in south-central China. As a result, the water depth upstream of the dam had increased by 66 m since the Three Gorges Reservoir (TGR) began storing water in late 2003. The formerly narrow channel characterized by torrential water flow has been converted to an extensive stagnant water body similar to the limnetic zone of a large lake system, with corresponding increases in the water depth, alternations of the flow rate, dissolved oxygen, light availability, and temperature in the TGR (Table 1).

Sample collection and processing

Detail methods for sample collection and processing can be found in Wang et al. (2014). Briefly, all samples were collected in two trips to the study sites in September 2004 and May 2005 which represent the wet and dry periods, respectively. The validity of this study is first based on the assumption that equally effective sampling effort was placed and resulted in similar probability of harvesting a fish species across sites. This is valid

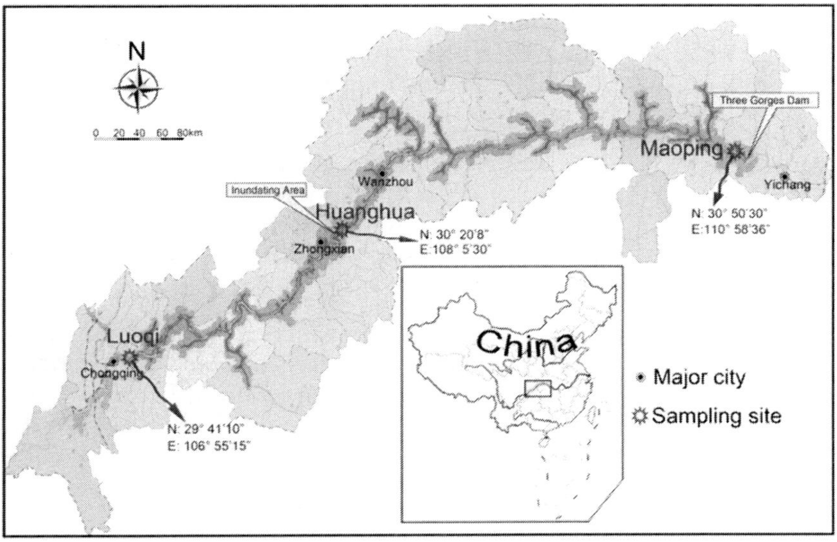

Figure. 1 Map of Three Gorges Reservoir area showing the three study sites and major cities

Table 1 Averages of selected environmental variables during the study period (2004–2005) at river sections near the three sampling sites along Yangtze River. Data are taken from Wang et al. (2014)

Variable	Upstream	Midstream	Downstream
Flow rate, m s^{-1}	1.75	0.36	0.17
TP, mg L^{-1}	0.216	0.145	0.138
TN, mg L^{-1}	1.665	1.494	1.357
pH	7.36	7.9	8.2
DO, mg L^{-1}	7.28	8.04	8.97
Conductivity, μs cm^{-1}	248	372	279
COD, mg L^{-1}	13.64	10.95	16.18
Chl a, mg m^{-3}	2.9	14.7	15.0
Number of fish species	22	12	14

because of the same fishing tools and similar time spent on each site. The middle Yangtze River hosted a high range of fish species, and our collection only reflected a portion of the common species present in this region.

Epiphytic algae were collected from substrate surface and rinsed repeatedly in deionized water to remove sediment particles and detritus. Particulate organic matter (POM) was collected from several depths with a water bottle along five cross-transects at each site. In the laboratory, water was passed through coarse (100 μm) sieves and then filtered onto precombusted Whatman GF/C (1.2 μm) glass fiber filters.

The invertebrates collected included zooplankton, snails, the stream crab (*Sinopotamon yangtsekiense*), and the river shrimp (*Macrobranchium nipponense*). Zooplankton were obtained from vertical hauls of a 100-μm plankton net from each of the five transects at each station. The snails, stream crabs, and river shrimp were hand-collected or captured with fish traps. Shells and gut contents were removed, and the individuals were rinsed with distilled water.

A total of 27 species of fish consisting of three to four individuals of each species (Appendix 1) were collected utilizing various fishing tools. For each fish captured, the total length was measured and recorded, and the dorsal white muscle tissue was removed and placed on ice in the field. In laboratory, all samples were dried at either 80 °C for 24 h (plant organic matter) or 60 °C for 48 h (consumers) and ground into fine power using a mortar and pestle, or an agate mill for vascular plant samples. All equipment were washed with ethanol and dried between samples.

Stable isotope analysis

The $^{13}C/^{12}C$ and $^{15}N/^{14}N$ ratios were determined using Thermo Finnigan MAT DELTA^{plus} XP isotope-ratio mass spectrometers at the Stable Isotope Laboratory for Ecological and Environmental Research at the Research Center for Plant Ecology, Beijing Institute of Botany, the Chinese Academy of Sciences. The isotope ratio is expressed in the conventional delta (δ) notation, defined as the per mil (‰) deviation from the isotope standard, Peedee Belemnite (PDB) formation and atmospheric N_2 for $^{13}C/^{12}C$ and $^{15}N/^{14}N$, respectively. Routine precision for replicate samples was ± 0.1 ‰ for $\delta^{15}N$ and ± 0.3 ‰ for $\delta^{13}C$.

Estimate of stable isotope enrichments between fish and diets

We investigated the diet compositions of yellow catfish (*Pelteobagrus fulvidraco*) and the Amur catfish (*Silurus asotus*) by gut content analysis.

The food components were parted into terrestrial plants, plant debris, phytoplankton, benthic algae, zooplankton, shrimp, mollusk, and fish, and the proportion in bulk diet and their stable isotope values were measured for each of them (Table 2). The stable isotope values of bulk diets could be calculated as follow:

Table 2 The dietary compositions of *P. fulvldraco and C. heterodon* revealed by gut content analysis and the stable isotope signatures

Food sources	Diets of *P. fulvldraco*			Diets of *C. heterodon*		
	%	$\delta^{13}C$	$\delta^{15}N$	%	$\delta^{13}C$	$\delta^{15}N$
Terrestrial plants	-	-	-	-	-	-
Plant debris	-	-	-	14.5	−25.1	1.7
Phytoplankton	8.5	−23.8	5.3	23.0	−23.2	4.2
Beneath algae	-	-	-	8.5	−22.9	10.3
Zooplankton	27.0	−22.9	6.7	21.5	−23.7	6.2
Shrimp	12.5	−21.5	8.7	-	-	-
Mollusk	32.5	−22.1	7.4	27.5	−22.6	6.9
Fish	19.5	−20.7	11.4	5.0	−21.5	9.2
Bulk diet		−22.1	8.2		−22.5	6.6
Discrimination (Δ)		+0.47	3.3		−0.51	2.9

"-" indicates that the food items were not found in the consumers' guts

$$\delta^{15}\mathrm{N}_{\mathrm{bulk-diet}} = \sum f_i \times \delta^{15} N_i \qquad (1)$$

where $\delta^{15}N_{\text{bulk-diet}}$ were the stable isotope values of fish; the subscript *i* represents one food source, *f* represents the fractional contribution of each food source to fish diets, and $\delta^{15}N_i$ represents the stable isotope values for each food source. The isotope enrichment ($\Delta\delta^{15}N$) between fish and their diets is calculated as follow:

$$\Delta\delta^{15}\mathrm{N} = \delta^{15}\mathrm{N}_{\mathrm{consumer}} - \delta^{15}\mathrm{N}_{\mathrm{bulk-diet}} \qquad (2)$$

There is a general agreement in that $\delta^{15}N$ increases by 2.8 to 3.8 ‰ per trophic level upward ($\Delta\delta^{15}N$) in freshwater systems (Minagawa and

Wada 1984; Post 2002). We calculated the isotope fractionation factor ($\Delta\delta^{15}N$) from yellow catfish and the Amur catfish and their diets, and ^{15}N-enrichment per trophic level is about 3.1 ‰.

Data analysis

Site-specific stable isotope data from the two collection events were pooled prior to further data analysis. This is done because no significant differences in $\delta^{13}C$ or $\delta^{15}N$ of fish between the two sample events for each site were found (ANOVA, all $p > 0.05$). Trophic position (λ) of each fish species was calculated with following equation (Post 2002):

$$\lambda = \left(\delta^{15}N_{consumer} - \delta^{15}N_{base}\right)/\Delta\delta^{15}N + 2 \quad (3)$$

where λ is trophic position of a species of fish. $\delta^{15}N_{consumer}$ is the average $\delta^{15}N$ of the fish, $\delta^{15}N_{base}$ is the average $\delta^{15}N$ of the baseline organism. We used average $\delta^{15}N$ of grass carp (*Ctenopharyngodon idellus*) and Wuchang bream (*Megalobrama amblycephala*) which are true herbivores as the baseline. $\Delta\delta^{15}N$ is the isotope enrichment factor. The trophic level of the consumers used to estimate the baseline is 2.

Six community-wide trophic metrics proposed by layman et al. (2007) were used to compare trophic niche between the study sites. The $\delta^{13}C$ range (CR) is the difference between the species with the most enriched and most depleted $\delta^{13}C$ ratio and is a measure of basal resource diversity. The $\delta^{15}N$ range (NR) is the difference between the species with the most enriched and most depleted $\delta^{15}N$ ratio and is a measure of trophic length within a food chain. The total area (TA) of the $\delta^{13}C$ and $\delta^{15}N$ bi-plot space (convex hull) describes the total space occupied by a community. The mean distance to the $\delta^{13}C$-$\delta^{15}N$ centroid (CD) provides a measure of trophic diversity. The mean nearest neighbor distance (MNND) is the mean of the Euclidean distances to each species' nearest neighbor in bi-plot space and thus a measure of the overall density of species packing. Small MNND suggests increased trophic redundancy. Finally, the standard deviation of nearest neighbor distance (SDNND) is a measure of the evenness of species packing in bi-plot space. Low SDNND values mean more even distribution of trophic niches. $\delta^{13}C$ vs. $\delta^{15}N$ biplots of each fish species in each site with standard ellipse areas (SEA) and convex hull (TA) were used to give the general figures for stable isotope distribution among sites. SEA was estimated using multivariate ellipse based metrics, which are bivariate equivalents to standard deviations in univariate analysis (Jackson et al. 2011). These analyses were performed using the R statistical computing packages (R Development Core Team, 2007).

RESULTS

Environmental variables and species composition of fish

Select hydrological, chemical, and biological variables from the three study sites are present in Table 1. The hydrological flow rate before the construction of the TGD was high across the study sites (Wang et al. 2014) but was greatly reduced after the completion of the dam to nearly stagnant at the downstream site. Nutrient (TP and TN) concentrations were high, indicating that all study sites were eutrophic, which is also supported by the moderately higher pH, DO, and COD and chlorophyll, a concentration in the midstream and downstream sites.

A total of 27 fish species were collected from the three study sites with species abundance highest at the upstream (22 species), moderate at downstream (14), and lowest at midstream (12). Chinese perch (*Siniperca* spp.) and largemouth bronze gudgeon (*Coreius guichenoti*) are on the national protected fish list which appear in all of the three study sites (Tables 3–5). However, ten species collected at the upstream site were not found at the midstream, and eight species were not found in the downstream sites and most of these fish are omnivores occupying the benthic habitat. The silver carp (*Hypophthalmichthys molitrix*) which is a planktivore living in the upper water column was collected at the midstream and downstream sites, but not at the upstream site. The longsout catfish (*Leiocassis longirostris*) and prenant's schizothoracin (*Schizothorax prenanti*) were collected from the downstream site only.

Table 3 Average, standard deviation (SD) of $\delta^{13}C$, $\delta^{15}N$ ratios, and trophic level for all consumer taxa at the upstream site (Luoqi) in Three Gorges Reservoir. Data were pooled from the wet and dry periods between 2004 and 2005

Species	$\delta^{13}C$	$\delta^{15}N$	Trophic level
Ctenopharyngodon idellus	−24.1 ± 0.5	5.5 ± 0.3	2.0 ± 0.2
Megalobrama amblycephala	−21.9 ± 0.6	8.7 ± 0.3	2.0 ± 0.1
Cyprinus carpio	−20.6 ± 0.6	7.2 ± 05	2.0 ± 0.1
Carassius auratus	−21.2 ± 2.1	8.1 ± 0.6	2.3 ± 0.2
Coreius guichenoti	−21.9 ± 0.5	8.6 ± 0.6	2.5 ± 0.3
Hemibarbus maculatus	−21.5 ± 0.4	9.5 ± 0.5	2.8 ± 0.1

Pelteobagrus fulvidraco	−21.6 ± 0.8	9.7 ± .15	2.8 ± 0.4
Distoechodon tumirostris	−21.1 ± 0.4	9.3 ± 0.4	2.7 ± 0.2
Hemiculterella sauvagei	−23.3 ± 0.8	8.4 ± 0.4	2.4 ± 0.1
Abbottina guentheri	−21.1 ± 0.6	8.3 ± 0.3	2.4 ± 0.1
Hemimzon sinensis	−22.3 ± 0.3	9.0 ± 0.5	2.6 ± 0.1
Leptobotia elongate	−21.9 ± 0.7	9.9 ± 0.7	2.9 ± 0.4
Misgurnus anguillicaudatus	−23.0 ± 0.5	9.3 ± 0.7	2.7 ± 0.1
Schizothorax prenanti	−22.6 ± 0.5	9.4 ± 0.4	2.7 ± 0.2
Botia superciliaris	−21.9 ± 0.4	10.9 ± 0.4	3.2 ± 0.1
Gobiobotia Xenophysogobio	−21.2 ± 0.9	8.9 ± 0.5	2.6 ± 0.3
Saurogobio dabryi	−22.6 ± 0.5	8.3 ± 0.3	3.0 ± 0.1
Rhinogobio typus	−12.0 ± 0.1	8.3 ± 0.2	2.4 ± 0.1
Silurus asotus	−21.1 ± 0.9	10.9 ± 1.8	3.2 ± 0.5
Ophiocephalus argus	−25.6 ± 0.9	10.3 ± 0.7	3.0 ± 0.3
Siniperca chuatsi	−24.0 ± 0.8	11.7 ± 0.4	3.5 ± 0.2
Culter erythropterus	−21.0 ± 0.7	11.9 ± 0.4	3.5 ± 0.1

Table 4 Averages, standard deviation (SD) of δ^{13}C and δ^{15}N ratios for all consumer taxa at the midstream site (Huanghua) in Three Gorges Reservoir. Data were pooled from the wet and dry periods between 2004 and 2005

Species	δ^{13}C	δ^{15}N	Trophic level
Ctenopharyngodon idellus	−23.3 ± 0.4	7.2 ± 0.3	2.0 ± 0.1
Megalobrama amblycephala	−23.3 ± 0.3	8.8 ± 0.4	2.0 ± 0.1
Hypophthalmichthys molitrix	−21.7 ± 0.6	7.7 ± 0.3	1.8 ± 0.1
Cyprinus carpio	−22.1 ± 0.6	8.1 ± 0.6	1.9 ± 0.2
Carassius auratus	−21.3 ± 0.8	10.4 ± 0.4	2.7 ± 0.1
Coreius guichenoti	−21.5 ± 0.5	9.5 ± 0.3	2.4 ± 0.1
Hemibarbus maculatus	−22.5 ± 0.7	9.8 ± 0.3	2.5 ± 0.1
Pelteobagrus fulvidraco	−21.2 ± 1.2	10.6 ± 1.4	2.8 ± 0.4
Distoechodon tumirostris	−22.0 ± 0.2	11.6 ± 0.3	3.1 ± 0.2
Hemiculterella sauvagei	−22.5 ± 0.6	9.1 ± 0.3	2.2 ± 0.1
Silurus asotus	−21.4 ± 0.6	12.7 ± 0.9	3.4 ± 0.2
Siniperca chuatsi	−23.5 ± 0.9	12.7 ± 0.7	3.4 ± 0.2

Table 5 Averages, standard deviation (SD) of δ^{13}C and δ^{15}N ratios for all consumer taxa at the upstream site (Maoping) in Three Gorges Reservoir. Data were pooled from the wet and dry periods between 2004 and 2005

Species	δ^{13}C	δ^{15}N	Trophic level
Ctenopharyngodon idellus	−23.1 ± 0.4	8.0 ± 0.4	2.0 ± 0.2
Megalobrama amblycephala	−21.5 ± 0.3	6.6 ± 0.4	2.3 ± 0.1

Hypophthalmichthys molitrix	−22.8 ± 0.6	7.6 ± 0.3	2.1 ± 0.1
Cyprinus carpio	−21.9 ± 0.3	9.5 ± 0.7	2.7 ± 0.2
Carassius auratus	−22.8 ± 1.1	10.1 ± 0.4	2.9 ± 0.2
Coreius guichenoti	−22.9 ± 0.4	10.4 ± 0.5	3.0 ± 0.1
Pelteobagrus fulvidraco	−23.7 ± 0.9	13.0 ± 1.1	3.8 ± 0.4
Hemiculterella sauvagei	−21.4 ± 0.6	7.8 ± 0.5	2.2 ± 0.1
Ophiocephalus argus	−23.2 ± 1.0	12.5 ± 0.5	3.7 ± 0.2
Schizothorax prenanti	−23.4 ± 0.6	13.0 ± 0.5	3.8 ± 0.1
Silurus asotus	−22.2 ± 1.0	12.6 ± 0.9	3.7 ± 0.2
Siniperca chuatsi	−22.4 ± 0.3	13.5 ± 0.5	4.0 ± 0.1
Culter erythropterus	−22.0 ± 0.4	12.4 ± 0.4	3.6 ± 0.1
Leiocassis longirostris	−22.2 ± 0.6	13.0 ± 0.5	3.9 ± 0.1

Appendix 1

Name, common name, and trophic role for all fish taxa at the three study sites of the middle Yangtze River

Scientific name	Common name	Trophic role
Ctenopharyngodon idellus	Grass carp	Herbivore
Megalobrama amblycephala	Wuchang bream	Herbivore

Hypophthalmichthys molitrix	Silver carp	Planktivore
Cyprinus carpio	Common carp	Omnivore
Carassius auratus	Crucian carp	Omnivore
Coreius guichenoti	Largemouth bronze gudgeon	Omnivore
Hemibarbus maculatus	Chinese minnow	Omnivore
Pelteobagrus fulvidraco	Yellowhead catfish	Omnivore
Distoechodon tumirostris	Round snout	Omnivore
Hemiculterella sauvagei	Spotted steed	Omnivore
Abbottina guentheri	Smallmouth gudgeon	Omnivore
Hemimzon sinensis	Chinese hillstream loach	Omnivore
Leptobotia elongate	Elongate loach	Omnivore
Misgurnus anguillicaudatus	Weather loach	Omnivore
Schizothorax prenanti	Prenant's schizothoracin	Omnivore
Botia superciliaris	Chinese sand loach	Omnivore
Xenophysogobio boulengeri	Ray finned carp	Omnivore
Saurogobio dabryi	Chinese lizard gudgeon	Omnivore
Rhinogobio typus	Longnose gudgeon	Omnivore
Silurus asotus	Amur catfish	Carnivore
Ophiocephalus argus	Snakehead	Carnivore

Siniperca chuatsi	Chinese perch	Piscivore
Culter erythropterus	Redfin culter	Piscivore
Leiocassis longirostris	Longsnout catfish	Piscivore

Trophic level

The diets of *P. fulvldraco* and *S. asotus* were identified and analyzed for stable isotope composition along with their muscle tissue. The trophic fractionation factor ($\Delta\delta^{15}N$) from *P. fulvldraco* and *S. asotus* over their diets were 3.3 and 2.9 ‰, respectively, and the average (3.1 ‰) is used as the $\Delta\delta^{15}N$ for the estimates of trophic position for each species. Although the $\Delta\delta^{15}N$ values were based on the instant stomach content analysis and apparently did not consider seasonal variations in dietary composition, the average is similar to those reported previously (Post 2002).

The fish collected during this study may be classified into one of five general trophic guilds: herbivore, planktivore, omnivore, carninvore, and piscivore but were dominated by benthic omnivorous fish as other rivers. (Tables 3, 4, and 5). The trophic level of the fish assemblages at the three study sties ranged from 1.8 to 4.5 with an average of 2.7, 2.5, and 3.0 for the upstream, midstream, and downstream, respectively (Fig. 2). The grass carp and Chinese bream which appeared in all study sites were the only true herbivorous fish relying on large aquatic plants, and they are assigned as trophic level 2 fish. Silver carp which was only found in the midstream and downstream sites had a trophic level about 2, indicating that they relied on primary producers. Most other fish were omnivores, typically positioned between trophic levels of 2 and 3.5. Species which were collected from all study sites often tended to increase in trophic level from upstream to downstream. The high trophic level was especially apparent at the fish collected from the downstream site which were often 0.5 trophic level higher than the same species from the upstream site. Yellow catfish was a full trophic level higher than those from the upstream and midstream sites. However, the trophic level of *Hemimzon sinensis* at the downstream site was slightly lower than those form the other sites.

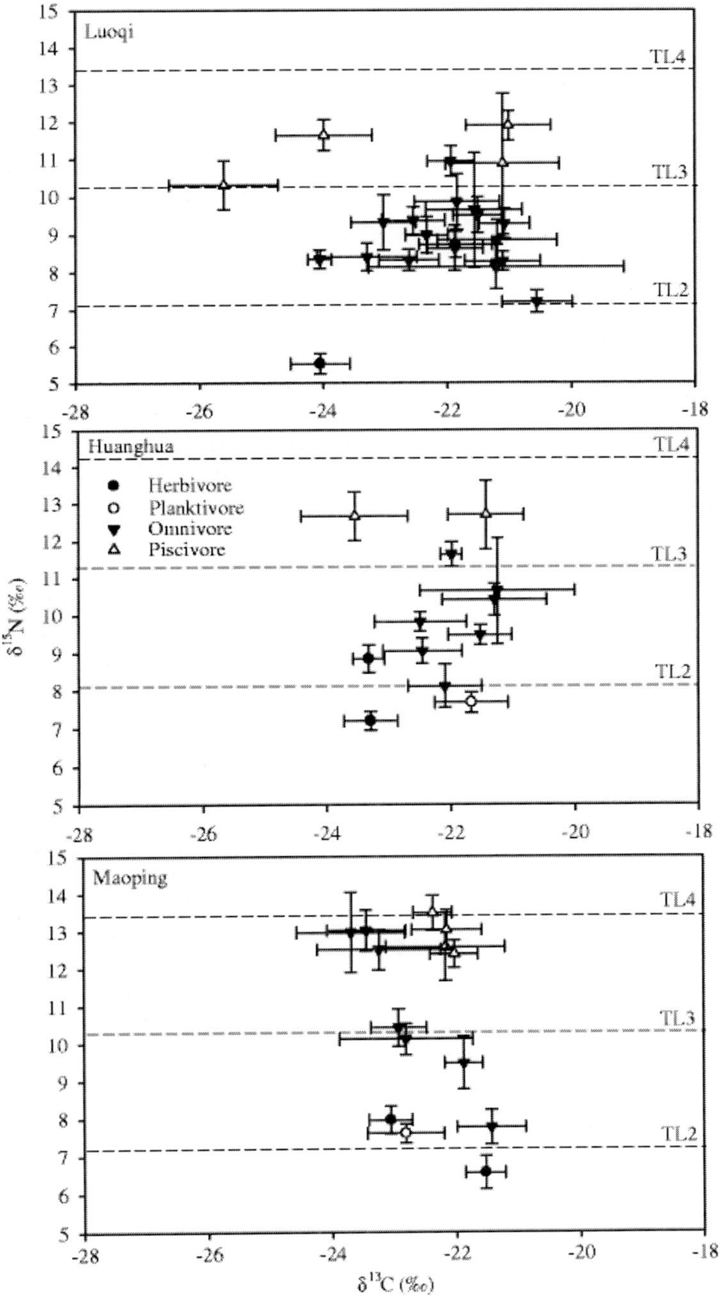

Figure. 2 Dual isotope plots of fish by trophic roles and trophic levels (TL) at the three study sites

Food chain length defined as the highest trophic level of the fish from a habitat was 3.5 at the upstream, 3.4 at the midstream, and 4.0 at the downstream. Chinese perch was positioned at the highest TL at all study sites while the Redfin culter and the Amur catfish shared the same TL as the Chinese perch from the upstream and midstream site. The TL (3.9) of longsnout catfish which was only collected from the downstream site was slightly lower than the Chinese perch. Except for Amur catfish which preys on invertebrates and small fish, other three fish are largely piscivores.

Trophic niche structure

The Bayesian results of Layman trophic niche metrics were displayed in Fig. 3. The nitrogen range (NR) values were similar at upstream and downstream, but lower at the midstream. By contrast, carbon range (CR) was highest at the upstream and drastically reduced at the midstream and downstream sites. Similarly, the standard ellipse area (SEA) size was greatest at upstream and decreased considerably at midstream and downstream. The mean distance to the $\delta^{13}C$-$\delta^{15}N$ centroid (CD) was nearly identical at both upstream and midstream and higher at downstream. Similar to CR and SEA, the mean nearest neighbor distance (MNND) and the standard deviation of nearest neighbor distance (SDNND) also displayed a trend of decreasing from upstream to downstream.

DISCUSSION

Before the construction of the TGD, hydrology and fish assemblage at the three study sites was similar with high flow rates and fish diversity (Wang et al. 2014). A number of studies reported decreases in fish species after the completion of TGD (Gao et al. 2010; Liu et al. 2012). The alterations of physical (flow rate and pattern, and river stage) and biological habitat (dietary sources and abundance) are attributed to the changes in fish composition and abundance (Gao et al. 2010; Dudgeon 2011; Zhang et al. 2014). However, information on intrinsic trophic niche as a function of change of hydrological and biological regimes has not been reported. This study provided new and confirmative data for the illustration of the changes in fish ecology following hydrological alterations of dam construction.

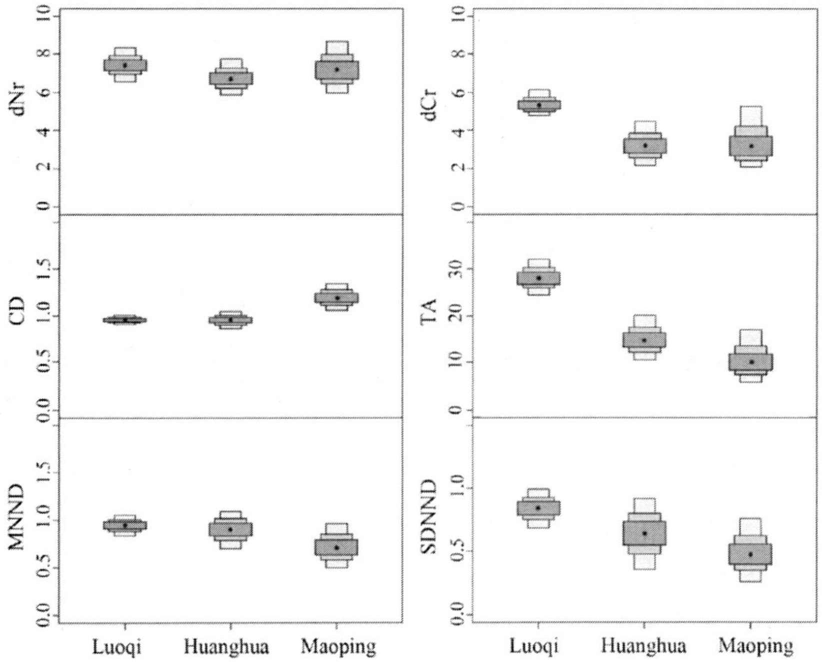

Figure. 3 Bar plots of the six Layman's niche metrics calculated using the Bayesian statistics based on resampling ($n = 10{,}000$) of uncertainty in sample mean estimates. *Black dots* represent means, and *shaded boxes* (dark to light) represent 50, 75, and 95 % credible intervals for mean estimates

Along with decreases in species diversity and appearance of the limnetic species, several major trophic niche metrics also pointed to the deterioration of the ecological niche at the downstream site since the completion of the TGD. Both $\delta^{15}N$ range (NR) and mean TL were greater at the downstream site than at the upstream site. Furthermore, the food chain length was also greatest at the downstream site with the Chinese perch occupying the highest trophic levels. The about 0.5 food chain length longer at the downstream than at other sites is likely attributed to more dietary organisms with higher trophic positions available at the more eutrophic downstream (Table 1).

The $\delta^{13}C$ range (CR) is a measure of basal resource diversity. Compared to the midstream and downstream sites, the high CR at the upstream site reflected higher degree of resource diversity and the utilization of these resources by similarly high diversity of the fish assemblage. The low CR at the midstream and downstream sites where hydrological flow rates were also dramatically reduced may be attributed to the decreases in natural resources. This view is supported by the $\delta^{13}C$ enrichment and depletion in common carp and snakehead collected at the upstream, respectively, which is over 1 ‰ and over 2 ‰ greater or less than those at the midstream and downstream. This might suggest a ^{13}C enriched and depleted basal resource, respectively, were absent at the lower reach of the river, likely as the result of decreasing availability of organic matter. The absence of snakehead with the most depleted $\delta^{13}C$ at the midstream directly contributed to the small CR in this habitat. By contrast, Layman et al. (2007) showed that compared to pre-restored period, CR did not change because similar basal resources were present (Layman et al. 2005). Abrantes et al. (2014) also found trophic diversity in an African estuary increased as greater input of terrestrial organic matter during the wet period.

The niche space (SEA) is influenced by the end members of $\delta^{13}C$-$\delta^{15}N$ for each consumer community. In this study, as discussed previously, the changes in SEA are mainly determined by CR which decreased considerably from upstream to downstream. The slight variation in NR from the upstream to downstream site does not affect the SEA as much as CR. Since the species which determined the trophic space basically appeared at all sites, we may conclude that it is resource availability which decreased along the river that affected fish trophic space of the community niche. This further supported findings from other measures in this study and other studies that alteration of hydrology impairs trophic niche in Yangtze and other lotic systems. The SEA is not affected by fish with intermediate $\delta^{13}C$-$\delta^{15}N$ values so decreases in fish species diversity do not necessarily affect trophic space (Layman et al. 2007).

The mean distance to the $\delta^{13}C$-$\delta^{15}N$ centroid (CD) provides a measure of average trophic diversity and is largely affected by the degree of species spacing (Layman et al. 2007). Although the trophic space (SEA) decreased from upstream to downstream, average trophic diversity (CD) increased from upstream to downstream, indicating that due to the absence of basal resources with extreme $\delta^{13}C$ values, fish trophic spacing was more aggregated and fish shared more similar basal resources at the downstream. This is supported by the decreases in both NND and SDNND at the midstream and downstream sites where trophic distribution became more even and redundant within the reduced trophic space. Cooper and Wissel (2012) show that with species redundancy increased in more productive waters. Layman et al. (2007) but showed that NND and SDNND decreased as the result of similar trophic niches shared by new immigrant species post-restoration.

CONCLUSIONS

Our results on species abundance and composition and trophic niche characteristics pointed to the negative impacts of the construction of SGD on fish community structure and trophic spacing. This is done likely through reduced hydrological flow and increased water stagnancy which are unfavorable to fish adapted to lotic water before the dam construction. The slight increase in food chain length at the impacted or more productive site seems to be positive but severe loss of species diversity and trophic space also occurred. Our data were collected only 1 year after the completion of the TGD and can be viewed as the immediate impact, if any, by TGD. Data are being collected to assess the longer term effect of TGD on trophic niche and the contribution of basal resources to the fish assemblages at similar sites along the Middle Yangtze River.

ACKNOWLEDGEMENTS

The authors are grateful to Leyi Li for his assistance in the stable isotopic measurements. The field and laboratory analyses were supported by two NSFCs (Grant Nos. 51179094 and 30700091) to J. Wang. Data analysis and manuscript preparation were supported by a NSFC (Grant No. 41376158) to B. Gu.

REFERENCES

1. Abrantes KG, Barnett A, Bouillon S (2014) Stable isotope-based community metrics as a tool to identify patterns in food web structure in east African estuaries. Funct Ecol 28:270–282
2. Bearhop S, Adams CE, Waldron S, Fuller RA, Macleod H (2004) Determining trophic niche width: a novel approach using stable isotope analysis. J Anim Ecol 73:1007–1012
3. Cabana G, Rasmussen J (1996) Comparison of aquatic food chains using nitrogen isotopes. Proc Natl Acad Sci 93:10844–10847
4. Cai S, Ni Z, Li Y, Shen Z, Xiong Z, Zhang Y, Zhou Y (2012) Metals in the tissues of two fish species from the rare and endemic fish nature reserve in the upper reaches of the Yangtze River, China. B Environ Contam Tox 6:922–927
5. Chen D, Xiong F, Wang K, Chang Y (2009) Status of research on Yangtze fish biology and fisheries. Environ Boil Fish 85:337–357
6. Chen S, Fath B, Chen B, Su M (2011) Evaluation of the changed properties of aquatic animals after dam construction using ecological network analysis. Procedia Environ Sci 5:114–119
7. Cooper RN, Wissel B (2012) Loss of trophic complexity in saline prairie lakes as indicated by stable-isotope based community-metrics. Aquat biosyst 8:6

8. Dudgeon D (2000) The ecology of tropical Asian rivers and streams in relation to biodiversity conservation. Annu Rev Ecol Syst 31:239–263
9. Dudgeon D (2011) Asian river fishes in the Anthropocene: threats and conservation challenges in an era of rapid environmental change. J Fish Boil 6:1487–1524View ArticleGoogle Scholar
10. Dynesius M, Nilsson C (1994) Fragmentation and flow regulation of river systems in northern third of world. Science 226:753–762
11. Finer M, Jenkins CN (2012) Proliferation of hydroelectric dams in the Andean Amazon and implications for Andes-Amazon connectivity. Plos one 7(4):e35126
12. Fu CZ, Wu JH, Chen JK, Wu QH, Lei GC (2003) Freshwater fish biodiversity in the Yangtze River basin of China: patterns, threats and conservation. Biodivers Conserv 12:1649–1685
13. Gao X, Zeng Y, Wang J, Liu H (2010) Immediate impacts of the second impoundment on fish communities in the Three Gorges Reservoir. Environ Biol Fish 2:163–173
14. Gu B, Schelske CL, Hoyer MV (1996) Stable isotopes of carbon and nitrogen as indicators of diet and trophic structure of the fish community in a shallow hypereutrophic lake. J Fish Biol 49:1233–1243
15. Hu M, Hua Q, Zhou H, Wu Z, Wu X (2014) The effect of dams on the larval abundance and composition of four carp species in key river systems in China. Environ Biol Fish 98:1201–1205
16. Jackson AL, Inger R, Parnell AC, Bearhop S (2011) Comparing isotopic niche widths among and within communities: SIBER–Stable Isotope Bayesian Ellipses in R. J Anim Ecol 3:595–602
17. Kling GW, Fry B, O'Brien WJ (1992) Stable isotopes and planktonic trophic structure in arctic lakes. Ecology 73:561–566
18. Layman CA, Winemiller KO, Arrington DA, Jepsen DB (2005) Body size and trophic position in a diverse tropical food web. Ecology 86:2530–2535
19. Layman CA, Arrington DA, Montana CG, Post DM (2007) Can stable isotope ratios provide for community-wide measures of trophic structure? Ecology 1:42–48
20. Liu J, Cao W (1992) Fish resources of the Yangtze River basin and the tactics for their conservation. Resources and Environment in the Yangtze Basin 1:17–23
21. Liu F, Wang J, Cao W (2012) Long-term changes in fish assemblage following the impoundments of the Three Gorges Reservoir in Hejiang, a protected reach of the upper Yangtze River. Knowl Manag Aquat Ec 407:06
22. McClain ME (2013) Balancing water resources development and environmental sustainability in Africa: a review of recent research findings and applications. Ambio 42:549–565
23. Minagawa M, Wada E (1984) Stepwise enrichment of $\delta^{15}N$ along food chains: further evidence and the relation between $\delta^{15}N$ and animal age. Geochim Cosmochim Ac 48:1135–1140
24. Newsome SD, Yeakel JD, Wheatley PV, Tinker MT (2012) Tools for quantifying isotopic niche space and dietary variation at the individual and population level. J Mammal 93:329–341
25. Park Y, Chang J, Lek S, Cao W, Brosse S (2003) Conservation strategies for endemic fish species threatened by the Three Gorges Dam. Conserv Biol 17:1748–1758
26. Perkins MJ, McDonald RA, van Veen FF, Kelly SD, Rees G, Bearhop S (2014) Application of nitrogen and carbon stable isotopes ($\delta^{15}N$ and $\delta^{13}C$) to quantify food chain length and trophic structure. PLoS One 3, e93281
27. Pinnegar JK, Polunin NVC (2000) Contributions of stable-isotope data to elucidating food webs of Mediterranean rocky littoral fishes. Oecologia 122:399–409
28. Post DM (2002) Using stable isotopes to estimate trophic position: models, methods, and assumptions. Ecology 83:703–718
29. R Development Core Team (2007) R: A Language and Environment for Statistical Computing. R Foundation for Statistical Computing, Vienna, Austria.
30. Salazar JG (2000) Damming the child of the ocean: the Three Gorges Project. J Environ Develop 9:160–1743

31. Sternberg R (2006) Damming the river: a changing perspective on altering nature. Renew Sust Energ Rev 10:165–197
32. Vander Zanden MJ (2001) Variation in $\delta^{13}C$ and $\delta^{15}N$ trophic fractionation: implications for aquatic food web studies. Limnol Oceanogr 8:2061–2066
33. Wang J, Gu B, Huang J, Han X, Lin G et al (2014) Terrestrial contributions to the aquatic food web in the middle Yangtze River. PLoS One 7, e102473. doi:10.1371/journal.pone.0102473
34. Wu J, Huang J, Han X et al (2003) Three-Gorges Dam—experiment in habitat fragmentation? Science 300:1239–1240
35. Xie P (2003) Three-Gorge Dam: risk to ancient fish. Science 302:1149
36. Xu X, Tan Y, Yang G (2013) Environmental impact assessments of the Three Gorges Project in China: issues and interventions. Earth-Sci Rev 124:115–125
37. Ye S, Li Z, Zhang T, Liu J, Xie S (2014) Assessing fish distribution and threats to fish biodiversity in the Yangtze River Basin, China. Ichthyol R 2:183–188
38. Yi YJ, Zhang SH (2012) Heavy metal (Cd, Cr, Cu, Hg, Pb, Zn) concentrations in seven fish species in relation to fish size and location along the Yangtze River. Environ Sci Pollut R 9:3989–3996
39. Zhang L, Xu J, Xie P, Zang X, Qiu G, Zeng J (2007) Stable isotope variations in particulate organic matter and a planktivorous fish in the Yangtze River. J Freshwater Ecol 3:383–386
40. Zhang G, Wu L, Li H, Liu M, Cheng F, Murphy BR, Xie S (2012) Preliminary evidence of delayed spawning and suppressed larval growth and condition of the major carps in the Yangtze River below the Three Gorges Dam. Environ Biol Fish 3:439–447
41. Zhang H, Wang CY, Yang DG, Du H, Wei QW, Kang M (2014) Spatial distribution and habitat choice of adult Chinese sturgeon (Acipenser sinensis Gray, 1835) downstream of Gezhouba Dam, Yangtze River, China. J Appl Ichthyol 6:1483–1491

CHAPTER 13

Distribution, Abundance, and Feeding Ecology of Baleen Whales in Icelandic Waters: Have Recent Environmental Changes had an Effect?

Gísli A.Víkingsson[1*], DanielG.Pike[2], HéðinnValdimarsson[3], AnnaSchleimer[4], ThorvaldurGunnlaugsson[1], TeresaSilva[3,5], BjarkiÞ.Elvarsson[1,6], BjarniMikkelsen[7], NilsØien[8], GenevièveDesportes[9], ValurBogason[1] and PhilipS.Hammond[4]

[1]*Marine Resources Section, Marine Research Institute, Reykjavík, Iceland*

[2]*Esox Associates, North Bay, ON, Canada*

[3]*Marine Environment Section, Marine Research Institute, Reykjavík, Iceland*

[4]*Sea Mammal Research Unit, Scottish Oceans Institute, University of St Andrews, St Andrews, UK*

[5]*Institute of Biology, University of Iceland, Reykjavík, Iceland*

[6]*Faculty of Physical Sciences, School of Engineering and Natural Sciences, University of Iceland, Reykjavík, Iceland*

[7]*Faroese Museum of Natural History, Torshavn, Faroe Islands*

[8]*Marine Mammal Research Group, Institute of Marine Research, Bergen, Norway*

[9]*GDnatur, Kerteminde, Denmark*

ABSTRACT

The location of Iceland at the junction of submarine ridges in the North-East Atlantic where warm and cold water masses meet south of the Arctic Circle contributes to high productivity of the waters around the island. During the last two decades, substantial increases in sea temperature and salinity have been reported. Concurrently, pronounced changes have occurred in the distribution of several fish species and euphausiids. The distribution and abundance of cetaceans in the Central and Eastern North Atlantic have been monitored regularly since 1987. Significant changes in the distribution and abundance of several cetacean species have occurred in this time period. The abundance of Central North Atlantic (CNA) humpback and fin whales has increased from 1800 to 11,600 and 15,200 to 20,600, respectively, in the period 1987–2007. In contrast, the abundance of minke whales on the Icelandic continental shelf decreased from around 44,000 in 2001 to 20,000 in 2007 and 10,000 in 2009. The increase in fin whale abundance was accompanied by expansion of distribution into the deep waters of the Irminger Sea. The distribution of the endangered blue whale has shifted northwards in this period. The habitat selection of fin whales was analyzed with respect to physical variables (temperature, depth, salinity) using a generalized additive model, and the results suggest that abundance was influenced by an interaction between the physical variables depth and distance to the 2000 m isobaths, but also by sea surface temperature (SST) and sea surface height (SSH), However, environmental data generally act as proxies of other variables, to which the whales respond directly. Overall, these changes in cetacean distribution and abundance may be a functional feeding response of the cetacean species to physical and biological changes in the marine environment, including decreased abundance of euphausiids, a northward shift in summer distribution of capelin and a crash in the abundance of sand eel.

INTRODUCTION

Cetaceans are important top predators in Icelandic waters with a total of 23 species recorded (Hersteinsson, 2004) of which 12–14 species are considered regular inhabitants. In terms of biomass and consumption, cetaceans play an important role in the Icelandic ecosystem. Sigurjónsson and Víkingsson (1997) estimated the total annual consumption by 12 cetacean species as 6 million tons corresponding to around four times the total Icelandic fishery landings. The diet composition of these cetaceans is poorly known except for common minke whales (*Balaenoptera acutorostrata*), fin whales (*Balaenoptera physalus*) and harbor porpoises

(*Phocoena phocoena*). Fin whales feed almost exclusively on euphausiids, mostly *Meganyctiphanes norvegica*, on the traditional whaling grounds in the Irminger Sea (Víkingsson, 1997). Fluctuations in environmental conditions affecting *per capita* prey availability have been shown to affect body condition and pregnancy rate in North Atlantic fin whales (Williams et al., 2013). Common minke whales have a much more varied diet ranging from euphausiids to large gadoid fish. Sand eel (*Ammodytes* sp.), herring (*Clupea harengus*) and capelin appear to be the most preferred prey (Víkingsson et al., 2014). The feeding ecology of blue (*Balaenoptera musculus*) and humpback whales (*Megaptera novaeangliae*) in Icelandic waters is not known. However, blue whales are known to feed exclusively on krill in the North Atlantic (Sears and Perrin, 2009) whereas humpback whales are reported to feed on both krill and pelagic schooling fish such as capelin and herring (Clapham, 2009).

Substantial warming and salinification has been observed in Icelandic and adjacent waters in the past two decades (Mortensen and Valdimarsson, 1999; Bersch, 2002; Malmberg and Valdimarsson, 2003; Hátún et al., 2005). Concurrently, pronounced changes have been reported in the distribution and abundance of several animal species in Icelandic waters. Some of these changes appear to be related to increased flow of Atlantic water masses into North Icelandic waters, including a northward expansion of the distribution of haddock (*Melanogrammus aeglefinus*), monkfish (*Lophius piscatorius*), and capelin (*Mallotus villosus*) (Astthorsson et al., 2007; Solmundsson et al., 2010). Sand eel abundance in southern and western Icelandic waters was apparently drastically reduced around 2005, with severe consequences for some seabird populations (Lilliendahl et al., 2013; Vigfusdottir et al., 2013). In terms of biomass, the most influential change is probably the recent invasion of Northeast Atlantic mackerel (*Scomber scombrus*) into Icelandic waters (Astthorsson et al., 2012). There are limited data available on temporal trends in meso- and macro zooplankton in Icelandic waters. Silva et al. (2014) found a decrease in euphausiid abundance in the oceanic (offshore) waters south and west of Iceland during 1958–2007, while in south Icelandic shelf waters they reported an increase in euphausiid larvae during 1990–2011. Recent and ongoing changes in the Icelandic marine environment affect several species that are important as prey for cetaceans. Depending on the ecological flexibility of the different species, these changes could be expected to affect the distribution and abundance of cetaceans in the area.

Large scale surveys, aimed at estimating and monitoring the distribution and abundance of cetaceans in the Central North Atlantic (CNA), have been conducted regularly since 1987 (Sigurjónsson, 1992; Lockyer and Pike, 2009; Víkingsson et al., 2009). Major changes in the distribution and abundance of several cetacean species have been observed over the past 20 years, particularly around the turn of the century. However, the potential effects on habitat use of cetaceans are poorly documented. Within the context of a changing environment, it is important to understand the link between baleen whales and their environment at various temporal and spatial scales in order to assess potential ecosystem-level effects, mediated by changing environmental conditions (Hátún et al., 2009). For future conservation and management it is of utmost importance to gain better understanding of these effects and the interactions among key species in the ecosystem.

In this paper we describe these changes in cetacean distribution and abundance and investigate to what extent they can be explained by recent biological (prey) and physical (oceanography) changes in the marine environment in this area. As part of this investigation, we present a first attempt at modeling the habitat use of fin whales in the Northeast Atlantic over a 20-year period, as a function of both physiographic and remotely sensed environmental variables. Fin whales were chosen for this exercise because of their wide distribution within the survey area and their spatial and temporal overlap with readily available environmental parameters.

MATERIALS AND METHODS

Hydrography

Iceland is located at the intersection between large submarine ridges along the Mid Atlantic Ridge in the northern North Atlantic resulting in complex circulation in Icelandic waters (Figure 1). In combination with energetic atmospheric circulation this leads to a highly variable environment on the boundary between colder and warmer waters. The East Greenland Current and East Icelandic Current bring colder water from the north while the North Atlantic Current and the Irminger Current carry warmer waters from the south. With the polar front lying through the Denmark Strait area and the Irminger Sea, this climatic boundary fluctuates with various proportions of Polar, Arctic and Atlantic water, especially north of Iceland but also in the Irminger Sea.

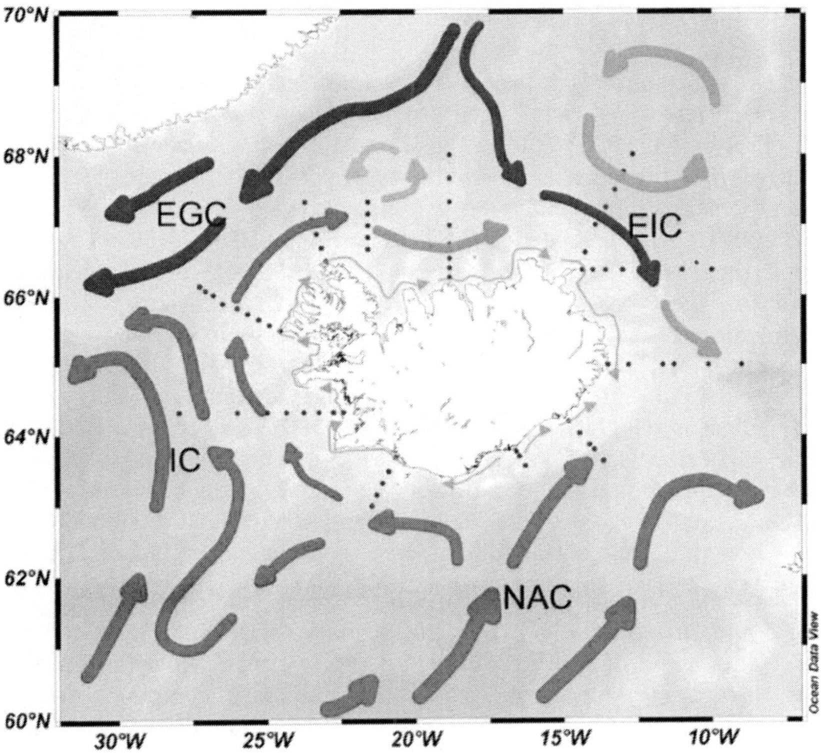

Figure 1. Oceanography of the Icelandic area. EGC, East Greenland Current; EIC, East Iceland Current; IC, Irminger Current; NAC, North Atlantic Current. The dotted lines represent sampling stations.

Oceanographic data to describe decadal changes in hydrographic conditions in Icelandic waters are based on long-term sampling conducted by the Marine Research Institute (MRI), Reykjavík Iceland (Malmberg and Valdimarsson, 2003; Valdimarsson et al., 2012). Temperature and salinity data used here were obtained with a CTD after 1990, mainly within the context of seasonal monitoring at stations around Iceland. Some data in the Irminger Sea were from an international redfish monitoring project coordinated through ICES. CTD data were processed according to internationally recommended standards. Data from before 1990 were obtained with Nansen bottles and reversing thermometers and then interpolated on 1 dbar intervals before being depth averaged in the same manner as CTD data.

Distribution and Abundance of Cetaceans

The North Atlantic Sightings Surveys (NASS and TNASS) constitute the largest series of cetacean surveys conducted to date in both spatial and temporal extent (Pike, 2009). These multi-national surveys have covered large parts of the summer distributions of North Atlantic baleen whales. To date, the series includes five large-scale surveys (1987, 1989, 1995, 2001, and 2007), the most recent one (TNASS 2007, CODA, SNESSA) extending across the whole North Atlantic from the east coast of North America to the European west coast (NAMMCO, 2009). In all these surveys, the CNA (Donovan, 1991) has been covered by Icelandic, Faroese and Norwegian research effort. The Icelandic continental shelf area was surveyed by air using cue-counting methods for abundance estimation (Donovan and Gunnlaugsson, 1989; Hiby and Hammond, 1989; Borchers et al., 2009; Pike et al., 2011a). Offshore waters of the CNA were covered by vessel surveys using conventional line transect methods (Sigurjónsson et al., 1989, 1991; Pike et al., 2009a, 2010b; Víkingsson et al., 2009). In addition to the above, full scale aerial surveys were conducted in Icelandic waters in 1986 and 2009 (Gunnlaugsson et al., 1988; Pike et al., 2011b).

The common minke whale was the primary target species of the aerial surveys covering Icelandic continental shelf waters, while the fin whales was the primary target species in the shipboard component covering Icelandic and adjacent offshore waters. The timing of the surveys was roughly the month of July, corresponding to the assumed peak abundance of the main target species, fin and common minke whales. The sei whale (*Balaenoptera borealis*) was a target species of the 1989 survey that had no aerial component and took place around 2 weeks later and extended further south than the other surveys (Sigurjónsson et al., 1991). The analysis presented here was largely confined to the CNA, an area that received a fairly systematic coverage by Iceland and the Faroes. To account for variable coverage within this area, post-stratification was applied for analysis of trends (Víkingsson et al., 2009). For comparisons within the shelf area, southern Icelandic waters were represented by aerial survey blocks 1, 2, 8, and 9 and northern Icelandic waters by blocks 4–7 (for delineation of aerial survey blocks Figure 6). Table 1 summarizes the aerial and shipboard surveys on which the analyses presented here are based.

Data on the catch distribution of fin whales were retrieved from the official databases of the MRI and the Icelandic Fisheries Directorate. Prior to the pause in fin whaling 1990–2005, positions were reported in squares (0.5° latitude × 1° longitude) used in the fisheries sector (Sigurjónsson, 1988) but, since 2006, more precise positions (to the nearest minute of latitude and longitude) have been reported.

Table 1. Survey data used in the analyses of cetacean abundance.

Species	Survey years	Area	Platform	Method	Figure	Source
Common minke whale	1986, 1987, 2001, 2007, 2009	Icelandic continental shelf	Aircraft	Cue counting (A)	5	Borchers et al., 2009; Pike et al., 2009b, 2011a
Common minke whale	1986, 1987, 1995, 2001, 2007, 2009	Icelandic continental shelf	Aircraft	Line transect densities (R)	6 and 7	Pike et al., 2009b, 2011b
Fin whale	1987, 1989, 1995, 2001, 2007	Central North Atlantic	Vessels	Line transect (A)	8	Sigurjónsson et al., 1989, 1991; Víkingsson et al., 2009, Pike et al., 2008
Fin whale	1987, 1989, 1995, 2001, 2007	Central North Atlantic	Vessels	Habitat modeling (R)	9	MRI database
Humpback whale	1986, 1987, 1995, 2001, 2007, 2009	Icelandic continental shelf	Aircraft	Sighting rates (whales/nm) (R)	12	Pike et al., 2009b, 2011b
Humpback whale	1987, 1995, 2001, 2007	Central North Atlantic	Vessels	Line transect (A)	12	Pike et al., 2010c

A, Estimated absolute abundance; R, Relative abundance; Figure, Figure numbers in this article. Source: References to reports giving details of data and methods.

Modeling of fin whale distribution

The large-scale link between fin whale distribution and its habitat in the Northeastern Atlantic was investigated by modeling the relationship between whale abundance and environmental covariates. Data on fin whale abundance were derived from the five Faroese and Icelandic line transect cetacean sighting surveys conducted from 1987 to 2007 (Lockyer and Pike, 2009). Transects were divided into effort segments, each of which ended when there was a change in survey conditions (sea conditions, weather) or a sighting. Data on cetaceans sighted included species, group size, and radial distance and angle to the group (Víkingsson et al., 2009).

In a first step, the detection probabilities, p^p^, of fin whales in each effort segment were estimated using the DISTANCE 6.2 software package (Thomas et al., 2010). Covariates available to improve precision and reduce bias in estimating detection probability included: sea conditions (Beaufort scale), weather index, visibility, vessel identity (nine vessels or three categories according to platform height), number of observers and group size. The weather index reflected the cloud cover as well as the amount of precipitation or fog and occasionally included wind direction. The final detection function was chosen by minimizing Akaike's Information Criterion (AIC) and by comparing the Cramér-von Mises goodness of fit test statistics (Thomas et al., 2010) and, all other things being equal, the precision of estimated average detection probability. Perpendicular distances were right truncated at 4000 m to improve detection function model fit; beyond this distance sightings were sporadic (Thomas et al., 2010). Only sightings with species identification of high and medium certainty were included in this analysis. Searching effort and sightings made in sea conditions greater than Beaufort scale 5 were excluded.

Abundance of fin whales, $\widehat{N}$, in each effort segment, *i*, was estimated using the Horvitz-Thompson estimator:

$$\widehat{N}_i = \sum_{j=1}^{n_i} \frac{1}{\widehat{p}_{ij}}$$

where n_i is the number of fin whales observed in the ith effort segment and $\widehat{p}_{ij}$ is the estimated probability of detection of the jth detected group in segment *i*.

Covariate data available to model the influence of environmental features on fin whale abundance included physiographic, remotely sensed and reconstructed data (Table 2). Some covariates of interest, especially salinity and chlorophyll a, could not be included because some survey years pre-dated the use of certain satellite sensors (e.g., SeaWIFS operated only since 1997). Satellite altimetry data date back only to 1993; however, a sea surface height (SSH) data set from 1950 to 2009, reconstructed based on *in situ* tide-gauge records, was made available by Meyssignac et al. (2012). This dataset had global coverage up to 70°N, which led to the exclusion of a minimal number of effort segments further north. Depth, slope, seabed aspect and distance covariates were linked to effort segment midpoints in Manifold 8.0. Sea surface temperature (SST) and SSH values were assigned to effort segment midpoints by interpolating the value from four surrounding cells based on coordinates and time using R 3.0.1 (R Core Team, 2013).

Multi-panel scatterplots and Pearson correlation coefficients were used to assess collinearity between pairs of covariates (Zuur et al., 2009). Collinearity between two covariates was considered significant when the Pearson correlation coefficient exceeded 0.6. During the model fitting process, only the covariate that explained most of the deviance was retained in the model for covariates identified as collinear.

The relationship between estimated abundance of fin whales and environmental covariates was modeled using Generalized Additive Models (GAMs) (Hastie and Tibshirani, 1990). GAMs are commonly used to relate distribution or abundance of animals to environmental characteristics due to their greater flexibility compared to Generalized Linear Models (GLMs) (Wood, 2001).

Table 2. List of physiographic and environmental covariates considered for inclusion in the model as explanatory variables.

Name	Description	Source
PHYSIOGRAPHIC VARIABLES		
Depth	Sea bed depth (m)	ETOPO1 1 arc-minute gridded global relief data. NOAA National Geophysical Data Center. http://www.ngdc.noaa.gov/
Slope	Slope(°) of the sea floor	Derived from ETOPO1 bathymetric data using transform surface dialog in Manifold 8.0
Aspect	Aspect (−180, 180) of the sea floor	Derived from ETOPO1 bathymetric data using transform surface dialog in Manifold 8.0
Distance to isobaths (Dist0, Dist200, Dist2000)	Distance of the effort segment midpoint to the coast, the 200 m deep isobath and the 2000 m deep isobath	Calculated in Manifold 8.0 using SQL queries
REMOTELY SENSED VARIABLES		
Monthly Sea surface temperature (SST)	Night-time monthly averages of SST (°C) from March to August	4 km Advanced Very High Resolution Radiometer (AVHRR) Pathfinder Version 5. Distributed by PO.DAAC and NODC http://podaac.jpl.nasa.gov/
Current Sea Surface temperature	Night-time current monthly average SST (°C) at the time of the cetacean survey	4 km Advanced Very High Resolution Radiometer (AVHRR) Pathfinder Version 5. Distributed by PO.DAAC and NODC http://podaac.jpl.nasa.gov/
RECONSTRUCTED VARIABLES		
Sea surface height (SSH)	1° × 1° annually averaged sea surface height (m)	Meyssignac et al., 2012

A GAM for abundance data with log link function was used with the general structure:

$$\widehat{N}_i = exp[\theta_0 + \sum_k f_k(X_{ik})]$$

where θ_0 is the intercept term, f_k are the smoothed functions of the explanatory covariates, and X_{ik} is the value of the kth explanatory covariate in the ith segment.

As is common with such data, the response variable, $\widehat{N}$ was overdispersed with a higher proportion of zeroes and of relatively large values than expected if the data were Poisson distributed. Models have been developed to deal with zero-inflated data (Barry and Welsh, 2002; Warton, 2005); however, a quasi-Poisson or negative binomial error distribution are appropriate when fitting models to overdispersed count data (Ver Hoef and Boveng, 2007). The Tweedie distribution has also been proposed because it has an additional parameter offering higher flexibility (Miller et al., 2013). These three error distributions, quasi-Poisson, negative binomial and Tweedie, were investigated when fitting the GAMs; quantile-quantile (QQ) plots, plots of Pearson residuals and response vs. fitted values plots were visually assessed to determine the most appropriate error distribution to use.

In models of count data, effort segments of variable length are typically accounted for by using an offset term, which assumes that the effort variable is a linear predictor with coefficient equal to 1 (Wood, 2006). However, in these data the relationship between the response variable and effort segment length did not meet this assumption and the latter was instead included as a smoothed covariate.

Data from surveys may be spatially auto-correlated and, if model covariates do not account for this, residuals may also be correlated leading to underestimated SEs of coefficients and possibly to unwarranted retention of covariates in model selection. A semi-variogram of model residuals from the final GAM plotted against distance between observations was inspected to assess whether there was spatial correlation in the model residuals (Zuur et al., 2009).

Smooth terms were modeled using penalized cubic regression splines with a shrinkage term, which allows smoothers to reduce to 0 degrees of freedom and to be dropped from the model (Wood, 2006). Interaction terms were fitted using tensor product interactions when the main effects were also included in the model. The package mgcv v. 1.8-1 for R was used to fit models (Wood, 2001).

Model selection used a forward stepwise procedure based on three evaluation criteria: (1) model GCV (Generalized Cross Validation) score, which is an approximation to AIC; (2) percentage of deviance explained; and (3) probability that a covariate was included in the model by chance using analysis of deviance tests (Wood, 2001).

For the predictive maps, a grid of 0.5° × 0.5° resolution populated with covariate values was generated in Manifold 8.0 for each survey year over the extent of survey coverage. Predictions of relative fin whale abundance in each grid cell for each survey year were generated in R based on the final GAM model. The point estimates of predicted relative abundance in each grid cell were smoothed into density surfaces using kriging with Gaussian distribution, implemented in Manifold 8.0, for presentation.

Cetacean Prey Species

Capelin

Capelin is among the most important commercially exploited pelagic fish species in Icelandic waters. The fishery for this species was initiated in the 1960s and continues to date. Acoustic monitoring of capelin distribution and abundance has been conducted by the MRI since 1980 by surveys in autumn and winter and has been reported annually in stock status reports (Marine Research Institute, 2014).

Sand eel

Sand eel have not been subject to commercial harvesting in Icelandic waters. However, they are important as a food source for other fish species, seabirds and marine mammals (Víkingsson et al., 2003, 2014; Lilliendahl and Solmundsson, 2006; Bogason and Lilliendahl, 2009; Vigfusdottir et al., 2013). Attempts to monitor sand eel abundance were not initiated until 2006, in relation to poor breeding success of seabirds in south Iceland. However, auxiliary information can be obtained from stomach content data from haddock sampled during MRI's annual groundfish surveys (Bogason and Lilliendahl, 2009; MRI, unpublished data).

Euphausiids (krill)

Data on euphausiid abundance were obtained from Sir Alistair Hardy Foundation for Ocean Science (SAHFOS) from their Continuous Plankton Recorder (CPR) program (Silva et al., 2014). The CPR collects plankton from vessels of opportunity crossing the North Atlantic along standard routes. The samples are processed, including species identification at SAFHOS (Batten et al., 2003). The northernmost routes of the CPR surveys cover the areas south and southwest of Iceland.

RESULTS

Hydrography

Changes in temperature and salinity in the waters south and west of Iceland in recent decades have been characterized by observations from the deepest station in the Faxaflói section west of Iceland (Figure 2A). This station is located in the core of the Atlantic water as it flows northwards along the shelf break in the Irminger Current. After the basin wide changes before the mid-1990s, temperature and salinity in the mixed layer started to increase rapidly, demonstrating the increase of water of Atlantic character in this part of the Sub Polar Gyre. Similar changes were observed in temperature further north on the Latrabjarg section (station Lb4 Northwest Iceland) in the northward flow of the branch of the Irminger Current which feeds the inflow of Atlantic water to the area north of Iceland, i.e., the North Icelandic Irminger Current (Figure 2B). The rise in temperature was particularly steep between 1995 and 2000 (Figure 2). In the following years, these changes are also apparent in the shelf area to the north of Iceland and most markedly as higher winter temperatures (Figure 3). Since 1998, winter temperature and salinity were higher than the average for the measurement period (1970–2014) in all years except

2002. In general, a similar but more variable rise in temperature was observed in other seasons. Figure 4 shows changes in summer temperatures at 200 m depth in the Irminger Sea between 1994 and 2003. Appreciable warming occurred in this period, particularly in the deep water area between the continental shelves of Iceland and Greenland.

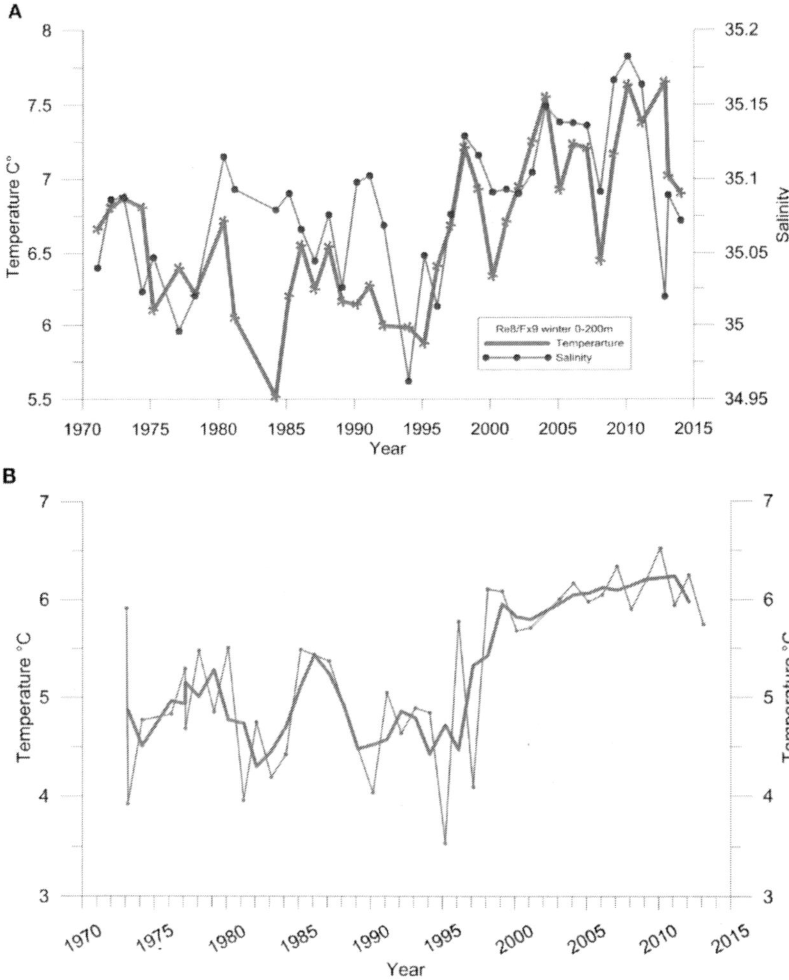

Figure 2. (A) Temperature (red) and salinity (blue) at a repeated hydrographic station, Fx9, in winter at the shelf slope west of Iceland. Dept average over 0–200 m depth. (B) Temperature at a repeated hydrographic station, Lb4, in winter at Latragrunn. Depth average over 80–180 m depth. Thick line shows 3-year running mean.

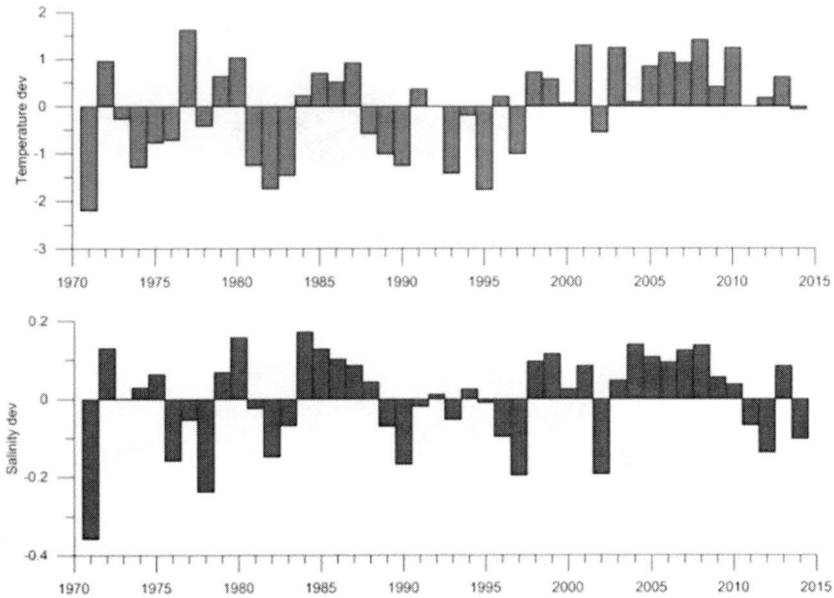

Figure 3. Temperature and salinity deviations from Siglunes section north of Iceland 1970–2013. Deviations are from depth average and mean from stations Si2–Si4.

Distribution and Abundance of Cetaceans

Common minke whale

Abundance estimates of common minke whales from aerial surveys conducted during 1987–2009 over the Icelandic continental shelf varied widely (Figure 5). Point estimates of abundance increased appreciably between 1987 and 2001 although the difference was only marginally significant (Borchers et al., 2009; Pike et al., 2009b). The survey in 2007 revealed a reduction in minke whale abundance to less than half that of 2001 and an extra aerial survey conducted in 2009 showed even further decline in numbers (Figure 5).

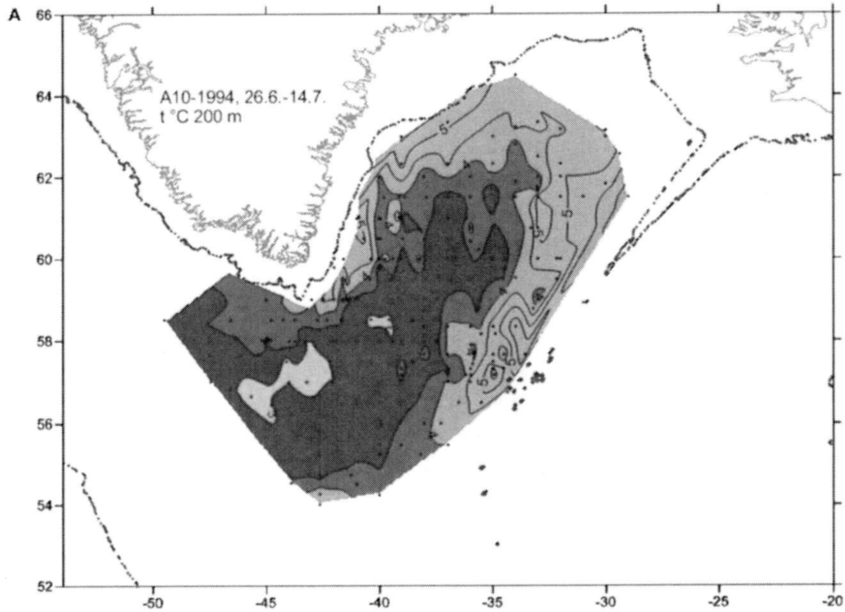

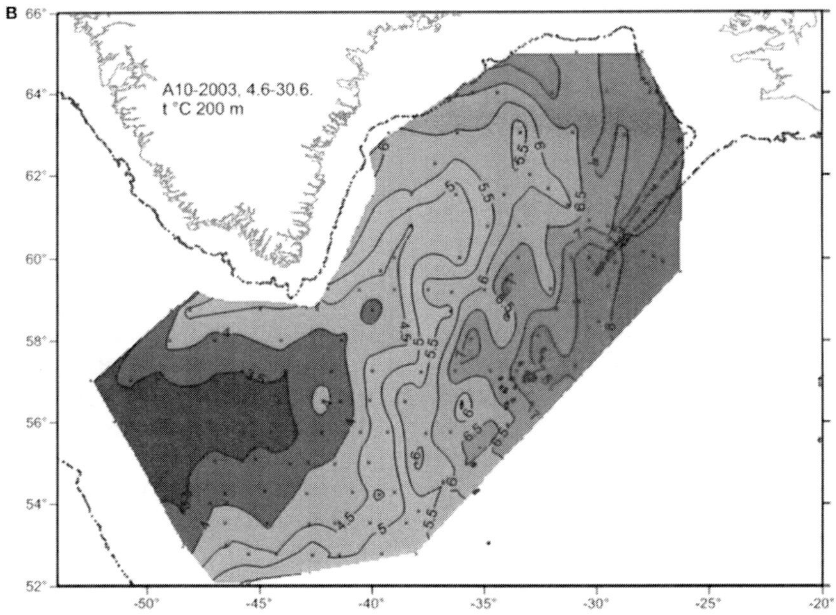

Figure 4. Sea temperatures at 200 m depth in the Irminger Sea in June 1994 (A) and 2003 (B).

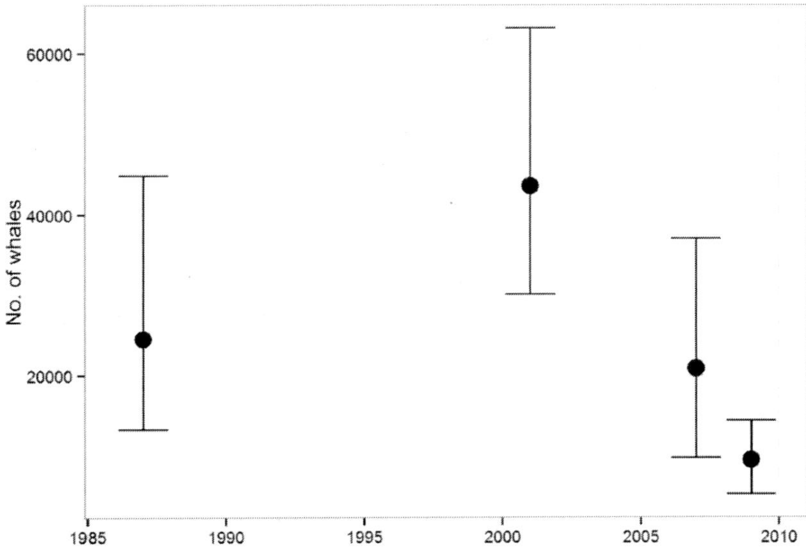

Figure 5. Abundance estimates (with 95% confidence intervals) of common minke whales in the Icelandic continental shelf area derived from aerial surveys conducted during 1987–2009.

Figure 6 shows concurrent changes in relative densities (number of whales/nm^2 at the surface) of common minke whales on the Icelandic continental shelf according to the aerial surveys conducted during 1986–2009. As for the estimated abundance, the estimates for the whole survey area show a slightly increasing, but non-significant trend within the range of 0.04 and 0.06 animals/nm^2 between 1986 and 2001. In the next survey (2007), the densities had decreased sharply to less than 0.02 animals/nm^2 and this drop in density was confirmed in the survey conducted in 2009 (Figure 6).

The decreased densities after 2001 were evident in most sub-areas of the aerial survey. However, the degree of the decline was highly variable within the continental shelf stratum with the steepest declines in southern and western coastal waters (Figure 6). Figure 7 compares the mean densities of the southern (blocks 1, 2, 8, and 9) and northern (blocks 4–7) survey blocks for the periods 1986–2001 and 2007–2009. In the earlier period, mean density in the southern/western area was more than double that of the northern area. In the later period, density in both areas declined to about the same level; thus the decline was much steeper in the southern area (79%) than in the northern area (46%).

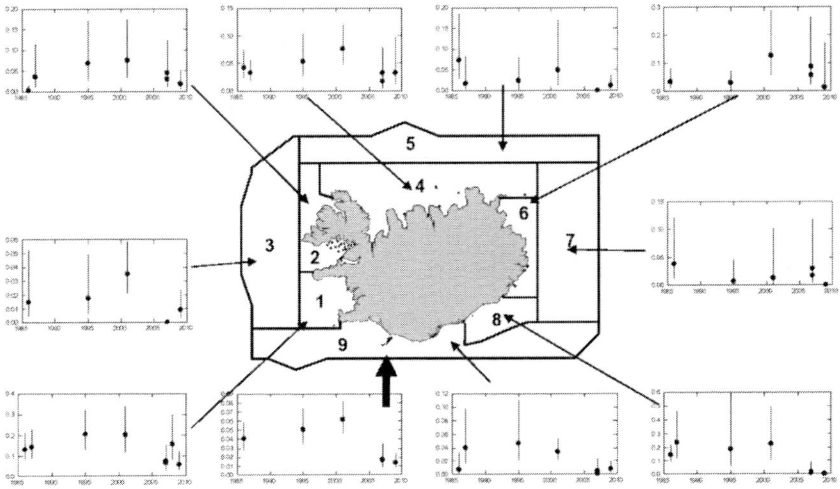

Figure 6. Trend in relative abundance (number of whales/nm2 at the surface with 95% confidence intervals) of common minke whales, 1986–2009, as indicated by line transect density in the Icelandic aerial survey area. Two estimates, using both primary observers, and a single primary observer, are shown for 2007 and 2009. The thick arrow refers to the total survey area.

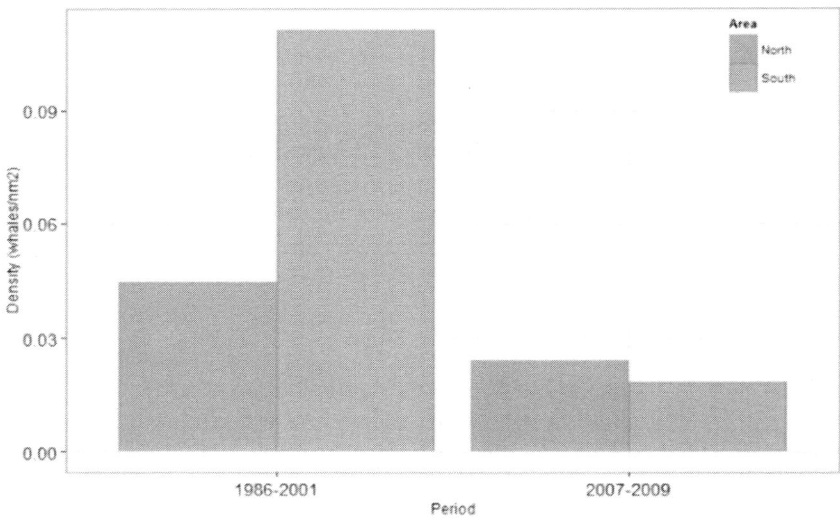

Figure 7. Comparison of changes in densities (Number of animals/nm2) of common minke whales in northern (blocks 4–7) and southern (blocks 1, 2, 8, and 9) coastal Icelandic waters. For delineation of aerial survey blocks see Figure 6.

Fin whale

The estimated abundance of fin whales increased during the survey period from 15,200 in 1987 to 20,600 in 2007 (Figure 8). During 1987–2001 the annual increase in the total CNA was estimated as 4%, while the annual growth rate was estimated as 10% in the Irminger Sea between Iceland and Greenland (Víkingsson et al., 2009).

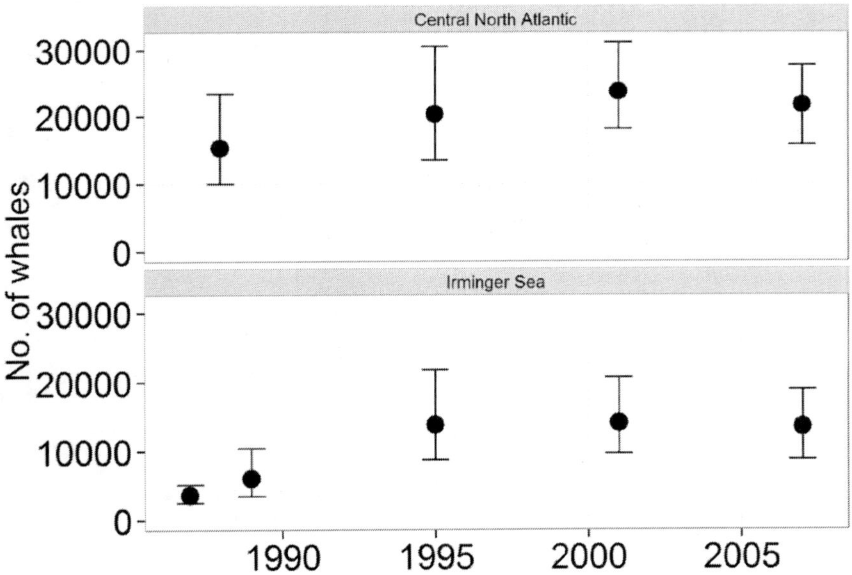

Figure 8. Abundance (with 95% confidence intervals) of fin whales according to the North Atlantic Sightings Surveys 1987–2001 in the total Central North Atlantic stock area (upper) and in the Irminger Sea west of Iceland (lower). For area delineations see Víkingsson et al. (2009).

In the habitat modeling, the covariates retained in the final GAM included smooth functions for SST, SSH and survey effort segment length, and an interaction term between depth and distance to the 2000 m isobath. A negative binomial error distribution with log link function was identified as the best description of the data. Pearson correlation coefficients showed high collinearity between all monthly SST covariates ($r = 1, p < 0.0001$), and between all combinations of Depth, Dist0, and Dist200 ($r > 0.7, p < 0.0001$). Accordingly, only one SST covariate and one of Dist0, Dist200, or Depth were included in the final model. The semi-variogram indicated that there was no spatial correlation in the model residuals (Supplementary Figure 1).

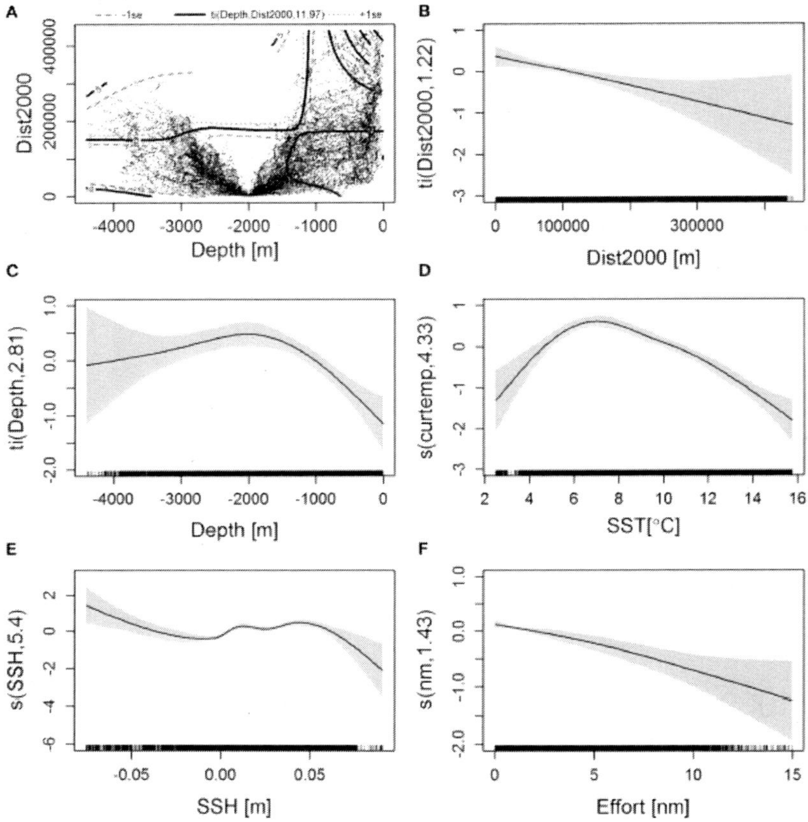

Figure 9. Smooth functions fitted in the final GAM for (A) 2-D interaction between Depth and Distance from the 2000 m isobath, (B) Distance from the 2000 m isobath, (C) Depth, (D) Sea surface temperature, (E) Sea surface height, and (F) survey effort. Positive values of the smoothed function indicate a positive effect on the response variable. In (A), the dots represent values of the response variable, and the black, green and red lines are contours of the fitted function ±1 SE. Given above the plot is the smoother term and the estimated degrees of freedom (see Supplementary Table 1). In (B–F), shaded areas represent the 95% confidence interval. Tick marks on the horizontal axis show the distribution of observations. The vertical axes describe the smoother term and estimated degrees of freedom (see Supplementary Table 1).

The results of the final model, which explained 15.7% of the deviance, are given in Supplementary Table 1. Figure 9 shows the fitted smooth function for each covariate included in the final model. The probability of sighting fin whales decreased with increasing segment length, supporting the decision to include effort as a smoothed covariate rather than an offset term (Figure 9F). SST in the study area at the time of the surveys ranged from 2.5°C to 15.7°C and the highest probability of encountering fin whales was predicted to be in waters with surface temperatures between 5 and 11°C peaking around 6–7°C (Figure 9D). Fin whales were less likely to be observed at greater distances from the 2000 m isobath and in shallower waters (Figures 9B,C). This pattern was also reflected in the interaction between these two covariates (Figure 9A). Sightings were mainly made at intermediate SSH of 0.02–0.04 m elevation as well as in waters with SSH depressions greater than 0.05 m (Figure 9E).

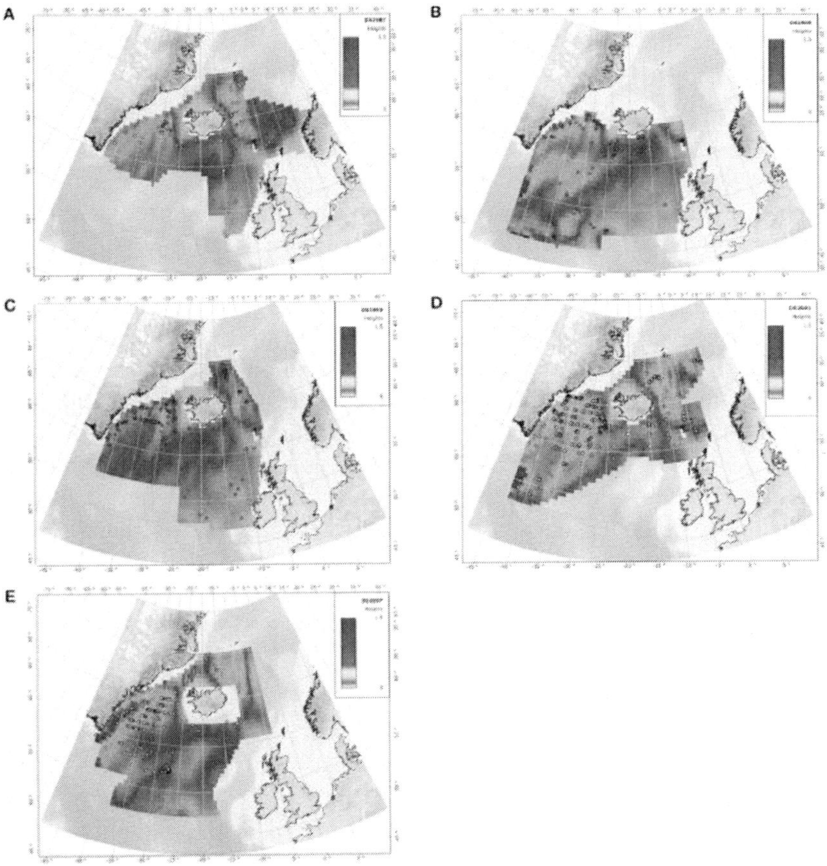

Figure 10. Predicted density surface maps derived from the final GAM with an average segment length of 1.75 nm and SST from July for the survey years 1987 (A), 1989 (B), 1995 (C), 2001 (D), and 2007 (E). Warmer colors indicate higher predicted relative abundance. Dots represent observations of fin whale groups during the surveys. Predicted relative abundance in each grid cell was smoothed using kriging.

The Irminger Sea and the area northeast of Iceland were predicted to hold the highest abundance of fin whales in all years (Figures 10A–E), except in 1989 when the area north of Iceland was not surveyed (Figure 10B). Within the Irminger Sea, fin whales were predicted to be mostly concentrated along the 2000 m isobath. Predicted relative abundance was high along the East Greenland shelf particularly in 2001 and 2007. The area west of the British Isles had consistent low predicted abundance of fin whales. These predictions were overall in agreement with the observations of fin whales.

As noted previously (Pike et al., 2008; Víkingsson et al., 2009), the distribution of fin whales has changed in the Irminger Sea. Whales aggregated over the continental slope in 1987, 1989, and 1995, but were dispersed throughout the Irminger Sea in 2001 and 2007. The model failed to explain the high number of sightings south of Iceland in 1989 (Figure 10B).

Since the resumption of fin whaling in 1948 the operation has been limited to a single land station in Hvalfjörður, West Iceland (Sigurjónsson, 1988). Because of factory regulations concerning freshness of products, catches have been limited to approximately 24 h cruising distance from the station. The distribution pattern of the fin whale catches has been rather stable and largely confined to areas close to the continental slopes of the Irminger Sea west of Iceland (Figure 11). The catch distribution of sei whales has been somewhat wider including areas further east along the southern coast of Iceland. The catch distribution of fin whales in 2014 differed appreciably from all the previous seasons since 1948. During 16–26 June, 26 fin whales were caught within the traditional fin whaling grounds west and southwest of Iceland. In the following 2 weeks only two fin whales were caught in this area despite considerable search effort. After widening the search area fin whales were found in high densities in offshore south Icelandic waters. During the rest of the season, 13 July-23 September 2014, 104 fin whales were taken in these "new" fin whaling grounds and four in the traditional grounds (Figure 11).

Humpback whale

Figure 12 shows the estimated relative abundance of humpback whales from the 1986–2009 aerial surveys and estimated absolute abundance from the 1987–2007 shipboard surveys. Both indices show an increase in the point estimates from 1986/1987 to 2001 with a slight decrease thereafter. However, confidence intervals are wide and the estimates since 1995 are not significantly different from each other.

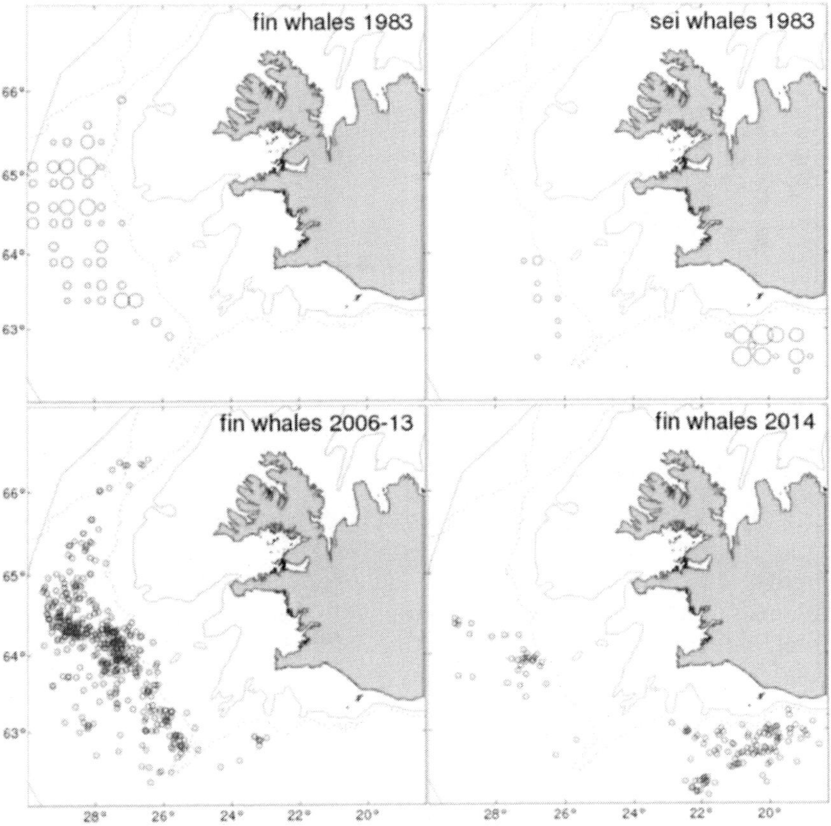

Figure 11. Catch positions of fin whales 1983, 2006–2013, and 2014, and sei whales 1983. Circles for individual whales 2006–2014, but in 1983 whales were reported by squares. Circle area is in proportion to catch numbers in square.

Cetacean Prey Species

Sand eel

Figure 12 shows the average number of sand eel in haddock stomachs sampled during the annual Icelandic groundfish surveys 1997–2010. Sand eel numbers decreased sharply from around 3/stomach in 1997 to 0 in 2004. In all subsequent years except 2007, the mean number of sand eel per stomach was less than 0.25 (Figure 13).

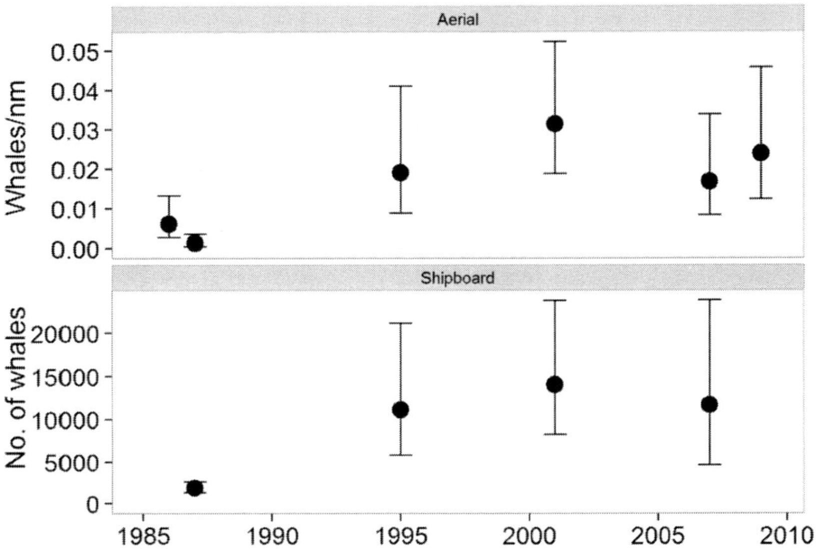

Figure 12. Humpback whale abundance 1986–2009. Sightings rates (animals/nm) in the Icelandic continental shelf area (upper) Estimated total abundance in the Central North Atlantic (lower). Ninety five percent confidence intervals are shown.

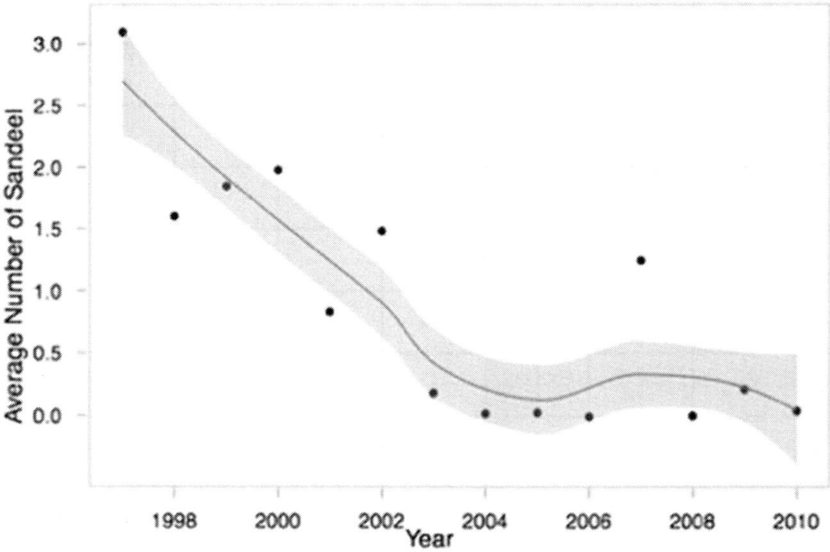

Figure 13. Mean number of sand eel in haddock stomachs analyzed in the Icelandic groundfish survey 1997–2010. A loess smoother has been fitted to the time series, shown with 95% confidence intervals.

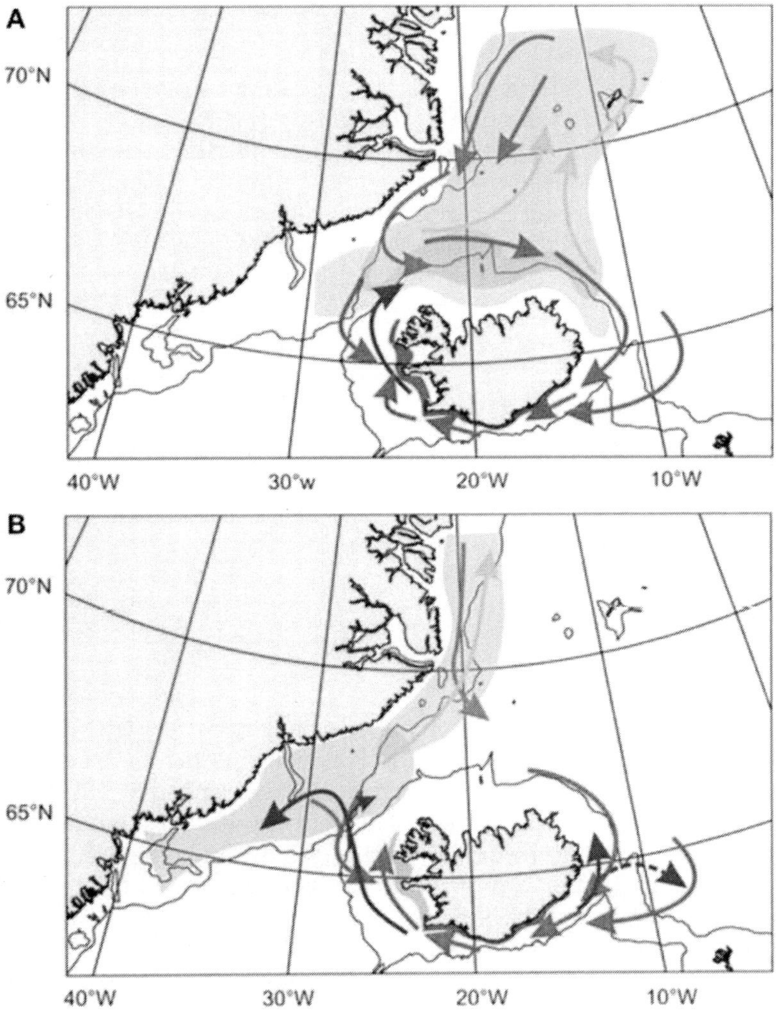

Figure 14. (A) Main patterns in distribution and migration of capelin until the turn of the last century. Light blue, Nursery area; Green, Feeding grounds of adult capelin. Blue arrows indicate feeding migrations and green arrows autumn return migrations toward Iceland waters. Red, Spawning migrations and areas; Violet arrows, 0-group drift (Modified after Vilhjalmsson, 2007). (B) A hypothesis for changed patterns of distribution and migration of capelin in recent years. Blue: Nursery area. Green: Feeding grounds of adult capelin, Spring and summer feeding migrations are indicated by blue arrows, but autumn return migrations by green arrows. Red: Spawning migrations (arrows) and main spawning areas. Violet arrows: 0-group drift. (After Pálsson et al., 2012b).

Capelin

A major shift in distribution and migration pattern of capelin appears to have taken place around the turn of the century (Figure 14). A northward shift in distribution of 0-group capelin away from the Icelandic continental shelf waters north off Iceland was first observed in 2003, coinciding with the onset of a greatly reduced recruitment to the capelin stock (Pálsson et al., 2012a,b, 2014). This situation seems to have prevailed ever since (Marine Research Institute, 2014). The distribution of adult capelin has also shifted westwards toward the coast of Greenland (Figure 14).

Euphausiids

Figure 15 shows total abundance of euphausiids in South Icelandic waters (area A6 in Silva et al., 2014) during 1958–2007 as recorded by the CPR. A significant decreasing trend was observed in this period (simple linear regression $r^2 = 0.25$, $p < 0.001$).

DISCUSSION

During the past two decades, pronounced changes have occurred in the marine environment around Iceland. Following 1995, an increase in temperature and salinity was observed in the waters south and west of Iceland (Figure 2), which prevailed for almost two decades and has been reflected in the biota (see below). Generally, salinity and temperature have also been higher in the waters north of Iceland since the late 1990s (Figure 3). These changes have been linked with the changed distribution and recruitment of capelin north of Iceland (Pálsson et al., 2012a,b, 2014). Concurrent with increasing temperatures and salinity and a northward shift of fronts following increased flow of Atlantic water to the areas north of Iceland, marked changes have occurred in the distribution of several fish species, including a northward shift in haddock and monkfish (Astthorsson et al., 2007; Solmundsson et al., 2010; Valdimarsson et al., 2012). At the same time an expansion in mackerel distribution westwards and northwards into Icelandic waters took place (Astthorsson et al., 2012). Due to the large size of the stock, this "mackerel invasion" into Icelandic waters represents a substantial change in the ecosystem of the Icelandic continental shelf. In addition, 0-group fish of the cold-water species capelin retreated northwards on the Icelandic shelf while the mature component has progressively spread along the shelf break off East Greenland (Vilhjalmsson, 2007; Pálsson et al., 2012a). The sand eel population off southern and western Iceland has been in decline and was drastically reduced around 2005 because of recruitment failure (Bogason and Lilliendahl, 2009). The oceanographic and ecological changes are both likely to have contributed to the changes observed in distribution and abundance of several cetacean populations.

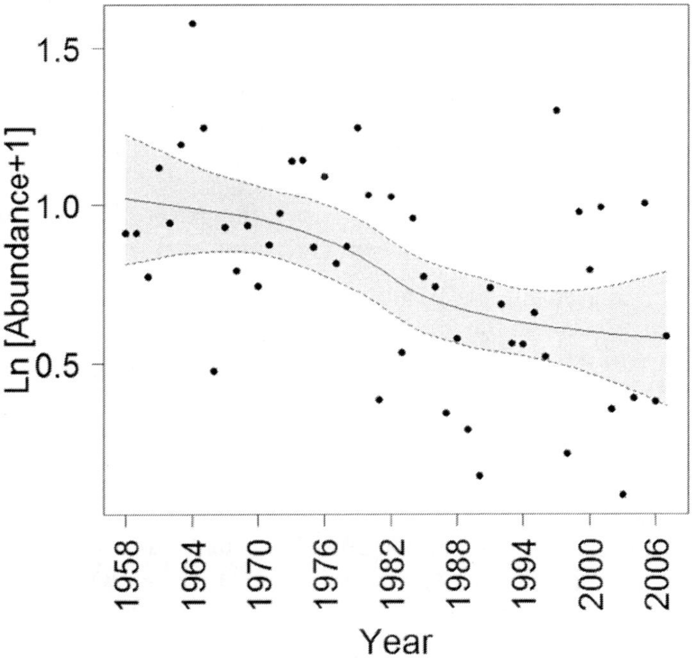

Figure 15. Total euphausiid abundance (Ind m−3) in area A6 from 1958 to 2007 as recorded by the Continuous Plankton recorder (CPR). Abundance values indicate annual averages and a loess curve (span = 0.71) with 95% confidence intervals (shaded gray area) is fitted to the data. Refer to Figure 1 in Silva et al. (2014) for location of CPR area.

Common Minke Whales

Densities of common minke whales in the Icelandic continental shelf area decreased markedly between 2001 and 2007 and the estimated abundance in 2009 was less than a quarter of that in 2001 (Figure 5). The shipboard component of the 2007 survey failed to detect a corresponding increase in abundance of minke whales in the offshore regions of the CNA stock area, although large areas north of Iceland and along the coast of East Greenland received no or poor coverage due to unfavorable weather and ice conditions (Pike et al., 2010a). Therefore, the details of a potential shift in distribution are unknown.

Investigations of stomach contents of minke whales have shown pronounced changes in diet composition in recent decades. During 1977–1997, sand eel was the predominant prey in southern and western Iceland while capelin occurred in 42% and euphausiids in 58% of the stomachs in the northern areas (Sigurjónsson et al., 2000). During 2003–2007 a significant change had occurred in the minke whales' diet composition with the contribution of euphausiids and capelin decreasing from 45 to 17% and 36 to 22%, respectively in comparable areas (Víkingsson et al., 2014). The proportions of large gadoids [mainly haddock and cod (*Gadus morhua*)] and herring increased coincidently between these two periods. The overall percentage of sand eel in minke whale stomachs was still high in the latter period, but decreased appreciably within the period from 90% in 2003 to 20% in 2007 in the southern areas (Víkingsson et al., 2014).

These dietary changes coincide well with the documented changes in the local abundance of the prey species sand eel, capelin and euphausiids (Astthorsson et al., 2007; Marine Research Institute, 2014). The massive decrease in sand eel abundance during 1997–2003 indicated by the analyses of haddock stomachs is supported by decreased breeding success of sand eel-dependent seabirds in southern Iceland (Lilliendahl et al., 2013; Vigfusdottir et al., 2013). The subsequent low levels of sand eels in minke whale stomachs are in agreement with data from a sand eel monitoring program initiated in 2006 (Bogason and Lilliendahl, 2009). The decreased proportion of capelin in the minke whale diet is also consistent with lower availability of capelin in north Icelandic waters during summer (Figure 14) and the increased contribution of haddock and herring reflects growth in these stocks (Marine Research Institute, 2014). These diet studies, together with the series of abundance estimates, suggest that minke whales appear to have responded to the environmental changes both by changes in diet and by leaving the Icelandic continental shelf area. While it seems likely that some of the whales followed the capelin to the coast of East Greenland, this cannot be verified because of the lack of coverage of that area by the 2007 survey.

Fin Whales

The habitat use modeling indicated that fin whale summer feeding areas are characterized by water depth, distance to the 2000 m isobath, SST and SSH (Table 2, Figure 9). Inter-annual variations in the predicted distribution of fin whales showed that increased use of the Irminger Sea in 2001 and 2007 (Figures 10D,E) coincided with an increase in SSH in the area. Sea-level changes are often related to ocean heat-content changes, as water expands with increasing temperatures (Church et al., 2013). The observed changes in SST and SSH over the years have previously been reported to have ecosystem consequences (Ottersen et al., 2001; Häkkinen and Rhines, 2004; Hátún et al., 2009).

While recovery following the pause in whaling in 1990 could be part of the explanation of the increase of fin whale abundance in Icelandic waters, the catch history of the 20th Century and stock assessments suggest that the stock had recovered to a large extent before the period being considered here (Víkingsson et al., 2009). The increase in fin whale abundance in the Irminger Sea between 1989 and 2001 coincided with a marked increase in sea temperature in the area (Figures 4, 10). In particular, the expansion of fin whale distribution into deeper waters (>1000 m) of the Irminger Sea seems to be related to appreciable warming of this area between 1994 and 2003 (Figure 4) as noted by Víkingsson and Valdimarsson (2006). Assuming that the diet of fin whales in this area consists overwhelmingly (> 90%) of euphausiids, mainly *Meganyctiphanes norvegica* (Víkingsson, 1997), it is tempting to conclude that the increased temperatures have facilitated growth in euphausiid abundance and thereby increased carrying capacity for fin whales (Víkingsson et al., 2009). However, there has been a decrease in euphausiids in shelf and oceanic habitats southwest, south and southeast of Iceland from 1958 to 2007 with the lowest numbers in each region in 2007 (Figure 14), despite increasing primary production in terms of phytoplankton in those areas (Silva et al., 2014).

Results from the habitat modeling presented here support a link between environmental conditions and fin whale abundance. However, environmental data generally act as proxies of other variables, to which the whales respond directly (Pershing et al., 2009). Prey availability often forms the missing link between oceanographic variables and higher-trophic level predators (Croll et al., 2005; Friedlaender et al., 2006; Anderwald et al., 2012). The key to understanding this link is to understand how fin whales respond to changes in prey availability, as well as how prey itself is affected by changing environmental conditions. For instance, the predicted highest abundance of fin whales around the 2000 m isobaths and virtual absence from shallow coastal waters is in accordance with the preference of the fin whale's main prey *M. norvegica* for deeper waters. The observed decrease in *M. norvegica* after 1995 could be an indication of a change in fin whale diet in recent years, similar to the diet shift in minke whales (Víkingsson et al., 2014). Such a shift would explain a less concentrated distribution along the shelf edge and higher observed and predicted numbers of fin whales throughout the Irminger Sea. Increased predation from the growing fin whale population might also have contributed to the decrease in euphausiid abundance.

Unfortunately, euphausiid monitoring by the extensive long-term Continuous Plankton Recorder (CPR, SAHFOS) series only reaches as far north as the waters south and southwest of Iceland and recent Icelandic research has mostly been confined to continental shelf waters where different euphausiid species prevail (Astthorsson and Gislason, 1997). Therefore, these studies are not necessarily representative of the main fin whale feeding grounds west of Iceland. It is noteworthy that while the CPR data show a decrease in adult euphausiid abundance during 1960–2007 (Figure 14), there was an increase in the density of euphausiid larvae during 1990–2010 in southwestern Icelandic waters (Silva et al., 2014). While this apparent discrepancy might possibly be explained by the different time frames of these two data series or by a weakened temporal synchrony between the development of young euphausiids and the phytoplankton bloom it would also be consistent with increased predation pressure on adult euphausiids, e.g., by mackerel and fin whales. The warming of the deep waters of the Irminger Sea might also have facilitated growth of other potential prey for fin whales. Updated studies on diet composition of fin whales might clarify this.

Overall, the habitat modeling results suggest that the weakening of the Subpolar Gyre, with associated changes in SST and salinity, had bottom-up effects on primary and secondary production. The poor spatial and temporal overlap between the CPR data and NASS cetacean data precluded the inclusion of krill abundance as a covariate. The explanatory power of the GAM might have been improved if data on prey biomass or primary production had been available because baleen whale distribution has been reported to be tightly coupled with zooplankton density (Friedlaender et al., 2006). However, the dynamic nature of krill requires close spatio-temporal proximity in whale sightings and measurements of krill to detect this trophic link; temporal lags as a result of independently collected prey and cetacean data can weaken the association (Laidre et al., 2010). Further work is therefore required to clarify possible changes in prey availability and diet composition of fin whales.

The fin whale catch distribution in 2014 was highly unusual with an unprecedented pattern relative to the history of the whaling operation dating back to 1948. Historically, sei whales have been caught in considerable numbers in this area, and these catches would certainly have included fin whales had they been encountered. Therefore, the unusual catch distribution in 2014 seems to indicate a sudden shift in distribution of fin whales between 2013 and 2014. According to the whalers, the traditional western area looked lifeless with no visible krill patches at the sea surface and unusually few seabirds. A notable exception to this was the widespread presence in the area of large schools of mackerel that were clearly visible at the surface. These observations from the whalers have been confirmed by a 2014 ecosystem survey that recorded high abundance of mackerel in the Irminger Sea (Nøttestad et al., 2014).

Humpback Whales

Catch statistics from the first period of modern whaling in Icelandic waters, the so-called Norwegian whaling period 1883–1915, indicate that humpback whales were hunted down to very low levels. Only six humpback whales were caught after whaling was resumed in 1948 until they received total protection again in 1955. Sigurjónsson and Gunnlaugsson (1990) reported an annual increase in relative abundance of 14.8% during 1979–1988 west of Iceland. This growth rate, coupled with the small absolute abundance estimate of around 2000 humpback whales in 1987 (Gunnlaugsson and Sigurjónsson, 1990) indicates that the number of animals in the area must still have been very small prior to 1970. Pike et al. (2009b) reported annual rates of increase in abundance of 12% for the period 1987–2001.

The most recent data presented here suggest that this rapid growth may have leveled off around the turn of the century (after the 1995 survey). Feeding habits of humpback whales in Icelandic waters are not well corroborated by stomach contents data. However, visual observations have indicated that capelin and euphausiids likely constitute important parts of the diet (MRI unpublished information). In particular numerous reports from capelin fishermen and capelin research cruises have documented substantial numbers of humpback whales following the migration of capelin in Icelandic waters during winter (Gunnlaugsson and Víkingsson, 2014). In recent years, humpback whale songs associated with mating behavior have been recorded in North Icelandic waters (Magnúsdóttir et al., 2014). Although this acoustic activity does not demonstrate that breeding actually takes place in Icelandic waters, an establishment of a new breeding area, in addition to the two known southern breeding areas, could help explain the large population increase in Icelandic waters and the apparent contradiction between abundance estimates at feeding and breeding grounds of the North Atlantic, as discussed by Smith and Pike (2009).

The increase in humpback whale abundance in Icelandic waters, leveling off around the year 2000, follows a pattern similar to the rise in temperature (Figure 2). However, it is hard to relate this growth to any known biological changes in the marine environment because the presumed most important prey species (capelin, euphausiids, and possibly sand eel) have not followed the same trends in abundance. In recent decades, high rates of increase have also been documented for humpback whales in several other areas without a clear explanation (Bannister, 1994; Stevick et al., 2003; Findlay et al., 2011; Heide-Jørgensen et al., 2012).

Blue Whales

Blue whale populations have not recovered to the same extent as the closely related fin whales or humpback whales, with estimated abundance in the CNA of around 1000 animals (Pike et al., 2009c). However, a significant increase in abundance has been reported for the period 1969–2001 (Sigurjónsson and Gunnlaugsson, 1990; Pike et al., 2009c). During 1987–2001 the increase rate was higher in Northeastern Icelandic waters than west of Iceland (Pike et al., 2009c) indicating a northward shift in relative distribution. Such a shift is consistent with anecdotal evidence from whale watching operations in Icelandic waters. A whale watching company specializing on blue whales was operated from the Snæfellsnes peninsula, West Iceland during 1996–2004. From around 2000, the number of encounters with blue whales decreased appreciably leading the company to give up their blue whale tours in 2004 due to scarcity of blue whales (Pétur Ágústsson, pers. comm). During this period of decline in West Iceland, blue whale sightings increased in the whale watching area Skjálfandi Bay in Northeast Iceland. Photo-identification matches have shown that at least some of the whales previously frequenting West Icelandic waters now occur in Northeastern Icelandic waters during midsummer (MRI, unpublished data).

CONCLUSIONS

Pronounced oceanographic changes have occurred in Icelandic waters since the mid-1990s, including a rise in sea water temperature and increased flow of warm Atlantic water into the waters north of Iceland. Although the exact mechanisms remain unclear, these changes appear to have caused a northward shift in the distribution of several fish species, a decrease in krill abundance and a collapse in the sand eel population off southern and western Iceland. Considerable changes in distribution and abundance of several cetacean species are apparent from the series of cetacean surveys dating back to 1986. Some of these changes appear to be related to these oceanographic and biological changes while others are harder to explain. Thus, the decrease in common minke whale abundance in the Icelandic continental shelf area seems to be related to the decrease in the abundance of the preferred prey species, sand eel in the southern part and capelin in the northern part. The apparent shift of blue whales from southwestern to northeastern Icelandic waters may be related to the decrease in euphausiid abundance in the waters south and southwest of Iceland.

Concurrently with increasing sea temperature in the deep waters of the Irminger Sea, the distribution of fin whales expanded into this area and the total abundance of fin whales in the Irminger Sea increased. However, the relationship between the rise in temperature and fin whale abundance is unclear. A potential increase in euphausiids, the fin whale's dominant prey species in this area, is not supported by CPR data and could instead indicate that fin whales have switched prey. However, data from the main fin whale feeding grounds are lacking because the CPR data series covers only the southernmost part of that area.

Humpback whales were rare in Icelandic waters throughout most of the 20th century but a significant increase occurred between 1970 and 2001 in the CNA. From 1995, abundance has been rather stable at around 10–15,000 whales. Feeding ecology of humpback whales in Icelandic waters is poorly documented but capelin and euphausiids are known to be among their prey species. It is hard to relate the increased abundance of humpback whales since the 1970s to any known biological changes in the marine environment.

Continued monitoring of the distribution and abundance of cetaceans as well as further studies into their feeding ecology are essential for better understanding of the recent and ongoing changes documented here.

ACKNOWLEDGMENTS

The authors thank Ásta Guðmundsdóttir, Ólafur K. Pálsson, Ástþór Gíslason, Sverrir D. Halldórsson, and Valerie Chosson at the MRI for fruitful discussions and for help in preparing the data and providing figures. Thanks are also extended to Priscilla Licandro and SAHFOS for providing CPR data for total euphausiids. Funding for Teresa Silva was provided by the EURO-BASIN: European Union Basin-scale Analysis, Synthesis and Integration (FP7 Contract No. 264933).

REFERENCES

1. Anderwald, P., Evans, P., Dyer, R., Dale, A., Wright, P., and Hoelzel, A. (2012). Spatial scale and environmental determinants in minke whale habitat use and foraging. *Mar. Ecol. Prog. Ser.* 450, 259–274. doi: 10.3354/meps09573
2. Astthorsson, O. S., and Gislason, A. (1997). Biology of euphausiids in the subarctic waters north of Iceland. *Mar. Biol.* 129, 319–330. doi: 10.1007/s002270050172
3. Astthorsson, O. S., Gislason, A., and Jonsson, S. (2007). Climate variability and the Icelandic marine ecosystem. *Deep Sea Res. Part II Top. Stud. Oceanogr.* 54, 2456–2477. doi: 10.1016/j.dsr2.2007.07.030
4. Astthorsson, O. S., Valdimarsson, H., Gudmundsdottir, A., and Óskarsson, G. J. (2012). Climate-related variations in the occurrence and distribution of mackerel (*Scomber*

scombrus) in Icelandic waters. *ICES J. Mar. Sci. J. Cons.* 69, 1289–1297. doi: 10.1093/icesjms/fss084

5. Bannister, J. L. (1994). Continued increase in Group IV humpbacks off Western Australia. *Rep. Int. Whal. Comm.* 44, 309–310.
6. Barry, S., and Welsh, A. (2002). Generalized additive modelling and zero inflated count data. *Ecol. Modell.* 157, 179–188. doi: 10.1016/S0304-3800(02)00194-1
7. Batten, S., Clark, R., Flinkman, J., Hays, G., John, E., John, A., et al. (2003). CPR sampling: the technical background, materials and methods, consistency and comparability. *Prog. Oceanogr.* 58, 193–215. doi: 10.1016/j.pocean.2003.08.004
8. Bersch, M. (2002). North Atlantic Oscillation–induced changes of the upper layer circulation in the northern North Atlantic Ocean. *J. Geophys. Res.* 107, 3156. doi: 10.1029/2001JC000901
9. Bogason, V., and Lilliendahl, K. (2009). Rannsóknir á sandsíli. *Hafrannsóknir* 145, 36–41. Available online at: http://www.hafro.is/Bokasafn/Timarit/fjolrit-145.pdf
10. Borchers, D. L., Pike, D. G., Gunnlaugsson, T., and Víkingsson, G. A. (2009). Minke whale abundance estimation from the NASS 1987 and 2001 aerial cue–counting surveys taking appropriate account of distance estimation errors. *NAMMCO Sci. Publ.* 7, 95–110. doi: 10.7557/3.2708
11. Church, J., White, N., Domingues, C., Monselesan, D., and Miles, E. (2013). Sea-level and ocean heat-content change. *Int. Geophys. Ocean Circ. Clim.* 103, 697–725. doi: 10.1016/B978-0-12-391851-2.00027-1
12. Clapham, P. J. (2009). "Humpback whale, *Megaptera novaeangliae*," in *Encyclopedia of Marine Mammals*, 2nd Edn., eds W. F. Perrin, B. Würsig, and J. G. M. Thewissen (Amsterdam; Boston; Heidelberg; London; New York; Oxford; Paris; San Diego; San Fransisco; Singapore; Sydney; Tokyo: Academic Press; Elsevier), 582–585.
13. Croll, D., Marinovic, B., Benson, S., Chavez, F., Black, N., Ternullo, R., et al. (2005). From wind to whales: trophic links in a coastal upwelling system. *Mar. Ecol. Prog. Ser.* 289, 117–130. doi: 10.3354/meps289117
14. Donovan, G. P. (1991). A review of IWC stock boundaries. *Rep. Int Whal. Commn.* 13, 39–68.
15. Donovan, G. P., and Gunnlaugsson, T. (1989). North Atlantic Sightings Survey 1987: preliminary report of the aerial survey off Iceland. *Rep. Int. Whal. Commn.* 39, 437–441.
16. Findlay, K., Best, P., and Meÿer, M. (2011). Migrations of humpback whales past Cape Vidal, South Africa, and an estimate of the population increase rate (1988–2002). *Afr. J. Mar. Sci.* 33, 375–392. doi: 10.2989/1814232X.2011.637345
17. Friedlaender, A., Halpin, P., Qian, S., Lawson, G., Wiebe, P., Thiele, D., et al. (2006). Whale distribution in relation to prey abundance and oceanographic processes in shelf waters of the Western Antarctic Peninsula. *Mar. Ecol. Prog. Ser.* 317, 297–310. doi: 10.3354/meps317297
18. Gunnlaugsson, T., and Sigurjónsson, J. (1990). NASS-87: estimation of whale abundance based on observations made onboard Icelandic and Faroese survey vessels. *Rep. Int. Whal. Commn.* 40, 571–580.
19. Gunnlaugsson, T., Sigurjonsson, J., and Donovan, G. P. (1988). Aerial survey of cetaceans in the coastal waters off Iceland, June-July 1986. *Rep. Int. Whal. Commn.* 38, 489–500.
20. Gunnlaugsson, T., and Víkingsson, G. A. (2014). *Winter Occurrence of Whales in Waters Around Iceland.* IWC SC/65b/RMP06. Available online at: https://iwc.int/index.php?cID=sc-documents&page=1
21. Häkkinen, S., and Rhines, P. B. (2004). Decline of subpolar North Atlantic circulation during the 1990s. *Science* 304, 555–559. doi: 10.1126/science.1094917
22. Hastie, T., and Tibshirani, R. (1990). *Generalized Additive Models.* Boca Raton, FL: Chapman Hall; CRC, 43.
23. Hátún, H., Hansen, B., Sandø, A. B., Drange, H., and Valdimarsson, H. (2005). De–stabilization of the North Atlantic thermohaline circulation by a gyre mode. *Science* 309, 1841–1844. doi: 10.1126/science.1114777

24. Hátún, H., Payne, M. R., Beaugrand, G., Reid, P. C., Sandø, A., Drange, H., et al. (2009). Large bio-geographical shifts in the north-eastern Atlantic Ocean: from the subpolar gyre, via plankton, to blue whiting and pilot whales. *Prog. Oceanogr.* 80, 149–162. doi: 10.1016/j.pocean.2009.03.001
25. Heide-Jørgensen, M., Laidre, K., Hansen, R., Burt, M., Simon, M., Borchers, D., et al. (2012). Rate of increase and current abundance of humpback whales in West Greenland. *J. Cetacean Res. Manag.* 12, 1–14.
26. Hersteinsson, P. (2004). *Íslensk Spendýr (Icelandic Mammals).* Reykjavík: Vaka-Helgafell.
27. Hiby, A. R., and Hammond, P. S. (1989). Survey techniques for estimating abundance of cetaceans. *Rep. Int. Whal. Commn.* 11, 47–80.
28. Laidre, K., Heide-Jørgensen, M., Heagerty, P., Cossio, A., Bergström, B., and Simon, M. (2010). Spatial associations between large baleen whales and their prey in West Greenland. *Mar. Ecol. Prog. Ser.* 402, 269–284. doi: 10.3354/meps08423
29. Lilliendahl, K., Hansen, E. S., Bogason, V., Sigursteinsson, M., Magnúsdóttir, M. L., Jónsson, P. M., et al. (2013). Viðkomubrestur lunda og sandsílis við Vestmannaeyjar (Recruitment failure of puffin and sand eel in Vestmannaeyjar). *Náttúrufræðingurinn* 83, 65–79.
30. Lilliendahl, K., and Solmundsson, J. (2006). Feeding ecology of sympatric European shags *Phalacrocorax aristotelis* and great cormorants, *P. carbo* in Iceland. *Mar. Biol.* 149, 979–990. doi: 10.1007/s00227-006-0259-7
31. Lockyer, C., and Pike, D. G. (eds.). (2009). *North Atlantic Sightings Surveys 1987–2001. Counting Whales in the North Atlantic.* Tromsø: NAMMCO Sci. Publ. 7. North Atlantic Marine Mammal Commission.
32. Magnúsdóttir, E. E., Rasmussen, M. H., Lammers, M. O., and Svavarsson, J. (2014). Humpback whale songs during winter in subarctic waters. *Polar Biol.* 37, 427–433. doi: 10.1007/s00300-014-1448-3
33. Malmberg, S.-A., and Valdimarsson, H. (2003). Hydrographic conditions in Icelandic waters, 1990–1999. *ICES Mar. Sci. Symp.* 219, 50–60.
34. Marine Research Institute. (2014). *Ástand nytjastofna á Íslandsmiðum 2013/2014 Aflahorfur fiskveiðiárið 2014/2015 (State of Marine Stocks in Icelandic Waters 2013/2014—Prospects for the Quota Year 2014/2015).* Reykjavík: Marine Research Institute.
35. Meyssignac, B., Becker, M., Llovel, W., and Cazenave, A. (2012). An assessment of two-dimensional past sea level reconstructions over 1950–2009 based on tide-gauge data and different input sea level grids. *Surv. Geophys.* 33, 945–972. doi: 10.1007/s10712-011-9171-x
36. Miller, D. L., Burt, M. L., Rexstad, E. A., and Thomas, L. (2013). Spatial models for distance sampling data: recent developments and future directions. *Methods Ecol. Evol.* 4, 1001–1010. doi: 10.1111/2041-210X.12105
37. Mortensen, J., and Valdimarsson, H. (1999). "Thermohaline changes in the irminger sea," in *International Council for the Exploration of the Sea*, CM Doc. 1999/L16, 11 (Copenhagen).
38. NAMMCO. (2009). "Report of NAMMCO Scientific Committee on Abundance Estimates," in *NAMMCO Annual Report 2007–2008* (Tromsø: North Atlantic Marine Mammal Commission).
39. Nøttestad, L., Salthaug, A., Johansen, G. O., Anthonypillai, V., Tangen, Ø., Utne, K., et al. (2014). *Cruise Report from the Coordinated Ecosystem Survey (IESSNS) with M/V "Brennholm," M/V "Vendla," M/V "Finnur Fríði" and R/V "Árni Friðriksson" in the Norwegian Sea and surrounding waters.* ICES. Available online at: http://www.hafro.is/images/frettir/2014/IESSNS_2014_Cruise_Report_Ecosystem_Survey_July-August.pdf
40. Ottersen, G., Planque, B., Belgrano, A., Post, E., Reid, P., and Stenseth, N. (2001). Ecological effects of the North Atlantic Oscillation. *Oecologia* 128, 1–14. doi: 10.1007/s004420100655

41. Pálsson, Ó. K., Gislason, A., Guðfinnsson, H. G., Gunnarsson, B., Ólafsdóttir, S. R., Petursdottir, H., et al. (2012a). Ecosystem structure in the Iceland Sea and recent changes to the capelin (*Mallotus villosus*) population. *ICES J. Mar. Sci. J. Cons.* 69, 1242–1254. doi: 10.1093/icesjms/fss071
42. Pálsson, Ó. K., Gíslason, Á., Gunnarsson, B., Guðfinnsson, H. G., Valdimarsson, H., Pétursdóttir, H., et al. (2014). Meginþættir í vistkerfi Íslandshafs og breytingar á lífsháttum loðnu. *Náttúrufræðingurinn* 85, 4–18.
43. Pálsson, Ó. K., Sveinbjörnsson, S., Valdimarsson, H., Gíslason, Á., and Vilhjálmsson, H. (2012b). Lífshættir loðnu í Íslandshafi. *Hafrannsóknir* 164, 119–132. Available online at: http://www.hafro.is/Bokasafn/Timarit/fjolrit-164.pdf
44. Pershing, A., Record, N., Monger, B., Mayo, C., Brown, M., Cole, T., et al. (2009). Model-based estimates of right whale habitat use in the Gulf of Maine. *Mar. Ecol. Prog. Ser.* 378, 245–257. doi: 10.3354/meps07829
45. Pike, D. G. (2009). North Atlantic Sightings Surveys. Introduction. *NAMMCO Sci. Publ.* 7, 7–18. doi: 10.7557/3.2702
46. Pike, D. G., Gunnlaugsson, T., Elvarsson, B., and Víkingsson, G. A. (2011a). *Correcting Perception Bias for Icelandic Aerial Surveys, 2007 and 2009.* NAMMCO SC18AESP08, 12. Available online at: https://iwc.int/index.php?cID=sc-documents&page=1
47. Pike, D. G., Gunnlaugsson, T., and Víkingsson, G. A. (2010b). *Icelandic Aerial Survey 2009: Survey Report and a Preliminary Abundance Estimate for Minke Whales.* NAMMCO SC/17/AS/07, 29. Available online at: https://iwc.int/index.php?cID=sc-documents&page=1
48. Pike, D. G., Gunnlaugsson, T., and Víkingsson, G. A. (2011b). *Icelandic Aerial Survey 2009: Survey Report and a Preliminary Abundance Estimate for Minke Whales.* IWC SC/60/PFI12, 29. Available online at: https://iwc.int/index.php?cID=sc-documents&page=1
49. Pike, D. G., Gunnlaugsson, T., Víkingsson, G. A., Desportes, G., and Bloch, D. (2009a). Estimates of the abundance of minke whales (*Balaenoptera acutorostrata*) from Faroese and Icelandic NASS shipboard surveys. *NAMMCO Sci. Publ.* 7, 81–93. doi: 10.7557/3.2707
50. Pike, D. G., Gunnlaugsson, T., Víkingsson, G. A., Desportes, G., and Mikkelsen, B. (2010c). *Estimates of the Abundance of Humpback Whales (Megaptera novaeangliae) from the T-NASS Icelandic and Faroese Ship Surveys Conducted In 2007.* IWC, SC/62/O13, 15. Available online at: https://iwc.int/index.php?cID=scdocuments&page=1
51. Pike, D. G., Gunnlaugsson, T., Víkingsson, G. A., and Mikkelsen, B. (2008). *Estimates of the Abundance of Fin Whales (Balaenoptera physalus) from the T-NASS Icelandic and Faroese Ship Surveys Conducted in 2007.* IWCSC/60/PFI13-revised. Available online at: https://iwc.int/index.php?cID=sc-documents&page=1
52. Pike, D. G., Gunnlaugsson, T., Víkingsson, G. A., and Mikkelsen, B. (2010a). *Estimates of the Abundance of Minke Whales (Balaenoptera acutorostrata) from the T-NASS Icelandic and Faroese Ship Surveys Conducted in 2007.* IWC SC62RMP5. Available online at: https://iwc.int/index.php?cID=scdocuments&page=1
53. Pike, D. G., Paxton, C. G. M., Gunnlaugsson, T., and Víkingsson, G. A. (2009b). Trends in the distribution and abundance of cetaceans from aerial surveys in Icelandic coastal waters, 1986-2001. *NAMMCO Sci. Publ.* 7, 117–142. doi: 10.7557/3.2710
54. Pike, D. G., Víkingsson, G. A., Gunnlaugsson, T., and Øien, N. (2009c). A note on the distribution and abundance of blue whales (*Balaenoptera musculus*) in the Central and Northeast North Atlantic. *NAMMCO Sci. Publ.* 7, 19–29. doi: 10.7557/3.2703
55. R Core Team. (2013). *R: A Language and Environment for Statistical Computing.* Vienna: R Foundation for Statistical Computing. Available online at: http://www.R-project.org/
56. Sears, R., and Perrin, W. (2009). "Blue whale, *Balaenoptera musculus*," in *Encyclopedia of Marine Mammals*, 2nd Edn., eds W. F. Perrin, B. Würsig, and J. G. M. Thewissen (Amsterdam; Boston; Heidelberg; London; New York; Oxford; Paris;

San Diego; San Fransisco; Singapore; Sydney; Tokyo: Academic Press; Elsevier), 120–124.

57. Sigurjónsson, J. (1988). Operational factors of the Icelandic large whale fishery. *Rep. Int. Whal. Comm.* 38, 327–333.

58. Sigurjónsson, J. (1992). "Recent studies on abundance and trends in whale stocks in Icelandic and adjacent waters," in *Royal Academy of Overseas Sciences*, ed J. J. Symoens (Brussels: Royal Academy of Overseas Sciences), 77–111.

59. Sigurjónsson, J., Galan, A., and Vikingsson, G. A. (2000). A note on stomach contents of minke whales (*Balaenoptera acutorostrata*) in Icelandic waters. *NAMMCO Sci. Publ.* 2, 82–90. doi: 10.7557/3.2973

60. Sigurjónsson, J., and Gunnlaugsson, T. (1990). Recent trends in abundance of blue (*Balaenoptera musculus*) and humpback whales (*Megaptera novaeangliae*) off west and southwest Iceland, with a note on occurrence of other cetacean species. *Rep. Int. Whal. Commn* 40, 537–551.

61. Sigurjónsson, J., Gunnlaugsson, T., Ensor, P., Newcomer, M., and Víkingsson, G. (1991). North Atlantic sightings survey 1989 (NASS 89). Shipboard surveys in Icelandic and adjacent waters July-August 1989. *Rep. Int. Whal. Commn.* 41, 559–572.

62. Sigurjónsson, J., Gunnlaugsson, T. H., and Payne, M. (1989). Nass 87. Shipboard sightings surveys in Icelandic and adjacent waters June-July 1987. *Rep. Int. Whal. Commn.* 39, 395–409.

63. Sigurjónsson, J., and Víkingsson, G. A. (1997). Seasonal abundance of and estimated food consumption by cetaceans in Icelandic and adjacent waters. *J. Northwest Atl. Fish. Sci.* 22, 271–287. doi: 10.2960/J.v22.a20

64. Silva, T., Gislason, A., Licandro, P., Marteinsdóttir, G., Ferreira, A. S. A., Gudmundsson, K., et al. (2014). Long-term changes of euphausiids in shelf and oceanic habitats southwest, south and southeast of Iceland. *J. Plankton Res.* 36, 1262–1278. doi: 10.1093/plankt/fbu050

65. Smith, T. D., and Pike, D. G. (2009). The enigmatic whale: the North Atlantic humpback. *NAMMCO Sci. Publ.* 7, 161–178. doi: 10.7557/3.2712

66. Solmundsson, J., Jonsson, E., and Bjornsson, H. (2010). Phase transition in recruitment and distribution of monkfish (*Lophius piscatorius*) in Icelandic waters. *Mar. Biol.* 157, 295–305. doi: 10.1007/s00227-009-1317-8

67. Stevick, P. T., Allen, J., Clapham, P. J., and Lien, J. (2003). North Atlantic humpback whale abundance and rate of increase four decades after protection from whaling. *Mar. Ecol. Prog. Ser.* 258, 263–273. doi: 10.3354/meps258263

68. Thomas, L., Buckland, S. T., Rexstad, E. A., Laake, J. L., Strindberg, S., Hedley, S. L., et al. (2010). Distance software: design and analysis of distance sampling surveys for estimating population size. *J. Appl. Ecol.* 47, 5–14. doi: 10.1111/j.1365-2664.2009.01737.x

69. Valdimarsson, H., Astthorsson, O. S., and Palsson, J. (2012). Hydrographic variability in Icelandic waters during recent decades and related changes in distribution of some fish species. *ICES J. Mar. Sci. J. Cons.* 69, 816–825. doi: 10.1093/icesjms/fss027

70. Ver Hoef, J., and Boveng, P. (2007). Quasi-Poisson vs. negative binomial regression: how should we model overdispersed count data? *Ecology* 88, 2766–2772. doi: 10.1890/07-0043.1

71. Vigfusdottir, F., Gunnarsson, T. G., and Gill, J. A. (2013). Annual and between-colony variation in productivity of Arctic Terns in West Iceland. *Bird Study* 60, 289–297. doi: 10.1080/00063657.2013.811214

72. Víkingsson, G. A. (1997). Feeding of fin whales (*Balaenoptera physalus*) off Iceland—diurnal and seasonal variation and possible rates. *J. Northw. Atl. Fish. Sci.* 22, 77–89. doi: 10.2960/J.v22.a7

73. Víkingsson, G. A., Elvarsson, B. T., Ólafsdottir, D., Sigurjonsson, J., Chosson, V., and Galan, A. (2014). Recent changes in the diet composition of common minke whales (*Balaenoptera acutorostrata*) in Icelandic waters. A consequence of climate change? *Mar. Biol. Res.* 10, 138–152. doi: 10.1080/17451000.2013.793812

74. Víkingsson, G. A., Ólafsdóttir, D., and Sigurjónsson, J. (2003). Geographical and seasonal variation in the diet of harbour porpoises (*Phocoena phocoena*) in Icelandic coastal water. *NAMMCO Sci. Publ.* 5, 243–270. doi: 10.7557/3.2829
75. Víkingsson, G. A., Pike, D. G., Desportes, G., Öien, N., Gunnlaugsson, T., and Bloch, D. (2009). Distribution and abundance of fin whales (*Balaenoptera physalus*) in the Northeast and Central Atlantic as inferred from the North Atlantic Sightings Surveys 1987–2001. *NAMMCO Sci. Publ.* 7, 49–72. doi: 10.7557/3.2705
76. Víkingsson, G. A., and Valdimarsson, H. (2006). Hvalir og umhverfisþættir (Whales and the environment). *Hafrannsoknir. Fjölrit* 125, 30–33.
77. Vilhjalmsson, H. (2007). Impact of changes in natural conditions on ocean resources. *Law Sci. Ocean Manage.* 11, 225. doi: 10.1163/ej.9789004162556.i-0.59
78. Warton, D. I. (2005). Many zeros does not mean zero inflation: comparing the goodness-of-fit of parametric models to multivariate abundance data. *Environmetrics* 16, 275–289. doi: 10.1002/env.702
79. Williams, R., Vikingsson, G. A., Gislason, A., Lockyer, C., New, L., Thomas, L., et al. (2013). Evidence for density-dependent changes in body condition and pregnancy rate of North Atlantic fin whales over four decades of varying environmental conditions. *ICES J. Mar. Sci.* 70, 1273–1280. doi: 10.1093/icesjms/fst059
80. Wood, S. (2001). mgcv: GAMs and generalized ridge regression for R. *R. News* 1, 20–25.
81. Wood, S. N. (2006). *Generalized Additive Models: An Introduction With R*. Boca Raton, FL: Chapman & Hall.
82. Zuur, A. F., Ieno, E. N., Walker, N., Saveliev, A. A., and Smith, G. M. (2009). Mixed effects models and extension in ecology with R. *Stat. Biol. Heal.* 413, 579. doi: 10.1007/978-0-387-87458-6_1

CHAPTER 14

Influence of Floodplain Area on Fish Species Richness in Waterbodies of the Chao Phraya River Basin, Thailand

Wataru Tanaka[1], Rottapon Wattanasiriserekul[2], Yuta Tomiyama[1], Tomomi Yamasita[1], Wikit Phinrub[2], Tosapol Chamnivikaipong[2], Apinun Suvarnaraksha[2], Yukihiro Shimatani[1]

[1]*Department of Civil Engineering, Faculty of Engineering, Kyushu University, Fukuoka, Japan*

[2]*Faculty of Fisheries Technology and Aquatic Resources, Maejo University, Chiang Mai, Thailand*

ABSTRACT

River-floodplain ecosystems are in delicate balance and are impacted by even minor changes in water availability. In this study, we surveyed fish assemblages and investigated environmental and landscape parameters in a total of 135 floodplain waterbodies (rivers, diversion canals, ponds, irrigation ditches, paddy fields, and wetlands) in the Chao Phraya River Basin in rainy (September 2014) and dry (March 2015) seasons. Factors affecting fish species richness in each type of waterbody were analyzed using generalized linear mixed models. Floodplain area around each surveyed waterbody was a major factor determining fish species richness in rivers, diversion canals, and ponds in the region. The contribution of floodplain area was equivalent to that of hydrology (current velocity, water depth) and water quality (dissolved oxygen, turbidity) in the waterbodies. The population of juvenile fishes was increased in temporarily connected floodplain waterbodies to main rivers compared with isolated waterbodies, and fluvial and lacustrine fishes were observed in the temporary inundated floodplain waterbodies during the rainy season. The high dependence of fish species richness on floodplain area in the region appeared to be a result of the use of inundated floodplains by fish species to forage and breed. Our results highlight the impact of flood control measures that reduces floodplain area. These measures must be reviewed to ensure the conservation of fish biodiversity in the Chao Phraya River Basin, one of the world's most threatened floodplain systems.

Keywords: Floodplain, Fish Species Richness, Lateral Fish Migration, Chao Phraya River Basin, Flood Pulse Concept

1. INTRODUCTION

Floodplains are riparian ecotones in which there is a transition between terrestrial and aquatic communities [1] . These areas feature high biodiversity and are a highly productive part of the aquatic ecosystem [2] [3] . Previous studies on river-floodplain ecosystems have demonstrated the importance of flooding, which allows aquatic organisms to migrate into inundated floodplains and to use resources from the submerged terrestrial landscape for feeding and breeding. Therefore, the area of floodplains affects the recruitment of aquatic organisms [2] [4] [5] . Changes in or disturbance of the natural flooding cycle creates habitat heterogeneity and gradients in the hydrochemical environment in floodplain waterbodies, which in turn affect the distribution and species richness of aquatic organisms [6] - [8] .

Many studies have shown that the species richness of various biota is affected by floods in large river systems around the world, for example, fishes and phytoplankton in the Danube River in Europe [7] [9] , hydrochorous plants in the Twentekanaal Canal in the Netherlands [10], zooplankton and phytoplankton in the La Plata River in South America [11] [12] , and fishes and zooplankton in the Amazon River in South America [13] [14] . Although Southeast Asia is home to some of the world's most threatened floodplains, particularly as a result of human activities, there have been few studies conducted on the impact of flooding on species diversity in this region [15] .

The Chao Phraya River is one of the major rivers of Southeast Asia, and its basin has a vast floodplain system. The floodplains are maintained by regular and sustained flood events, and the life cycles of many fish species in the basin depend on these floods [16] . However, flood control measures have been implemented in the Chao Phraya River Basin following the devastation caused by the 2011 Thailand flood. Understanding of the impact of floodplain area and flood events on fishes in the basin is essential to ensure that the flood control measures are adapted to minimize negative effects on the river-floodplain ecosystem. Previous studies about fishes in the basin have focused on taxonomy and biodiversity (e.g., [17] - [19]), whereas ecological aspects have received little attention [16] [20] . In particular, the importance of floodplain area and flood events for fish species richness remains largely unknown. In the present study, fish assemblages and the hydrochemical environments of the main tributaries and floodplain water bodies in the mid-Chao Phraya River Basin were investigated. General linear mixed model (GLMM) analysis was used to determine the importance of flood events and the area of floodplains for the fishes in the region.

2. MATERIALS AND METHODS

2.1. Study Area

The study was conducted in the floodplains of the mid-Chao Phraya River Basin (Figure 1(a)) where the three main tributary rivers of the Chao Phraya River meet. The tributaries are the Ping River with an average discharge of 158 $m^3 \cdot s^{-1}$ and a maximum discharge of 403 $m^3 \cdot s^{-1}$ (P.7A water level station 2014), the Yom River with an average discharge of 101 $m^3 \cdot s^{-1}$ and a maximum discharge of 443 $m^3 \cdot s^{-1}$ (Y.16 water level station 2014), and the Nan River with an average discharge of 201 $m^3 \cdot s^{-1}$ and a maximum discharge of 862 $m^3 \cdot s^{-1}$ (N.7A water level station 2014) [21] . There are no barriers to the dispersal of fishes, such as large dams, on the rivers in the study area. This region has three seasons: a rainy season, typically from mid-May to mid-October, and cool and hot dry seasons, typically from mid-October to mid-May [22] . In the rainy season, floods generally occur from September to October. Flood characteristics among the three sub-basins differ in terms of floodplain area. The floodplain area in the Ping River sub-basin is smaller than that in the Yom and Nan River sub-basins (Figure 1(b)).

2.2. Fish Sampling and Environmental Parameter Assessment

Fish sampling was conducted using cast nets with a 20 mm mesh size and scoop nets with a 2 mm mesh size. The fishing effort at all sampling sites was 20 man-minutes for cast nets and 10 man-minutes for scoop nets. In total, 135 waterbodies were sampled in the Ping River sub-basin, the Yom River sub-basin, and the Nan River sub-basin (Figure 1(a)) in September 2014 (the rainy season: 15 rivers, 19 diversion canals, 12 ponds, 5 irrigation ditches, 18 paddies, and six wetlands) and March 2015 (the dry season: 15 rivers, 18 diversion canals, 12 ponds, 5 irrigation ditches, 11 paddies, and one wetland). The different waterbody types are shown in Figure 2. For safety reasons, sampling from rivers and ponds with a water depth > 2 m was carried out from the edge of the waterbody. Captured fishes were identified at species level and their total length was measured. Most individuals were released at the same site where they were caught, but some were preserved in 4% formalin when further identification was required. The fishes were identified according to [16] , Apinun (unpublished data), and the online database of Fishes of Mainland Southeast Asia (http://ffish.asia/; [23]), their scientific names following [19] .

At each study site, the following environmental parameters were measured in the field: electric conductivity (EC, μS; PCS Tester 35, Oakton

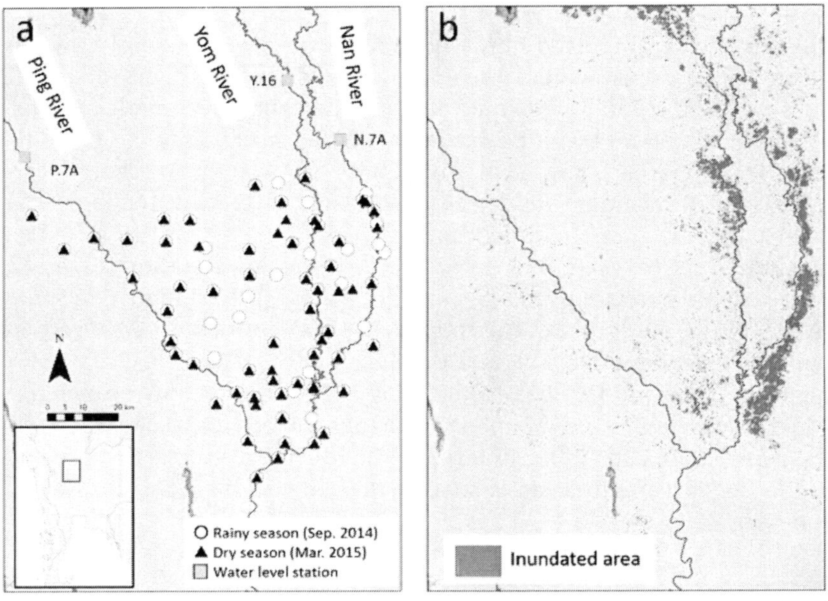

Figure 1. Maps of the study area. (a) Open circles indicate the sites surveyed in the rainy season (September 2014). Black triangles indicate the sites surveyed in the dry season (March 2015), (b) Inundated areas in 2014.

Figure 2. Photographs of the types of waterbodies surveyed. (a) River, (b) Diversion canal, (c) Pond, (d) Irrigation ditch, (e) Paddy field, and (f) Wetland.

Instruments, IL, USA), turbidity (Turb, NTU; HI 93703; Hanna Instruments Japan, Tokyo, Japan), dissolved oxygen (DO, $mg{\cdot}L^{-1}$; DO-5509; Mother Tool Co., Ltd., Ueda, Japan), water depth (Dep, m), and current velocity (Vel, $m{\cdot}s^{-1}$; CR-11; Cosmo-riken Co., Ltd., Kashiwara, Japan). These environmental parameters were measured once at the surface in the center of each study site. In addition to the above environmental parameters, waterbodies were classified as temporarily connected to main rivers by an annual flood or as isolated at the time of survey (Con; 1 for temporary connection, 0 for no connection).

2.3. Landscape Parameter Assessment

At each sampling location, the following landscape parameters were measured by geographic information system (GIS) manipulation: route distance of each waterbody via the channel network from the nearest river (Dist, km), and the floodplain area around each survey site (FA, km^2). The floodplain area was calculated within a buffered circle around each survey site according to [24] and [25] . Because the spatial scale at which the floodplain area should be extracted was not clear, buffers of a given radius (500 m, 1000 m, 2000 m, 3000 m, and 4000 m) were generated around the center of each study site. The floodplain GIS data were obtained from the Thailand Flood Monitoring System [26] . Data for the 2014 inundation range were used and areas that experienced flooding in that year were classified as floodplains. The landscape parameters were calculated using ArcGIS ver. 10.2 (ESRI Japan, Tokyo, Japan).

2.4. Data Analysis

To determine the size composition of fishes in the waterbodies connected to main rivers by an annual flood and in isolated waterbodies, total length histograms were constructed for the species with more than 50 individuals captured from both connected and isolated waterbodies during the rainy season survey (Esomus metallicus, Trichopsis vittata, Trichopodus trichopterus, and Trichopodus microlepis).

A GLMM [27] was used to identify the environmental and landscape parameters that influence fish species richness (Table 1). The analysis was performed for overall fish species richness and for each waterbody type surveyed at twenty or more locations (rivers, diversion canals, ponds, and paddy fields). A correlation matrix between environmental and landscape parameters was calculated to avoid multicollinearity. The Pearson's correlation coefficients (R) were <0.7, indicating that multicollinearity did not distort the GLMMs using these variables as predictors (Table 2) [28] .

In the GLMMs, the dependent values were the species richness of native fish species, migratory native fish species, and non-migratory native fish species in each waterbody type. The floodplain migratory behavior for each fish species was defined according to FishBase (http://www.fishbase.org/). The independent variables were EC, Turb, DO, Dep, Vel, Con, Dist, and FA as fixed factors, and season (rainy season, dry season) and waterbody type (river, diversion canal, pond, irrigation ditch, paddy field, and wetland) as random effect intercepts. A Poisson distribution with log link function was applied to the GLMMs, which were performed for all possible sets of independent variables. Model selection was made using Akaike's information criterion (AIC; [29]). The model with the lowest AIC was defined as the best model.

Table 1. Variables used during GLMM and model selection.

Variables	Description	Variable type	Unit
EC	Electric conductivity of survey site	Continuous	μS
Dep	Water depth of survey site	Continuous	m
Do	Dissolved oxygen of survey site	Continuous	$mg{\cdot}L^{-1}$
Turb	Turbidity of survey site	Continuous	NTU
Vel	Current velocity of survey site	Continuous	$m{\cdot}s^{-1}$
Con	Connected by an annual flood or not	Dummy	0, 1
Dist	Route distance from the nearest river	Continuous	km
FA	Floodplain area around each survey site	Continuous	km^2

It is possible that fish species richness showed spatial autocorrelation among the study sites (e.g., the fish species richness was less in the Ping River sub-basin study sites compared with the Yom and Nan River sub-basins). Species-environment models that have spatial autocorrelation in the residuals may be not appropriate because they can overestimate or underestimate the importance of environmental variables [30] -[32] . To test for spatial autocorrelation, Moran's I [33] was computed using the residuals from the best models. Moran's I is widely used to assess the existence and effect of spatial autocorrelation in species?environment regression models (e.g., [34]). Moran's I ranges from −1 to +1; a positive and high value indicates a spatial dispersion distribution pattern (positive autocorrelation), a zero value indicates a random distribution pattern, and a negative and low value indicates a spatial correlation distribution pattern (negative autocorrelation). The significance of Moran's I was tested by computing the Z score. According to the two-tailed Z test with a significance threshold (P value) of 0.05, P values <0.05 were considered significantly spatially autocorrelated. All statistical analyses were conducted using R statistical software (ver. 3.1.1; R Development Core Team) and its optional packages "lme4" for GLMM, "MuMIn" for model selection, and "spdep" for Moran's I test.

Table 2. Correlation matrix of variables used during GLMM and model selection.

	Total	1	2	3	4	5	6	7	8
1	EC	1							
2	Dep	−0.26	1						
3	Do	−0.02	−0.03	1					
4	Turb	0.03	−0.09	0.06	1				
5	Vel	−0.25	0.31	0.27	0.13	1			
6	Con	−0.15	0.00	−0.03	−0.11	−0.09	1		
7	Dist	0.14	−0.16	−0.05	0.20	−0.29	−0.14	1	
8	FA†	0.09	0.06	−0.16	−0.27	−0.06	0.33	−0.46	1
	River	**1**	**2**	**3**	**4**	**5**	**6**	**7**	**8**
1	EC	1							
2	Dep	−0.32	1						
3	Do	0.21	−0.27	1					
4	Turb	0.06	0.69	0.15	1				
5	Vel	−0.11	0.46	0.46	0.53	1			
6	Con	-	-	-	-	-	-		
7	Dist	-	-	-	-	-	-	-	
8	FA†	−0.36	0.2	−0.38	−0.21	−0.01	-	-	1
	Canal	**1**	**2**	**3**	**4**	**5**	**6**	**7**	**8**
1	EC	1							
2	Dep	−0.37	1						
3	Do	0.00	−0.15	1					
4	Turb	0.10	−0.22	0.07	1				
5	Vel	−0.12	0.48	−0.10	−0.07	1			
6	Con	-	-	-	-	-	-		
7	Dist	0.17	0.06	−0.15	0.16	0.26	-	1	
8	FA†	0.19	−0.08	0.1	−0.26	−0.1	-	−0.40	1
	Paddy	**1**	**2**	**3**	**4**	**5**	**6**	**7**	**8**
1	EC	1							
2	Dep	−0.34	1						
3	Do	−0.07	−0.02	1					
4	Turb	0.04	−0.31	0.06	1				
5	Vel	−0.10	0.29	−0.26	−0.08	1			
6	Con	−0.32	0.59	−0.09	−0.31	0.32	1		
7	Dist	0.10	−0.42	0.16	0.33	−0.17	−0.36	1	
8	FA†	0.02	0.48	−0.27	−0.41	0.27	0.54	−0.59	1
	Pond	**1**	**2**	**3**	**4**	**5**	**6**	**7**	**8**
1	EC	1							
2	Dep	−0.18	1						
3	Do	0.12	0.12	1					
4	Turb	0.07	−0.28	0.27	1				
5	Vel	−0.04	−0.16	−0.15	0.25	1			
6	Con	−0.32	−0.06	0.23	−0.07	−0.06	1		
7	Dist	−0.02	−0.16	0.52	0.23	−0.13	−0.16	1	
8	FA†	−0.03	0.13	−0.43	−0.13	0.12	0.21	−0.65	1

† Radius of buffer is 2000 m; -: Unused. Variable name abbreviation is indicated in Table 1 and text.

Table 3. Summary of fish sampling data.

Order	Family	Scientific name	Lateral migration[†]	Alien	Sites	*N*	Prevalence at each wtaerbody type: River	Diversion canal	Pond	Irrigation ditch	Paddy field	Wetland
Beloniformes												
	Adrianichthyidae	*Oryzias sp.*			22	108	○	○	○	○	○	○
	Belonidae	*Xenentodon cancila* (Hamilton 1822)	○		6	10	○		○		○	
		Xenentodon sp.			2	2	○					
	Hemiramphidae	*Dermogenys siamensis* (Fowler 1934)			27	103	○	○	○	○	○	○
Clupeiformes												
	Clupeidae	*Corica laciniata* (Fowler 1935)			1	1	○					
		Clupeoides borneensis (Bleeker 1851)			1	4	○					
		Clupeichthys aesarnensis (Wongratana 1983)			6	15	○					
		Clupeichthys sp.			2	4	○					
Cypriniformes												
	Cobitidae	*Acantopsis spp.*			16	42	○	○				
		Lepidocephalichthys hasselti (Cuvier & Valenciennes 1846)	○		4	7		○	○		○	
		Lepidocephalichthys (c.f. hasselti) (Cuvier & Valenciennes 1846)			2	10		○				
		Lepidocephalichthys (c.f. berdmorei) (Blyth, 1860)			1	1	○					
		Pangio anguillaris (Vaillant 1902)	○		3	3	○	○				
		Yasuhikotokia eos (Taki 1972)			1	11	○					
	Cyprinidae	*Amblypharyngodon chulabhornae* (Vidthayanon & Kottelat 1990)	○		22	75		○	○	○	○	○
		Barbichthys laevis (Cuvier & Valenciennes 1842)	○		1	1	○					
		Barbonymus altus (Günther, 1868)	○		15	98	○	○	○	○		
		Barbonymus gonionotus (Bleeker, 1849)	○		22	109	○	○	○	○	○	○
		Barbonymus schwanefeldii (Bleeker 1854)	○		10	78	○	○				
		Barbonymus sp.	○		2	6	○					
		Crossocheilus reticulates (Fowler 1934)	○		2	3	○					
		Cyclocheilichthys armatus (Cuvier & Valenciennes 1842)	○		13	82	○	○	○			
		Cyclocheilichthys apogon (Cuvier & Valenciennes 1842)	○		13	211	○	○	○	○		
		Cyclocheilichthys lagleri (Sontirat 1985)	○		7	15	○	○	○			
		Cyclocheilichthys repasson (Bleeker 1853)	○		1	2	○	○	○		○	○
		Cyclocheilichthys sp.			4	10		○				
		Cyclocheilos enoplos (Bleeker 1849)	○		9	11		○			○	
		Esomus metallicus (Ahl 1924)	○		72	1572	○	○	○	○	○	○

Gymnostomus siamensis (Sauvage 1881)	○	19	54	○	○	○		○	○
Henicorhynchus sp.	○	1	37	○					
Hypsibarbus malcolmi (Smith 1945)		6	73	○					
Hypsibarbus vernayi (Norman 1925)		5	14	○					
Hypsibarbus wetmorei (Smith 1931)		2	64	○					
Labeo chrysophekadion (Bleeker 1849)	○	1	1		○				
Labeo rohita (Hamilton 1822)	○	2	5		○	○			
Labeo sp.	○	1	3		○				
Labiobarbus siamensis (Sauvage 1881)		22	73	○	○	○		○	
Labiobarbus leptocheilus (Cuvier & Valenciennes 1842)		8	24	○	○	○			○
Labiobarbus sp.		4	13	○	○			○	
Mystacoleucus obtusirostris (Cuvier & Valenciennes 1842)		26	298	○	○				
Osteochilus vittatus (Cuvier & Valenciennes 1842)	○	13	19	○	○				
Osteochilus waandersii (Bleeker 1853)	○	1	1	○					
Osteochilus sp.		6	11	○	○				○
Paralaubuca typus (Bleeker 1864)	○	13	83	○	○		○	○	
Paralaubuca sp.		3	5		○			○	
Parachela oxygastroides (Bleeker 1852)	○	6	43	○	○				
Parachela siamensis (Günther 1868)	○	1	1	○					
Parachela williaminae (Fowler 1934)		1	1			○			
Parachela sp.	○	4	4	○	○			○	
Puntioplites proctozysron (Bleeker 1864)	○	18	45	○	○			○	
Puntius brevis (Bleeker 1849)	○	17	125	○	○	○	○	○	○
Puntius (cf. masyai) (Smith 1945)		3	4			○		○	
Puntius sp.		1	2					○	
Raiamas guttatus (Day 1870)	○	4	20	○					
Systomus rubripinnis (Cuvier & Valenciennes 1842)	○	1	3	○					
Thynnichthys thynnoides (Bleeker 1852)	○	1	10	○					
Rasbora aurotaenia (Tirant 1885)		5	15	○	○	○		○	
Rasbora borapetensis (Smith 1934)	○	10	20	○		○	○	○	○
Rasbora dusonensis (Bleeker 1850)		10	52	○		○			

	Rasbora paviana (Tirant 1885)		11	51	○	○	○		○	
	Rasbora sp.		4	14		○			○	
	Systomus rubripinnis (Cuvier & Valenciennes 1842)	○	2	6		○	○			
Gasterosteiformes										
Syngnathidae	*Doryichthys martensii* (Peters 1868)		1	1	○					
Osteoglossiformes										
Notopteridae	*Chitala ornate* (Gray 1831)		1	1	○					
	Notopterus notopterus (Pallas 1769)	○	6	8	○	○		○	○	
Perciformes										
Ambassidae	*Parambassis siamensis* (Fowler 1937)	○	40	418	○	○	○	○	○	○
	Parambassis wolffii (Bleeker 1850)	○	4	14	○	○			○	
Anabantidae	*Anabas testudineus* (Bloch 1792)	○	24	78	○	○	○	○	○	○
Channidae	*Channa striata* (Bloch 1793)	○	13	110	○	○	○		○	
Cichlidae	*Oreochromis niloticus* (Linnaeus 1758)	○	13	66	○	○	○	○	○	○
Eleotrididae	*Oxyeleotris marmorata* (Bleeker 1852)	○	5	8	○	○	○		○	
Gobiidae	*Gobiopterus chuno* (Hamilton 1822)		9	23	○	○	○		○	○
	Mugilogobius rambaiae (Smith 1945)		3	11	○	○				
	Mugilogobius sp.		1	1		○				
	Rhinogobius sp.		2	4		○			○	
	Brachygobius sp.		3	4		○	○		○	
Nandidae	*Pristolepis fasciata* (Bleeker 1851)		14	38	○	○	○		○	○
Osphronemidae	*Osphronemus goramy* La Cepède 1801)		2	2	○					
	Trichopsis pumila (Arnold 1936)		24	99	○	○	○	○	○	○
	Trichopsis vittata (Cuvier & Valenciennes 1831)	○	70	576	○	○	○	○	○	○
	Trichopodus microlepis (Günther 1861)		39	409	○	○	○	○	○	○

	Trichopodus pectoralis (Regan 1910)	○	8	23		○	○			○
	Trichopodus trichopterus (Pallas 1770)	○	67	682	○	○	○	○	○	○
Pleuronectiformes										
Soleidae	*Brachirus harmandi* (Sauvage 1878)		8	14	○					
Cynoglossidae	*Cynoglossus feldmanni* (Bleeker 1854)		1	1	○					
Siluriformes										
Bagridae	*Hemibagrus filamentus* (Chaux & Fang 1949)	○	4	6	○					
	Hemibagrus spilopterus (Ng & Rainboth 1999)	○	2	8			○		○	
	Mystus albolineatus (Roberts 1994)		11	22	○	○		○	○	
	Mystus multiradiatus (Roberts 1992)	○	11	40	○	○	○	○	○	○
	Mystus mysticetus (Roberts 1992)	○	9	43	○	○	○		○	
	Pseudomystus siamensis (Regan 1913)		3	11	○	○				
Clariidae	*Clarias macrocephalus* (Günther 1864)		1	1		○				
	Clarias batrachus (Linnaeus, 1758)		1	1			○			
Loricariidae	*Pterygoplichthys sp.*	○	3	5		○		○		
Pangasiidae	*Pangasius sanitwongsei* (Smith 1931)		2	5	○					
	Pangasius macronema (Bleeker 1850)	○	3	4	○					
	Pangasius sp		1	2	○					
Silurida	*Kryptopterus cryptopterus* (Bleeker 1851)		1	14	○					
	Phalacronotus bleekeri (Günther 1864)	○	2	3	○					
Sisoridae	*Bagarius yarrelli* (Sykes 1839)		1	1	○					
Synbranchiformes										
Mastacembelidae	*Mastacembelus armatus* (La Cepède 1800)	○	6	8	○					
Synbranchidae	*Monopterus javanensis* (La Cepède, 1800)		1	1			○			
Tetraodontiformes										
Tetraodontidae	*Pao cambodgiensis* (Chabanaud 1923)		1	1	○					

†: According to Fish Base (http://www.fishbase.org)

Table 4. General characteristics of environmental and landscape parameters at each waterbody type.

	Unit	River		Diversion canal		Irrigation ditch		Pond		Paddy field		Wetland	
		Mean (SD)		Mean (SD)		Mean (SD)		Mean (SD)		Mean (SD)		Mean (SD)	
		Rainy	Dry	Rainy	Dry	Rainy	Dry	Rainy	Dry	Rainy	Dry	Rainy	Dry
EC	µS	214 (39.4)	207 (61.2)	239 (45.7)	357 (156)	345 (143)	665 (257)	318 (104)	533 (328)	344 (163)	586 (549)	263 (91.0)	199 (0)
Dep	m	1.05 (0.367)	0.691 (0.311)	1.10 (0.505)	0.581 (0.356)	0.680 (0.232)	0.452 (0.142)	0.997 (0.328)	1.10 (0.498)	0.426 (0.317)	0.218 (0.126)	0.365 (0.0946)	0.47 (0)
Do	$mg \cdot L^{-1}$	5.61 (1.41)	4.76 (1.48)	3.40 (1.28)	3.66 (1.56)	3.35 (1.51)	3.68 (1.24)	3.68 (3.05)	3.66 (2.02)	3.96 (2.31)	4.11 (2.08)	4.19 (2.29)	0.300 (0)
Turb	NTU	130 (85.3)	12.6 (8.04)	70.9 (79.6)	141 (228)	108 (70.8)	51.2 (58.4)	47.8 (36.4)	19.6 (22.5)	156 (167)	36.9 (34.1)	21.9 (7.44)	0.640 (0)
Vel	ms^{-1}	0.824 (0.419)	0.38 (0.303)	0.183 (0.184)	0.0189 (0.0621)	0.0267 (0.0533)	0(0)	0.0139 (0.0461)	0(0)	0.0456 (0.167)	0(0)	0(0)	0(0)
Con	0, 1	0(0)	0(0)	0(0)	0(0)	0(0)	0(0)	0.167 (0.373)	0(0)	0.4444 (0.4969)	0(0)	0.167 (0.373)	0(0)
Dist	km	0.152 (0.132)	0.152 (0.132)	0.144 (0.165)	0.166 (0.185)	0.206 (0.127)	0.239 (0.0987)	0.428 (0.206)	0.387 (0.204)	0.397 (0.250)	0.333 (0.249)	0.325 (0.173)	0.0662 (0)
FA	km^2	3.90 (2.58)[†]	3.902 (2.58)[†]	2.80 (3.02)[†]	3.05 (3.18)[†]	3.41 (2.09)[†]	3.88 (1.79)[†]	6.05 (2.92)[†]	5.44 (2.82)[†]	5.17 (3.25)[†]	4.99 (3.47)[†]	5.23 (2.71)[†]	1.11 (0)[†]

[†] Radius of buffer is 2000 m; Variable name abbreviation is indicated in Table 1 and text.

3. RESULTS

3.1. Fish Species Richness and Environmental Characteristics

In total, 103 species (6555 individuals) belonging to 27 families were recorded (Table 3). Of the 135 study sites, 11 were inundated by the annual flood when the survey was conducted during the rainy season. The general characteristics of environmental and landscape parameters for each waterbody type during the rainy and dry seasons are shown in Table 4. No major differences were observed between the seasons except for Turb. Turb tended to be higher during the rainy season in all waterbody types except for diversion canal. Turb in diversion canal tended to be higher during the dry season. Figure 3 shows the total-length histograms for fishes captured in the study sites connected to main rivers by an annual flood and in isolated study sites during the rainy season survey. For all species, the proportion of small fish tended to be higher in connected study sites compared with isolated waterbodies.

3.2. GLMM Analysis and Model Selection

Table 5 shows the results of GLMM analysis with AIC model selection for fish species richness. The differences between the AIC value for the best models and null models (ΔAIC) were >4 except for the model for non-migratory fish species richness in canals, indicating substantial support for the best models [35] . No spatial autocorrelation was detected in the residuals of the best models. Figure 4 shows the relationship between floodplain area (2000 m buffer size; the best buffer size for GLMM analysis of the richness of all native fish in a waterbody) and richness of all native species in each waterbody type. The broken lines represent linear regression lines for significant correlations ($P < 0.05$). Significant positive correlations between native fish richness and floodplain area were found for rivers and ponds, but not for other waterbody types.

4. DISCUSSION

The most important finding of the present study is that the area of floodplains around each study site is a primary factor that determines fish species richness in rivers, diversion canals, and ponds in the study area. The contribution of floodplain area to fish species richness was equivalent to hydrology (Con, Vel, and Dep) and water quality (DO, Turb) parameters. This was supported by the fact that the best models for richness of all native fish species, floodplain migratory native fish species, and floodplain non-migratory native fish species showed a positive effect of floodplain area, and most were significant (Table 5; Figure 4), and by the fact that floodplain area was one of the top three parameters in all model selections (Table 5).

In paddy fields, the parameters EC and Con were the major determinants of fish species richness (Table 5). This may reflect the usage of fertilizers and pesticides, which increase electric conductivity, and the difficult access into isolated paddy field waterbodies because of migration barriers between paddy fields and ditches. The biological importance of submerged floodplains in fish ecology is well known. Shallow standing water and water with low dissolved oxygen provide refuge habitats from predators and flood disturbance [13] ; high nutrient inputs from rivers and high concentrations of feed organisms, such as phytoplankton [36] , zooplankton [11] , and fish [37] [38] , provide foraging and nursery habitats [37] [38] , and new connections between waterbodies and water level fluctuations caused by floods, trigger lateral spawning migration and egg deposition in some species [39] -[41] .

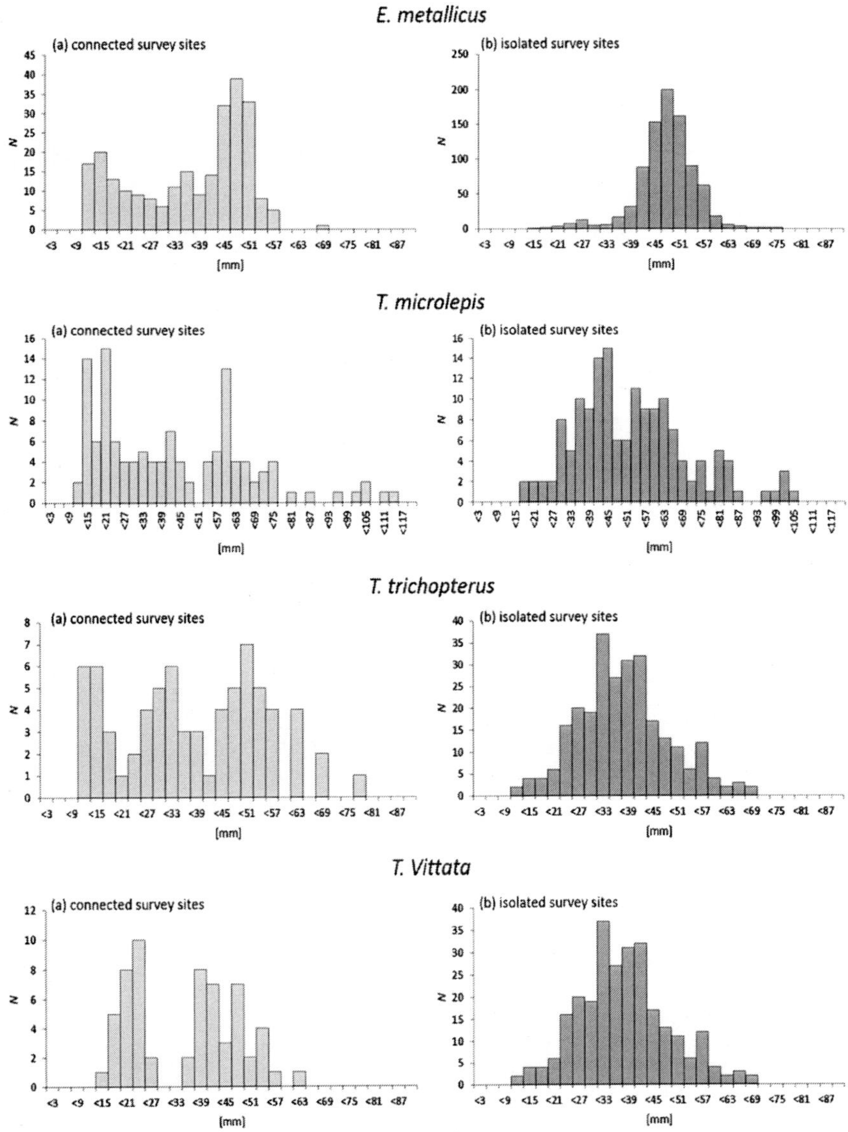

Figure 3. Total-length histograms for four common fish species (more than 50 individuals captured from both temporarily connected and isolated waterbodies) recorded during the rainy season survey. (a) Temporarily connected study sites and (b) Isolated study sites.

Table 5. Results of the general linear mixed model analysis of richness of all native species, floodplain migratory native species, and floodplain non-migratory native species in each waterbody type.

Waterbody type	Species richness	AIC	ΔAIC (from Null)	Wi	Moran's *I*	Intercept	Random intercept: Season		Waterbody Type					
							Rainy	Dry	River	Diversion Canal	Pond	Irrigation Ditch	Paddy	Wetland
Total	All native fish	765.8	96.6	0.296	0.0346^{NS}	1.35***	−0.0486	0.0486	0.600	0.178	−0.0743	0.0969	−0.651	−0.125
	Floodplain migratory	650.4	48.4	0.145	0.0596^{NS}	0.698**	−0.00203	0.00203	0.579	0.143	−0.135	0.171	−0.648	−0.0756
	Floodplain non-migratory	500	46.6	0.128	0.0390^{NS}	−0.0291	0	0	1.13	0.240	−0.350	−0.270	−0.609	−0.0659
River	All native fish	143.8	54.1	0.159	−0.0844^{NS}	3.14***	1.25e−17	−1.25e−17	-	-	-	-	-	-
	Floodplain migratory	133.5	40.3	0.177	−0.0136^{NS}	2.01***	0	0	-	-	-	-	-	-
	Floodplain non-migratory	119.4	14.7	0.117	0.0107^{NS}	2.72***	1.13e−18	−1.13e−18	-	-	-	-	-	-
Canal	All native fish	214.5	14.7	0.178	0.0707^{NS}	1.37***	0	0	-	-	-	-	-	-
	Floodplain migratory	182.3	12.2	0.161	0.0646^{NS}	0.904***	0	0	-	-	-	-	-	-
	Floodplain non-migratory	140.0	1.39	0.0479	−0.0106^{NS}	0.380^{+}	0	0	-	-	-	-	-	-
Paddy	All native fish	147.6	49.0	0.0730	0.105^{NS}	1.17**	0	0	-	-	-	-	-	-
	Floodplain migratory	120.0	19.6	0.0757	−0.107^{NS}	0.978*	0	0	-	-	-	-	-	-
	Floodplain non-migratory	85.4	33.8	0.0816	0.0137^{NS}	1.65*	0	0	-	-	-	-	-	-
Pond	All native fish	107.8	8.59	0.0765	−0.216^{NS}	1.02***	0	0	-	-	-	-	-	-
	Floodplain migratory	97.2	4.55	0.0651	−0.205^{NS}	0.835**	0	0	-	-	-	-	-	-
	Floodplain non-migratory	85.4	4.92	0.0519	−0.0878^{NS}	−0.475^{NS}	0	0	-	-	-	-	-	-

Waterbody type	Species richness	Coefficient of independent variables: EC (appearance)†	Dep (appearance)†	Do (appearance)†	Turb (appearance)†	Vel (appearance)†	Con (appearance)†	Dist (appearance)†	FA [Buffer Size] (appearance)†	
Total	All native fish	−0.000399^{+}			0.000955***	−0.347*	0.567***	0.00709*	0.106***	[2000]
		(50%)	(100%)	(33.3%)	(100%)	(100%)	(100%)	(66.7%)	(100%)	
	Floodplain migratory	−0.000454^{+}	0.695^{NS}		0.00119***	−0.285^{NS}	0.764***	0.00877*	0.118***	[2000]
		(50%)	(50%)	(50%)	(100%)	(50%)	(100%)	(100%)	(100%)	
	Floodplain non-migratory		0.504**		0.000689^{NS}	−0.576**	1.04***		0.0758*	[1000]
		(37.5%)	(100%)	(37.5%)	(43.8%)	(93.8%)	(100%)	(37.5%)	(93.8%)	
River	All native fish		−0.412*	−0.245***	0.00180*		ND	-	0.166***	[2000]
		(50%)	(42.9%)	(100%)	(57.1%)	(42.9%)			(100%)	
	Floodplain migratory			−0.202**			ND	-	0.202***	[2000]
		(42.9%)	(50%)	(100%)	(50%)	(42.9%)			(100%)	
	Floodplain non-migratory			−0.252***			ND	-		
		(28.6%)	(38.1%)	(95.2%)	(38.1%)	(38.1%)			(47.6%)	
Canal	All native fish				0.00116***		ND	0.0102***	0.0957**	[500]
		(36.4%)	(36.4%)	(36.4%)	(100%)	(36.4%)		(100%)	(100%)	
	Floodplain migratory				0.0031***		ND	0.0121**	0.0872*	[500]
		(35.7%)	(28.6%)	(28.6%)	(100%)	(28.6%)		(92.9%)	(78.6%)	
	Floodplain non-migratory				0.00997^{+}	1.17^{+}	ND		0.102*	[1000]
		(25.8%)	(40.4%)	(21.2%)	(50%)	(51.9%)		(25%)	(65.4%)	
Paddy	All native fish	−0.00108**		0.0684^{NS}	0.00133*		1.27***			
		(100%)	(38.9%)	(63.9%)	(58.3%)	(36.1%)	(100%)	(27.8%)	(47.2%)	
	Floodplain migratory	−0.00160*			0.00182*		1.12***			
		(100%)	(44.8%)	(37.9%)	(65.5%)	(31%)	(100%)	(24.1%)	(27.6%)	
	Floodplain non-migratory	−0.00432**					1.83***		−0.171^{+}	[1000]
		(100%)	(28.6%)	(42.9%)	(28.6%)	(32.1%)	(100%)	(25%)	(57.1%)	
Pond	All native fish			0.0793*	−0.00468^{NS}				0.144**	[2000]
		(23.3%)	(26.7%)	(86.7%)	(56.7%)	(36.7%)	(23.3%)	(23.3%)	(100%)	
	Floodplain migratory								0.130*	[2000]
		(23.3%)	(23.3%)	(23.3%)	(23.3%)	(23.3%)	(23.3%)	(26.7%)	(96.7%)	
	Floodplain non-migratory			0.171**	−0.00739^{NS}				0.148*	[1000]
		(24.4%)	(31.1%)	(100%)	(51.1%)	(26.7%)	(26.7%)	(42.2%)	(68.9%)	

†Appearance rate of containing model in the models with ΔAIC < 4 from the best model [%]; Level of significance: $^{***}P < 0.001$, $^{**}P < 0.01$, $^{*}P < 0.05$, $^{+}P < 0.1$, $^{NS}P > 0.1$; ND: No data, -: Unused; Variable name abbreviation is indicated in Table 1 and text.

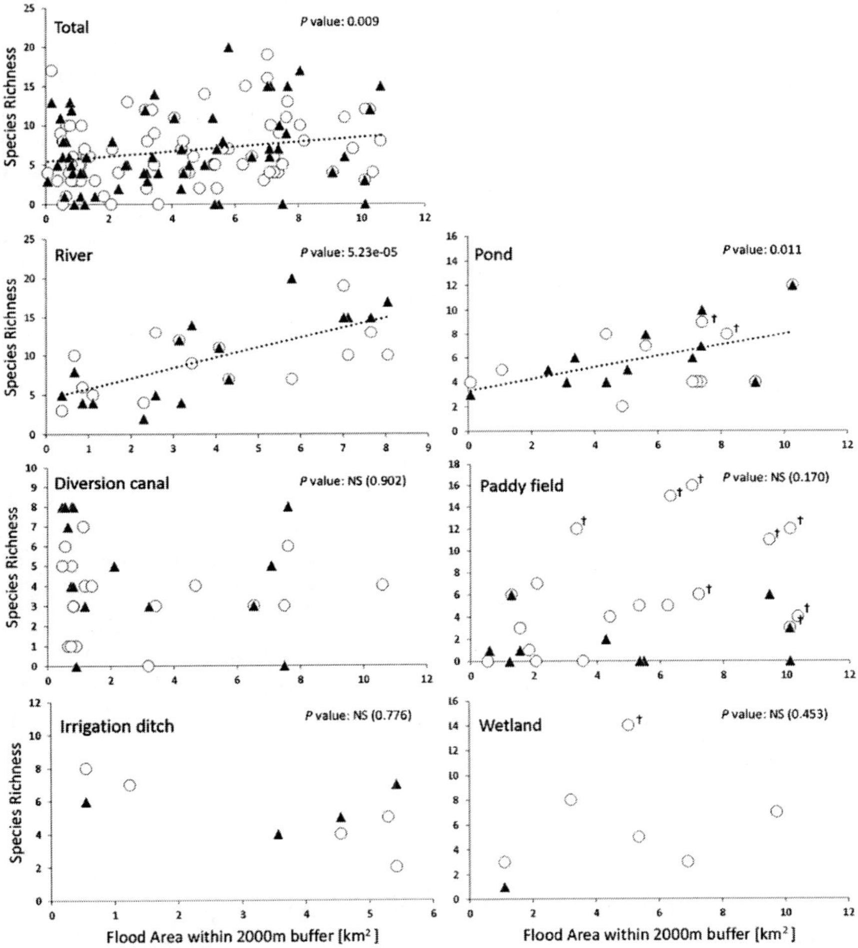

Figure 4. The relationship between floodplain area (2000 m buffer size) and richness of all native fish species for each waterbody type. The broken lines represent significant linear regressions ($P < 0.01$). P values are shown in the top right of each plot. Open circles indicate sites surveyed in the rainy season. Black triangles indicate sites surveyed in the dry season. Crosses to the upper right open circles represent the study sites temporarily connected to main rivers by floodwaters in the rainy season survey.

In the present study, increasing fish richness with increasing floodplain area probably reflected a direct association with floodplain utilization by migratory fish species. The parameter FA was retained with relatively higher frequency in model selections for floodplain migratory species compared with those for non-migratory species in every waterbody type except paddy fields (Table 5). In addition, the proportion of small size fishes (<20 mm total length) of four common species (Esomus metallicus, Trichopsis vittata, Trichopodus trichopterus, and Trichopodus microlepis), classified as 0+ juveniles, was increased in temporarily connected survey sites compared with isolated survey sites (Figure 3). Furthermore, although numbers were less than for the four common species mentioned above, some fluvial and lacustrine fish species, such as Xenentodon cancila, Barbonymus gonionotus, Cyclocheilichthys apogon, Cyclocheilichthys lagleri, Puntius brevis, and Amblypharyngodon chulabhornae (Table 3), were captured in temporarily inundated study sites. This finding indicates that individuals of these species moved into floodplains from permanent waterbodies in the early part of the rainy season, even before maximum flood levels occurred.

This study revealed the importance of floodplain area for fishes in the mid-region of the Chao Phraya River basin. The conservation of floodplains and fishes in the region is important for many reasons. The Chao Phraya River Basin is known for its high fish biodiversity. More than 690 freshwater fish species have been recorded and many utilize the floodplains during their life cycle [16] [18] . Moreover, inland fisheries have long been a part of Thai culture and are an important source of protein, especially for rural populations. Based on catch, the Thailand inland fishery ranked 13th in the world in 2012 [42] . Floodplains are maintained by dynamic interactions between flooding and landscape, and floodplains are disappearing at an accelerating rate in Southeast Asia, primarily as a result of changing hydrology caused by large-scale irrigation schemes and dams [15] . [43] investigated the impact of the operation of large-scale dams since the 2011 Thailand flood and found that the inundated area in the Chao Phraya River Basin was reduced by 40%. In the present study, the main stream of the Ping River, which has a smaller floodplain area compared with the other two rivers, showed a relatively low fish species richness (Figure 5). The reduction in floodplain area secondary to flood control measures (e.g., the Bhumibol Dam) may affect species richness, especially of migratory species in the Ping River Basin. On the other hand, flooding has the potential to cause serious damage to the economy of Thailand (e.g., the 2011 Thailand flood). Further studies are needed to assess the respective advantages and disadvantages of flood control measures in the Chao Phraya River Basin and to identify the best approach to protect both the economy and biodiversity. Future research should consider the detailed relationship between aquatic organisms and the characteristics of flood events, such as velocity, duration, timing, and the degree of fluctuation.

5. CONCLUSION

Fish assemblages and environmental, landscape parameters in floodplain waterbodies (rivers, diversion canals, ponds, irrigation ditches, paddy fields, and wetlands) in the mid-Chao Phraya River Basin were investigated in rainy (September 2014) and dry (March 2015) seasons. Factors affecting fish species richness in each type of waterbody were analyzed using generalized linear mixed models. The analysis showed that floodplain area around each surveyed waterbody was a major factor determining fish species richness in rivers, diversion canals, and ponds in the region. The observation of the study also showed that the population of juvenile fishes was increased in temporarily connected floodplain waterbodies to main rivers compared with isolated waterbodies. The high dependence of fish species richness on floodplain area in the region appeared to be a result of the use of inundated floodplains by fish species to forage and breed.

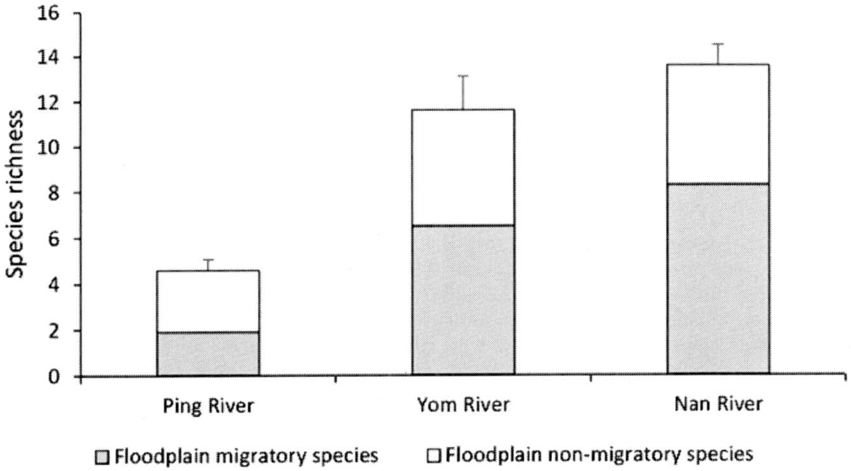

Figure 5. Native fish species richness and composition for each of the three rivers surveyed. Bars indicate the standard error.

ACKNOWLEDGEMENTS

We are grateful for the assistance provided by Y. Kano, T. Yahara, T. Pomun, N. Yamaguchi, and M. Shinohara. This work was supported by Program for Leading Graduate Schools (Graduate Education and Research Training Program in Decision Science for a Sustainable Society), JSPS, Japan and the Environment Research and Technology Development Fund (S9) of the Ministry of the Environment, Japan.

REFERENCES

1. Kolding, J. and van Zwieten, P.A.M. (2012) Relative Lake Level Fluctuations and Their Influence on Productivity and Resilience in Tropical Lakes and Reservoirs. Fisheries Research, 115, 99-109. http://dx.doi.org/10.1016/j.fishres.2011.11.008
2. Moss, B., Booker, I., Balls, H. and Manson, K. (1989) Phytoplankton Distribution in a Temperate Floodplain Lake and River System. I. Hydrology, Nutrient Sources and Phytoplankton Biomass. Journal of Plankton Research, 11, 813-838. http://dx.doi.org/10.1093/plankt/11.4.813
3. Junk, W.J. (1996) Ecology of Floodplains—A Challenge for Tropical Limnology. In: Schiemer, F. and Boland, K.T., Eds., Perspectives in Tropical Limnology, SPB Academic Publishng, Amsterdam, 255-265.
4. King, A.J., Humphries, P. and Lake, P.S. (2003) Fish Recruitment on Floodplains: The Roles of Patterns of Flooding and Life History Characteristics. Canadian Journal of Fisheries and Aquatic Sciences, 60, 773-786. http://dx.doi.org/10.1139/f03-057
5. Arthington, A.H., Balcombe, S.R., Wilson, G.A., Thoms, M.C. and Marshall, J. (2005) Spatial and Temporal Variation in Fish-Assemblage Structure in Isolated Waterholes during the 2001 Dry Season of an Arid-Zone Floodplain River, Cooper Creek, Australia. Marine and Freshwater Research, 56, 25-35. http://dx.doi .org/ 10.1071/MF04111
6. Junk, W.J. (1989) Flood Tolerance and Tree Distribution in Central Amazonian Floodplains. In: Tropical Forests: Botanical Dynamics, Speciation and Diversity, Academic Press, New York, 47-64. http://dx.doi.org/10.1016/b978-0-12-353550-4.50012-5
7. Ward, J.V., Tockner, K. and Schiemer, F. (1999) Biodiversity of Floodplain River Ecosystems: Ecotones and Connectivity. River Research and Applications, 15, 125-139. http://dx.doi.org/10.1002/(sici)1099-1646(199901/06)15:1/3<125::aid-rrr523>3.0.co;2-e
8. Tockner, K., Malard, F. and Ward, J.V. (2000) An Extension of the Flood Pulse Concept. Hydrological Processes, 14, 2861-2883. http://dx.doi.org/10.1002/1099-1085(200011/12)14:16/17<2861::AID-HYP124>3.0.CO;2-F
9. Mihaljevic, M., Stevic, F., Spoljaric, D. and Pfeiffer, T.Z. (2015) Spatial Pattern of Phytoplankton Based on the Morphology-Based Functional Approach along a River-Floodplain Gradient. River Research and Applications, 31, 228-238. http://dx.doi.org/10.1002/rra.2739
10. Boedeltje, G., Bakker, J.P., Brinke, A.T., Van Groenendael, J.M. and Soesbergen, M. (2004) Dispersal Phenology of Hydrochorous Plants in Relation to Discharge, Seed Release Time and Buoyancy of Seeds: The Flood Pulse Concept Supported. Journal of Ecology, 92, 786-796. http://dx.doi.org/10.1111/j.0022-0477.2004.00906.x
11. Simoes, N.R., Dias, J.D., Leal, C.M., Louizi, S.M.B., Lansac-Toha, F.A. and Bonecker, C.C. (2013) Floods Control the Influence of Environmental Gradients on the Diversity of Zooplankton Communities in a Neotropical Floodplain. Aquatic Sciences, 75, 607-617. http://dx.doi.org/10.1007/s00027-013-0304-9
12. Izaguirre, I., O'Farrell, I. and Tell, G. (2001) Variation in Phytoplankton Composition and Limnological Features in a Water-Water Ecotone of the Lower Paraná Basin (Argentina). Freshwater Biology, 46, 63-74.
13. Petry, P., Bayley, P.B. and Markle, D.F. (2003) Relationships between Fish Assemblages, Macrophytes and Environmental Gradients in the Amazon River Floodplain. Journal of Fish Biology, 63, 547-579. http://dx.doi.org/10.1046/j.1095-8649.2003.00169.x
14. Bozelli, R.L., Thomaz, S.M., Padial, A.A., Lopes, P.M. and Bini, L.M. (2015) Floods Decrease Zooplankton Beta Diversity and Environmental Heterogeneity in an Amazonian Floodplain System. Hydrobiologia, 753, 233-241. http://dx.doi.org/10.1007/s10750-015-2209-1

15. Tockner, K. and Stanford, J.A. (2002) Riverine Flood Plains: Present State and Future Trends. Environmental Conservation, 29, 308-330. http://dx.doi.org/10.1017/S037689290200022X
16. Rainboth, W.J. (1996) Fishes of the Cambodian Mekong. FAO Species Identification Field Guide for Fishery Purposes. FAO, Rome.
17. Zakaria-Ismail, M. (1994) Zoogeography and Biodiversity of the Freshwater Fishes of Southeast Asia. Hydrobiologia, 285, 41-48. http://dx.doi. org/ 10.1007/BF00005652
18. Kottelat, M. and Whitten, T. (1996) Freshwater Biodiversity in Asia: With Special Reference to Fish. World Bank Technical Paper, 343. http://dx.doi.org/10.1596/0-8213-3808-0
19. Kottelat, M. (2013) The Fishes of the Inland Waters of Southeast Asia: A Catalogue and Core Bibliography of the Fishes Known to Occur in Freshwaters, Mangroves and Estuaries. Raffles Bulletin of Zoology, 27, 1-663.
20. Vidthayanon, C. (2002) Peat Swamp Fishes of Thailand. Office of Environmental Policy and Planning, Bangkok. (In Thai)
21. Hydro and Agro Informatics Institute (2015). http://www.thaiwater.net/web/
22. Thai Meteorological Department (2015). http://www.tmd.go.th/ en/ archive/thailand_climate.pdf
23. Kano, Y., Adnan, M.S.B., Grudpan, C., Grudpan, J., Magtoon, W., Musikasinthorn, P., Natori, Y., Ottomanski, S., Praxaysonbath, B., Phongsa, K., Rangsiruji, A., Shibukawa, K., Shimatani, Y., So, N., Suvarnaraksha, A., Thach, P., Thanh, P.N., Tran, D.D., Utsugi, K. and Yamashita, T. (2013) An Online Database on Freshwater Fish Diversity and Distribution in Mainland Southeast Asia. Ichthyological Research, 60, 293-295. http://dx.doi.org/10.1007/s10228-013-0349-8
24. Kano, Y., Kawaguchi, Y., Yamashita, T. and Shimatani, Y. (2010) Distribution of the Oriental Weatherloach, Misgurnus anguillicaudatus, in Paddy Fields and Its Implications for Conservation in Sado Island, Japan. Ichthyological Research, 57, 180-188. http://dx.doi.org/10.1007/s10228-009-0146-6
25. Tanaka, W., Kano, Y., Yamasita, T., Saitou, K., Kawaguchi, Y. and Shimatani, Y. (2011) Determinants of the Misgurnus anguillicaudatus Population and Its Application to Conservation Planning in Sado Island, Japan. Ecology and Civil Engineering, 14, 1-9. (In Japanese with English Abstract) http://dx.doi.org/10.3825/ece.14.1
26. GISTDA (2015). http://flood.gistda.or.th/
27. Pinheiro, J.C. and Bates, D.M. (2000) Mixed-Effects Models in S and S-Plus. Springer, New York. http://dx.doi.org/10.1007/978-1-4419-0318-1
28. Dormann, C.F., Elith, J., Bacher, S., Buchmann, C., Carl, G., Carré, G., Marquéz, J.R.G., Gruber, B., Lafourcade, B., Leitao, P.J., Münkemüller, T., Mcclean, C., Osborne, P.E., Reineking, B., Schroder, B., Skidmore, A.K., Zurell, D. and Lautenbach, S. (2013) Collinearity: A Review of Methods to Deal with It and a Simulation Study Evaluating Their Performance. Ecography, 36, 27-46. http://dx.doi.org/10.1111/j.1600-0587.2012.07348.x
29. Akaike, H. (1973) Information Theory and an Extension of the Maximum Likelihood Principle. In: Petran, B.N. and Csari, F., Eds., International Symposium on Imformation Theory, 2nd Edition, Akademiai Kiado, Budapest Hungary, 267-281.
30. Keitt, T.H., Bjornstad, O.N., Dixon, P.M. and Citron-Pousty, S. (2002) Accounting for Spatial Pattern When Modeling Organism-Environment Interactions. Ecography, 25, 616-625. http://dx.doi.org/10.1034/j.1600-0587.2002.250509.x
31. Overmars, K.P., De Koning, G.H.J. and Veldkamp, A. (2003) Spatial Autocorrelation in Multi-Scale Land Use Models. Ecological Modelling, 164, 257-270. http://dx.doi.org/10.1016/S0304-3800(03)00070-X
32. Ferrer-Castán, D. and Vetaas, O.R. (2005) Pteridophyte Richness, Climate and Topography in the Iberian Peninsula: Comparing Spatial and Nonspatial Models of Richness Patterns. Global Ecology and Biogeography, 14, 155-165. http://dx.doi.org/10.1111/j.1466-822X.2004.00140.x

33. Moran, P.A.P. (1950) Notes on Continuous Stochastic Phenomena. Biometrika, 37, 17-23. http://dx.doi.org/10.1093/biomet/37.1-2.17
34. Lichstein, J.W., Simons, T.R., Shriner, S.A. and Franzreb, K.E. (2002) Spatial Autocorrelation and Autoregressive Models in Ecology. Ecological Monographs, 72, 445-463. http://dx.doi.org/10.1890/0012-9615(2002)072[0445:SAAAMI]2.0.CO;2
35. Burnham, K.P. and Anderson, D.R. (2002) Model Selection and Multimodel Inference: A Practical Information-Theoretic Approach. 2nd Edition, Springer, New York.
36. Keruzoré, A.A., Willby, N.J. and Gilvear, D.J. (2013) The Role of Lateral Connectivity in the Maintenance of Macrophyte Diversity and Production in Large Rivers. Aquatic Conservation: Marine and Freshwater Ecosystems, 23, 301-315. http://dx.doi.org/10.1002/aqc.2288
37. Gordon, H.C. (1989) The Habitat Diversity and Fish Reproductive Function of Floodplain Ecosystems. Environmental Biology of Fishes, 26, 1-27. http://dx.doi.org/10.1007/BF00002472
38. Pander, J., Mueller, M. and Geist, J. (2015) Succession of Fish Diversity after Reconnecting a Large Floodplain to the Upper Danube River. Ecological Engineering, 75, 41-50. http://dx.doi.org/10.1016/j.ecoleng.2014.11.011
39. Merron, G.S. and Mann, B.Q. (1995) The Reproductive and Feeding Biology of Schilbe Intermedius Rüppell in the Okavango Delta, Botswana. Hydrobiologia, 308, 121-129. http://dx.doi.org/10.1007/BF00007397
40. Agostinho, A.A., Gomes, L.C., Veríssimo, S. and Okada, E.K. (2004) Flood Regime, Dam Regulation and Fish in the Upper Paraná River: Effects on Assemblage Attributes, Reproduction and Recruitment. Reviews in Fish Biology and Fisheries, 14, 11-19. http://dx.doi.org/10.1007/s11160-004-3551-y
41. Bonvechio, T.F. and Allen, M.S. (2005) Relations between Hydrological Variables and Year-Class Strength of Sportfish in Eight Florida Waterbodies. Hydrobiologia, 532, 193-207. http://dx.doi.org/10.1007/s10750-004-1388-y
42. Moffitt, C.M. and Cajas-Cano, L. (2014) Blue Growth: The 2014 FAO State of World Fisheries and Aquaculture. Fisheries, 39, 552-553. http://dx.doi.org/10.1080/03632415.2014.966265
43. Mateo, C.M., Hanasaki, N., Komori, D., Tanaka, K., Kiguchi, M., Champathong, A., Sukhapunnaphan, T., Yamazaki, D. and Oki, T. (2014) Assessing the Impacts of Reservoir Operation to Floodplain Inundation by Combining Hydrological, Reservoir Management, and Hydrodynamic Models. Water Resources Research, 50, 7245-7266. http://dx.doi.org/10.1002/2013WR014845

CHAPTER 15

Analysis of Long-Term Changes in a Mediterranean Marine Ecosystem Based on Fishery Landings

[1]ECHO Group, Division of Oceanography, National Institute of Oceanography and Experimental Geophysics, Trieste, Italy

[2]Italian National Institute of Environmental Protection and Research, Chioggia, Italy

[3]Institute of Marine Sciences, National Research Council, Ancona, Italy

[4]Department of Environmental Sciences, Informatics and Statistics, Ca' Foscari University, Venice, Italy

[5]The Abdus Salam International Centre for Theoretical Physics, Trieste, Italy

ABSTRACT

In the Mediterranean Sea, structured and standardized monitoring programs of marine resources were set only in the last decades, so the analysis of changes in marine communities over longer time scale has to rely on other sources. In this work, we used seven decades (1945–2014) of disaggregated landings statistics for the Northern Adriatic Sea (Mediterranean) to infer changes in the ecosystem. Analysis of landings composition was enriched with the application of a suite of ecological indicators (e.g., trophodynamic indicators, such as the primary production required to sustain the catches—PPR; size-based indicators, such as the large species indicator—LSI; other indicators, such as the elasmobranchs-bony fish ratio—E/B ratio). Indicators were further compared with main ecosystem drivers, i.e., fishing capacity, nutrient loads and climate change. Species most vulnerable to fishing (i.e., elasmobranchs and large-sized species) dramatically declined at the beginning of the industrialization of fishery that occurred right afterwards World War II, as can be inferred by the negative drop of LSI and E/B ratio in the mid-1950s. However, until the mid-1980s landings and PPR increased due to improvements in fishing activities (e.g., the introduction of more efficient fishing gears) increasing fishing capacity, high productivity of the ecosystem. Overall, the effects of fishing were buffered by an increase in productivity in the period of high nutrient discharge (up to mid-1980s), while significant changes in fish community structure were already occurring. From the mid-1980s, a reduction in nutrient load caused a decline in productivity but the food-web structure was already modified and unable to support, or recover from, such unbalanced situation, resulting in the collapse of landings. This collapse is coherent with alternative stable states hypothesis, typical of complex real systems, that implies drastic interventions that go beyond fisheries management and include regulation of nutrient release for recovery. The work highlights that, despite poor capabilities to track species dynamics, landings and applied indicators might help to shed light on the long-term dynamics of marine communities, thus contributing to place current situation in an historical framework with potential for supporting management.

Keywords: landings, marine historical ecology, ecological indicators, long-term changes, ecosystem drivers, Adriatic Sea

INTRODUCTION

Long-term analyses of the interactions between human society and the oceans are necessary for understanding the processes that brought the marine ecosystems as we see today, avoiding the so-called "shifting the baseline syndrome" and understanding the magnitude and causes of change (Pauly, 1995; Jackson et al., 2001). In this framework, marine historical ecology (MHE) can bring a significant contribution to present-day management of marine ecosystems, both for conservation and for sustainable exploitation of resources (Engelhard et al., 2015). However, often the analyses of historical changes in marine communities cannot be based on results of structured and standardized monitoring programs, since in most of the cases those were set in the very last decades. The need to bridge the gap between request of knowledge on past status of ecosystems and the available monitoring data, triggered the uses of different approaches to extract information from data coming from other sources, including paleontological, archeological, and historical sources, as well as the use of landings (e.g., Rosenberg et al., 2005; Sàenz-Arroyo et al., 2005; Lotze et al., 2006; Fortibuoni et al., 2010, 2016; Van Beveren et al., 2016).

In this context, detailed and disaggregated fishery statistics represent an important source of information that can be used as proxies to evaluate long-term changes in marine fisheries and communities. Changes in the composition of landings, evaluated through opportune weighting metrics (indicators), showed to reflect changes in the structure of underlying fish communities due to anthropogenic impacts and environmental changes (e.g., Caddy, 1993; Pauly et al., 1998; de Leiva Moreno et al., 2000; Pinnegar et al., 2002; Libralato et al., 2004; Pauly and Watson, 2005; Baeta et al., 2009; Munyandorero and Guenter, 2010; Kleisner et al., 2013). Moreover, catch statistics are recognized to be linked to fishing and environmental pressures and respond selectively to management action (Coll et al., 2016).

Nevertheless, the intrinsic limitations of fishery-dependent data—such as landings—include the lack of standardization, the dependence from fishing fleet activity features as well as market preferences of products that usually change across time and space. All this imposes caution in deriving marine population densities directly from catch statistics (e.g., Essington et al., 2006; Pauly et al., 2013).

The capability to connect modification in landings composition to changes in the community at sea is more robust when local disaggregated landings result from multi-target and multi-gear fisheries (i.e., several distinct métiers), and when changes in fishing activities (e.g., introduction of new technologies, shift from one fishing gear to another, shift in fishing grounds, fishing capacity as number and tonnage of boats) are traceable. In this context an important aspect to be considered is the trend in indices over time, rather than the absolute values they assume, being reference points or limit values for many indicators not yet been established (Shin et al., 2010).

The Northern Adriatic Sea (Mediterranean Sea) represents a valuable case study for MHE, due to the long history of exploitation (Botter et al., 2006; Fortibuoni, 2010), human-induced changes (Lotze et al., 2011), as well as documented long-term modifications of the physical and chemical oceanographic characteristics of the basin due to anthropogenic impact (Mozetič et al., 2009; Solidoro et al., 2009) and temperature change (Russo et al., 2002).

In this study, a long-term time-series (1945–2014) of landings disaggregated by species (Mazzoldi et al., 2014) for the Northern Adriatic Sea was analyzed to detect changes in total yields and variations in landings composition by functional groups over time. A suite of ecological indicators was applied to landings data to integrate responses to multiple stressors (Fu et al., 2015; Coll et al., 2016). They include trophodynamic indicators (e.g., mean trophic level, primary production required), climatic indicators (mean temperature of the catch), and other indicators, such as elasmobranchs-bony fish ratio. These indicators were compared with independent data describing main ecosystem drivers, in order to corroborate findings.

Our analysis falls within the general need of taking into account ecological processes when considering long-term changes in fishery resources where fishery-independent data are lacking, by testing several ecological indicators, and their responsiveness to fishery and environmental drivers. The approach is suitable to be applied for the purposes of the Ecosystem Approach to Fishery Management (EAFM) and contributes to establishing historical baselines to be used to compare current and future ecosystem status, for instance in the context of the Marine Strategy Framework Directive (MSFD) implementation.

MATERIALS AND METHODS

Area of Study

The Northern Adriatic Sea is the shallowest (average depth of 33.5 m) and northern-most area of the Adriatic and Mediterranean Seas (Figure 1). This area is characterized by strong riverine outflows from the Po River (the largest Italian river) that is a primary source of nutrients and organic matter to the basin (Giani et al., 2012). Water circulation is dominated by two counter-clockwise gyres, which confine a large part of the nutrient-enriched riverine inputs along western coastal regions (Zavatarelli et al., 1998). In fact, the western coastal waters were long considered as ones of the most productive of the Mediterranean Sea (Hopkins et al., 1999), and occasionally local hypoxia/anoxia events have been reported especially close to the Po River mouth (Giani et al., 2012). The reduction in phosphorus loads in Italian rivers in the 1980s triggered reversal in the eutrophication trend and was indicated as the start of a (cultural) oligotrophication process for the basin (Mozetič et al., 2009; Solidoro et al., 2009).

The seabed is characterized by muddy and sandy bottoms, with the presence of few rocky outcrops. Due to the presence of a wide flat trawlable platform coupled with the high productivity, the Northern Adriatic Sea is the Italian basin with the highest fishery pressure, and one of the most exploited areas in the Mediterranean Sea (AdriaMed, 2004).

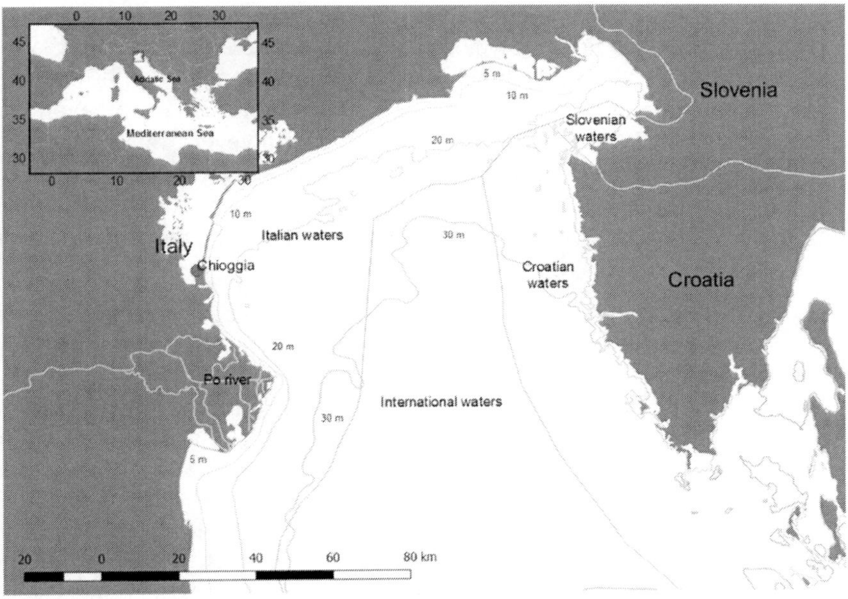

FIGURE 1. THE AREA OF STUDY (NORTHERN ADRIATIC SEA). Commercial fisheries landings derive from the area shown in the figure, Slovenian and Croatian waters excluded.

History of Fisheries in the Northern Adriatic Sea

The biological resources of the study area have been intensively exploited since centuries, and the port of Chioggia (Figure 1) hosts the most important fishing fleet of the Adriatic Sea and one of the most important of the entire Mediterranean basin (Botter et al., 2006; Fortibuoni, 2010; Mion et al., 2015).

Before and immediately after World War II (WWII) the Chioggia fishing fleet adopted a wide variety of artisanal gears, both passive and active and the industrialization of fisheries in the Adriatic Sea gradually became consolidated starting from the 1950s, having begun in the period between the two world wars with the first experiments with engines (Fortibuoni, 2010). The period following WWII was characterized by marked changes in fishing equipment and technologies. There was a great increase in both demand for fish products and technical innovations, which enabled fleets to expand considerably. Consequently, there was a continuous and substantial increase in the fishing capacity (Cataudella and Spagnolo, 2011).

The use of progressively larger ships and engines allowed fishing areas to be expanded as well as larger and heavier gears to be used. As an example, the bottom otter-trawl was previously towed by pair sailing boats (Botter et al., 2006), and after WWII it was increasingly adopted by single mechanized vessels. Most traditional fishing gears were abandoned (e.g., beach seines and longlines, commonly used throughout the nineteenth century) and new and more efficient ones were introduced. However, the geographical characteristics of the Northern Adriatic Sea (shallow and semi-closed basin) constrained the expansion of fisheries, thus fishery grounds exploited by Chioggia's fleets can be considered almost stable throughout the period analyzed (Mazzoldi et al., 2014).

The first of the major innovations in fishing equipment introduced was the "*saccaleva*" surrounding net aided with a light source for attracting fish. It became widespread in the 1940s in Chioggia and gradually replaced all other methods for fishing small pelagic fish, such as the traditional "*menaide*" drift net (Fortibuoni, 2010). Successively, the "*saccaleva*" was substituted in the late 1960s by the more efficient "*volante*" (mid-water pelagic trawl, towed by paired vessels) that since the 1970s is the main gear used to catch pelagic species (Cingolani et al., 1996). In the mid-1950s the "*rapido*" trawl (a sort of beam trawl rigged with 10 cm long iron teeth; Pranovi et al., 2001) was introduced targeting flatfish and shellfish (such as the Mediterranean scallop *Pecten jacobaeus*), but also demersal resources, as spottail mantis shrimp (*Squilla mantis*) and common cuttlefish (*Sepia officinalis*). In the early 1970s, the first hydraulic dredge came into use, substituting the traditional hand-maneuvered gears to harvest different species of clams (Romanelli et al., 2009). In relation to this, the clam fishery quickly became one of the most valuable.

Furthermore, in the mid-1980s, the introduction of LORAN (Long Range Navigation) and subsequently the video plotter and GPS (Global Positioning System) greatly improved navigation precision, allowing the exploitation of areas that were previously inaccessible because of their proximity to unsuitable trawling sites, as the presence of rocky outcrops.

As regard fisheries management, a comprehensive scheme in Italy was initiated with the Law 41/1982, establishing that all professional fishing vessels had to possess a license reporting the characteristics of the vessel (e.g., GT), limitations of fishing areas, gear use and spatial licensing (Piroddi et al., 2015). The Italian fisheries management system is actually based on fishing effort/capacity regulation systems, and technical measures. No quotas or TACs (total allowable catch) have been established, except for Atlantic bluefin tuna (*Thunnus thynnus*) and Striped venus clam (*Chamelea gallina*). From 1983, the national fishing fleet was subject to reduction constraints in relation to two reference parameters, fleet tonnage and engine power. A further incentive toward fleet capacity/effort reduction was provided by European Structural Funds that financed the voluntary removal of vessels. Moreover, in recent years there has been a voluntary departure from the sector due to the general old age of the fishing fleet and the fisheries crisis (Cataudella and Spagnolo, 2011). However, as regards the Chioggia's fishing fleet, fishing capacity (i.e., total GT) has increased until recent years (Barausse et al., 2011).

Data

Landings Dataset

Official landings data (1945–2014) from the Chioggia's wholesale fish-market were retrieved from the Clodia database (Clodia database, 2015). Landings of a wide variety of benthic and pelagic species represent the aggregated commercialized quantities caught by the highly diversified fishing gears employed by fishermen of Chioggia. Data do not include any estimate of the discard, and landings disaggregated by gear are not available. Landings refer only to fish and seafood caught by local fishermen from the Chioggia's fleet that operates in the Adriatic Sea. The database was validated by Mazzoldi et al. (2014) to show that landings composition provides reliable indication of fish abundance.

Information on the habitat (pelagic/demersal) and maximum length (L_{max}) for each species were taken from FishBase (Froese and Pauly, 2016). Moreover, the thermal preference (median temperature preference, T) was assigned according to Cheung et al. (2013). The trophic level of species (TL), specific for the Northern Adriatic Sea, was obtained from Fortibuoni et al. (2013), from other literature and estimates (e.g., Stergiou and Karpouzi, 2002) or retrieved from FishBase in the case of fish, and SeaLifeBase (Palomares and Pauly, 2016) in the case of invertebrates.

Ecosystem Drivers: Fishing Capacity, Nutrient Loads, and Climate Change

Data on Chioggia's fishing fleet capacity in terms of gross registered tonnage (GRT) were gathered from ISTAT (Italian National Institute of Statistics) for the period 1951–1989 and in terms of gross tonnage (GT) from the Community Fleet Register for the period 1990–2014. Total fishing capacity was expressed as GT for the entire time-series and used as a proxy for fishing pressure in the analyses because no long-term records of fishing effort were available. It was also not possible to include in fishing pressure any estimate of the technological creep due to the lack of data (e.g., CPUE from scientific surveys; Engelhard, 2016).

The monthly discharge of the Po river (m^3/s), measured at Pontelagoscuro (Ferrara, Italy) for the period 1945–2014 was provided by the Regional Environmental Protection Agency of Emilia Romagna and used to calculate monthly mean annual discharge. Nutrient yearly discharge, in terms of annual input of nitrogen (NO_3) and phosphorus (PO_4) load (t/y) from the Po river, was obtained from Ludwig et al. (2009) for the period 1960–2000, and from EU FP7 PERSEUS project for the period 2001–2014.

The winter (December-March) station based index of the North Atlantic Oscillation index (Hurrell and National Center for Atmospheric Research Staff, 2016) was used to represent the large-scale climatic variability in the area in the period 1945–2014. The NAO is based on the difference of normalized sea level pressure between Lisbon (Portugal) and Stykkisholmur/Reykjavik (Iceland). Positive values of the NAO index are typically associated with stronger-than-average westerlies over the middle latitudes and wetter/milder weather over western Europe.

Finally, monthly data on sea surface temperature (SST) were downloaded from the International Comprehensive Ocean–Atmosphere Data Set (ICOADS). SST data at 13°E–45°N, 15°E–45°N, 13°E–43°N, and 15°E–43°N for the period 1957–2014 were used to compute the annual mean values.

Ecological Indicators

Data were also used to compute ecological indicators, meant to provide a holistic description of the system. Disaggregated landings per species (landed quantities in kg per year) were integrated into metrics by applying a suite of ecological indicators (described in Table 1) based on TL, thermal preference, habitat and species' size.

TABLE 1. Ecological Indicators Calculated For The Time Series Of Landings.

Indicator (acronym)	Rationale	Formulation	References
Mean Trophic Level of landings (mTL)	Indicator of the impact of fishing on ecosystems, which selectively removes larger (and higher TL) species. It is expected to decrease with increasing fishing pressure.	$mTL = \frac{\sum_{i=1}^{n} TL_i Y_i}{\sum_{i=1}^{n} Y_i}$ where TL_i is the trophic level of species i, n is the total number of species, and Y_i is the landings of species i	Pauly et al. (1998)
Mean Temperature of the Catch (MTC)	Indicator of ocean warming on fish communities that leads to increased catches of warmer-water species and decreased catches of colder-water species. It is expected to increase with increasing ocean warming.	$MTC = \frac{\sum_{i=1}^{n} T_i Y_i}{\sum_{i=1}^{n} Y_i}$ where T_i is the median temperature preference of species i, n is the total number of species, and Y_i is the landings of species i	Cheung et al. (2013)
Primary Production Required to sustain fishery (PPR)	The PPR is a measure of the level of exploitation of the studied area, accounting for the fraction of Primary Production sequestrated by fisheries. The method is based on the trophic level of the caught species, the energy transfer efficiency between trophic levels, and on the primary productivity of the basin. It is expected to increase with increasing fishing pressure.	$PPR = \sum_{i=1}^{n} \frac{Y_i}{CR} \cdot \left(\frac{1}{TE}\right)^{(TL_i - 1)}$ where Y_i is the landings of species i, n is the total number of species, CR is the conversion rate of wet weight to carbon (fixed at 1:9), TE is the transfer efficiency (fixed at 10%), and TL_i is the trophic level of species i	Pauly and Christensen (1995)
Ratio between pelagic and demersal species biomass in landings (P/D ratio)	Highlights the influence of nutrient enrichment on fish communities, since nutrient availability has a differential effect on pelagic fish (positive) and on demersal stocks (negative). It is expected to increase with increasing nutrients availability.	$P/D\ ratio = \frac{\text{Landings of small pelagic species}}{\text{Landings of demersal fish plus benthic species}}$	de Leiva Moreno et al. (2000)
Large Species Indicator (LSI)	Quantifies relative changes in landings of larger fish species that are selectively removed by fishing. It is expected to decrease with increasing fishing pressure	$LSI = \frac{\sum_{i=1}^{n} Y_i \xi_{max} \geq 30}{\sum_{i=1}^{n} Y_i}$ where L_{max} is the maximum length of the species i, n is the total number of species, and Y_i is the landings of species i. Species are classified as large or small according to their L_{max} adopting a threshold value that was set at 30 cm in agreement to the method proposed by Greenstreet et al. (2011)	Shephard et al. (2012)
Ratio between elasmobranch and bony fish landings (E/B ratio)	Summarizes changes in the fish community structure between groups of species that are characterized by contrasting life-histories traits and by differential vulnerability to fishing activities. It is expected to decrease with increasing fishing pressure.	$E/B\ ratio = \frac{\text{Landings of elasmobranch}}{\text{Landings of bony fish}}$	Piet and Pranovi (2005)
Q-90 statistic	It is a variant on Kempton's Q index developed to measure the effects of mortality from fishing on species diversity. The statistic represents the slope of the cumulative functional groups abundance curve between the 10- and 90-percentiles. It is expected to decrease with increasing fishing pressure	$Q-90 = \frac{0.8 \times n}{\log(Y2/Y1)}$ where n is the total number of functional groups, $Y1$ and $Y2$ are landings of the 10th and 90th percentiles in the cumulative abundance distribution	Ainsworth and Pitcher (2006)

Metrics used included: the mean Trophic Level of landings, an indicator of food webs structure (mTL; Pauly et al., 1998); the Mean Temperature of the Catch (MTC; Cheung et al., 2013), which is the average inferred temperature preference of the species weighted by their annual catch used for evaluating the effect of sea warming on fish communities; the ratio of small pelagic fish to demersal and benthic landings (P/D ratio; de Leiva Moreno et al., 2000); the Large Species Indicator (LSI), i.e., the biomass proportion in landings of large-sized fish species (Shephard et al., 2012); the ratio between elasmobranch and bony fish in the landings (E/B ratio; Piet and Pranovi, 2005); the Primary Production Required to sustain fishery catches (PPR; Pauly and Christensen, 1995) that is the amount of energy exported from the system by landings and can be seen as the ecological footprint of fishing activities (Swartz et al., 2010); the Q-90 statistic (Ainsworth and Pitcher, 2006) that is a variant on Kempton's Q index to evaluate biodiversity (Kempton and Taylor, 1976). Landings per Unit of Fishing Capacity (LPUC) were also computed considering the yearly total gross tonnage of the fleet.

ANALYSIS OF DATA

Multivariate Analysis

Landings data were aggregated into 12 functional groups, according to their taxonomical and ecological features, to reduce missing values and restrict the number of variables in order to better elucidate main changes. Invertebrates were subdivided into three groups, i.e., bivalves, cephalopods, and crustaceans (no reliable data for gastropods were available in the database and thus this group was omitted). Pelagic bony fish species were grouped into three size classes according to fish species L_{max}(small: $L_{max} <$ 30 cm; medium: 30 cm $\leq L_{max} <$ 90 cm; large: $L_{max} \geq$ 90 cm). Analogously, three groups were used for aggregating demersal bony fish species by size. Flatfishes, sharks and "skates and rays" represented three further groups.

Functional groups' proportion in landings (%) was computed yearly in order to analyse landings composition and remove the effect of landings abundances. A fourth root transformation was applied to data to downweight the influence of predominant groups (Kaiser et al., 2000) and the similarity between every pair of years was computed using the Bray-Curtis similarity coefficient. Chronological clustering using the un-weighted pair-group average (Legendre and Legendre, 1998) was applied to identify periods with similar landings composition. Functional groups' proportion in landings was compared among periods by means of the non-parametric Kruskal-Wallis H-test (α = 0.05). Ecological indicators were compared between consecutive periods identified through cluster analysis by means of the non-parametric Mann-Whitney U-test ($\alpha = 0.05$). Analysis was done with PAST v. 3.14 (PAleontological STatistics; Hammer et al., 2001).

Ecological indicators and drivers relationship was analyzed for the period 1960–2014 (for which all drivers were available) through the statistical procedure BIO-ENV from PRIMER v. 6.1.5 (global BEST-test; Clarke et al., 2008). The method consists in the computation of the correlation coefficients between similarity matrices of ecological indicators and drivers, and identifies the combination of drivers that maximizes the correlation. Indicators and drivers were normalized prior to the construction of the Euclidean distance matrices, since they represented different units of measure. The skewness and the individual correlations between drivers were explored by constructing a draftsman plot and examining the resulting Spearman rank correlations to eventually reduce redundancy and dimensionality of the data. We included all drivers in the analysis after testing the absence of highly correlated drivers ($\rho \geq 0.95$).

Time-Series Analysis

Trends in total landings (Y), ecological indicators and drivers time-series were analyzed using the Mann–Kendall test and the non-parametric Sen's method was used to quantify the magnitude (slope) of the trend, using the Microsoft Excel template MAKESENS (Mann-Kendall test for trend and Sen's slope estimates) developed by Salmi et al. (2002).

Drivers time-series were further explored using the sequential *t*-test analysis of regime shift (STARS v. 3.4) first developed by Rodionov (2004) in order to detect potential abrupt changes in the mean. The algorithm was applied to the filtered time series calculated by removing red noise through a "pre-whitening" procedure based on the IP4 method (Rodionov, 2006) to take into account the effect of serial correlation on shift detection. The cut-off length was set at 10 years, the significance level at 0.05 and Huber's weight parameter at 1.

RESULTS

Multivariate Analysis

Chronological clustering allowed distinguishing seven periods with different landings composition at a similarity threshold of 0.94: 1945, 1946–1954, 1955–1961, 1962–1985, 1986–1993, 1994–2008, and 2009–2014 (Figure 2). Year 1945, which represented an autonomous cluster at this similarity threshold, was associated to the first period (then becoming 1945–1954).

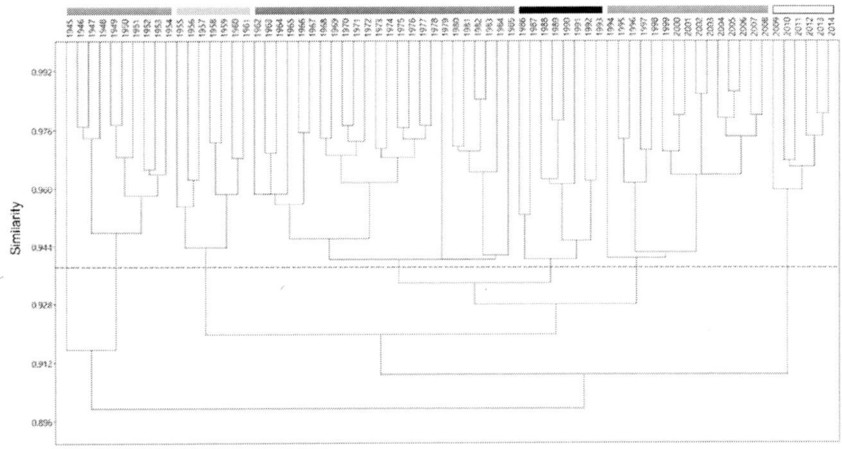

FIGURE 2. Chronological clustering (bray-curtis similarity, unweighted pair-group average) of the composition of landings for the years 1945–2014.

The difference among periods of functional groups' proportion in landings was statistically significant for all functional groups. Landings were dominated by small pelagics in all periods, with a percentage contribution ranging between 37% (*SD*: 8%; 1945–1954) and 68% (*SD*: 3%; 2009–2014). Cephalopods was the second most relevant group in the landings in all periods, with a contribution ranging between 10% (*SD*: 2%; 1955–1961) and 20% (*SD*: 5%; 1986–1993). In the first two periods, the third group in terms of biomass was medium pelagics that represented more than 10% of the landings [13% (*SD*: 4%) and 10% (*SD*: 6%), respectively], but then medium pelagics declined in the following periods down to 1.1% (*SD*: 0.4%) in recent years (2009–2014). Medium demersals ranged between 12% (*SD*: 2%; 1945–1954) and 6% (*SD*: 1%; 1955–1961). The other groups contributed with <10% in all periods. It is worth noting that skates and rays showed a sharp decline, from 3% (*SD*: 1%) in 1945–1954 to 0.04% (*SD*: 0.01%) in 2009–2014 (Figure 3). Another interesting dynamic was showed by bivalves that from 2% (*SD*: 1%) of the landings biomass in the first period increased to 8% (*SD*: 6%) in the period 1986–1993, and then sharply declined down to 1.9% (*SD*: 0.4%) in the most recent period (Figure 3).

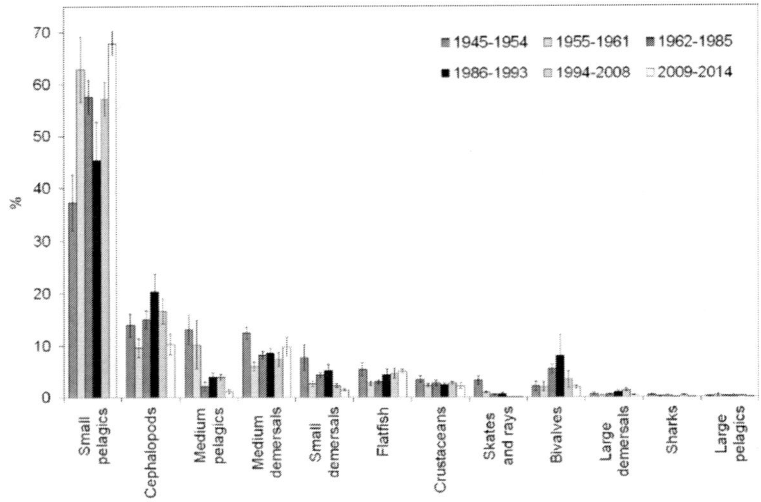

FIGURE 3. Percentage composition of the fish community in the periods identified through chronological clustering**.** The confidence interval (95%) is reported in the vertical bar.

Mann-Whitney *U*-test results are reported in Table 2. Y, MTC, LPUC, PPR, and P/D ratio significantly increased between the first two periods, while mTL, LSI and E/B ratio significantly decreased (Figures 4, 5). Between 1955–1961 and 1962–1985 only MTC significantly increased, while mTL, LSI, and P/D ratio significantly decreased (Figure 5). In the subsequent period, mTL and LSI significantly increased, while Y, LPUC, PPR, Q-90, and P/D ratio significantly decreased (Figures 4, 5). Between 1986–1993 and 1994–2008 a significant increase of MTC, Q-90, and P/D ratio was observed, while E/B ratio further decreased (Figure 5). Finally, by comparing the last two periods (1994–2008 and 2009–2014) a significant decrease of MTC, PPR, LSI, Q-90, and E/B ratio occurred, while P/D ratio significantly increased (Figure 5).

TABLE 2. Significant differences (mann-whitney u-test, $p < 0.05$) of total landings (*y*), landings per unit of capacity (lpuc), and ecological indicators between successive periods.

	1945–1954	1955–1961	1962–1985	1986–1993	1994–2008	2009–2014
Y (t)	3520 (3293–4174)	▲		▼		
LPUC (t)	0.99 (0.95–1.06)	▲		▼		
mTL	3.65 (3.12–3.18)	▼	▼	▲		
MTC	13.21 (12.42–13.55)	▲	▲		▲	▼
PPR (10^{12}gCy^{-1})	0.08 (0.07–0.11)	▲		▼		▼
LSI	0.37 (0.35–0.41)	▼	▼	▲		▼
E/T ratio	0.04 (0.03–0.06)	▼			▼	▼
Q-90	4.66 (4.48–4.96)			▼	▲	▼
P/D ratio	0.94 (0.80–1.19)	▲	▼	▼	▲	▲

It is reported the median value and the interquartile range (in brackets) for the first period and significant positive (green) and negative (red) changes for successive periods.

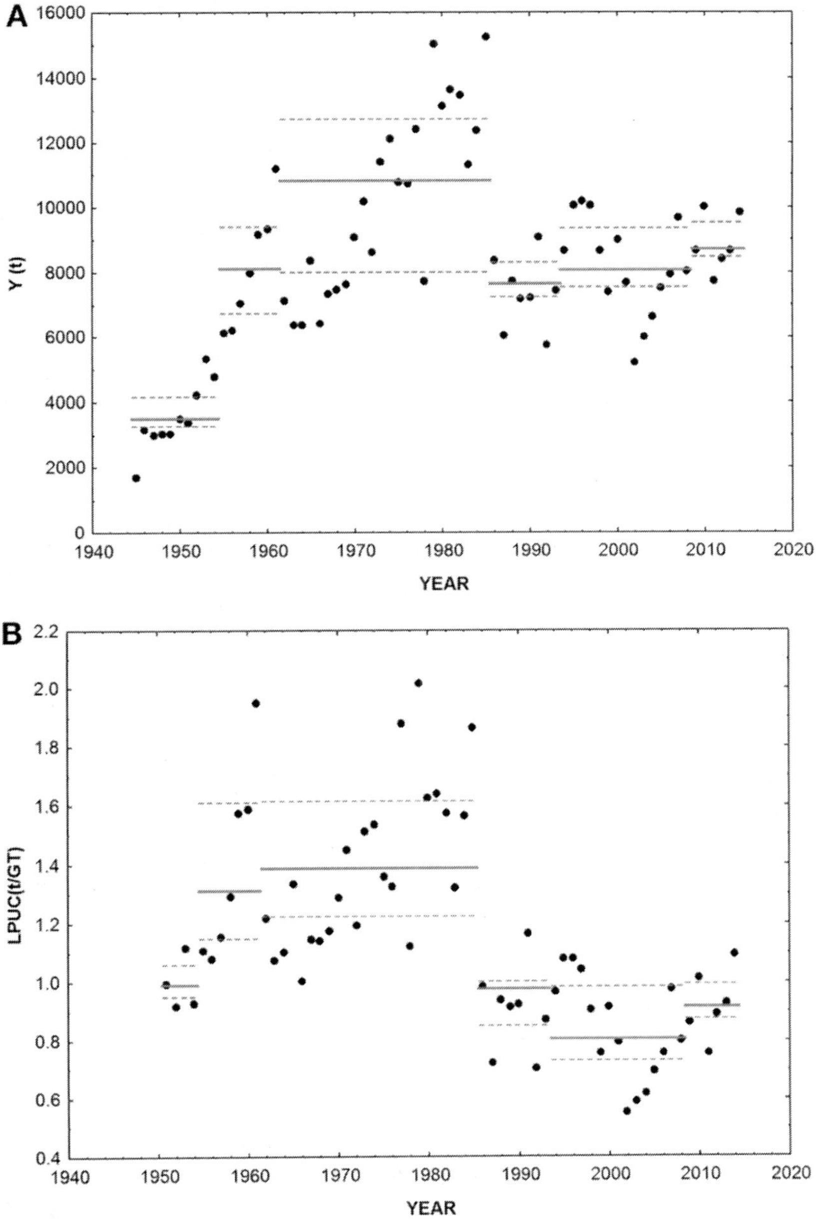

FIGURE 4. (A) Total landings (Y) and **(B)** landings per unit of fishing capacity (LPUC). Black dots, data; continuous red line, median value in each period identified through the cluster analysis; dotted red lines, interquartile range in each period.

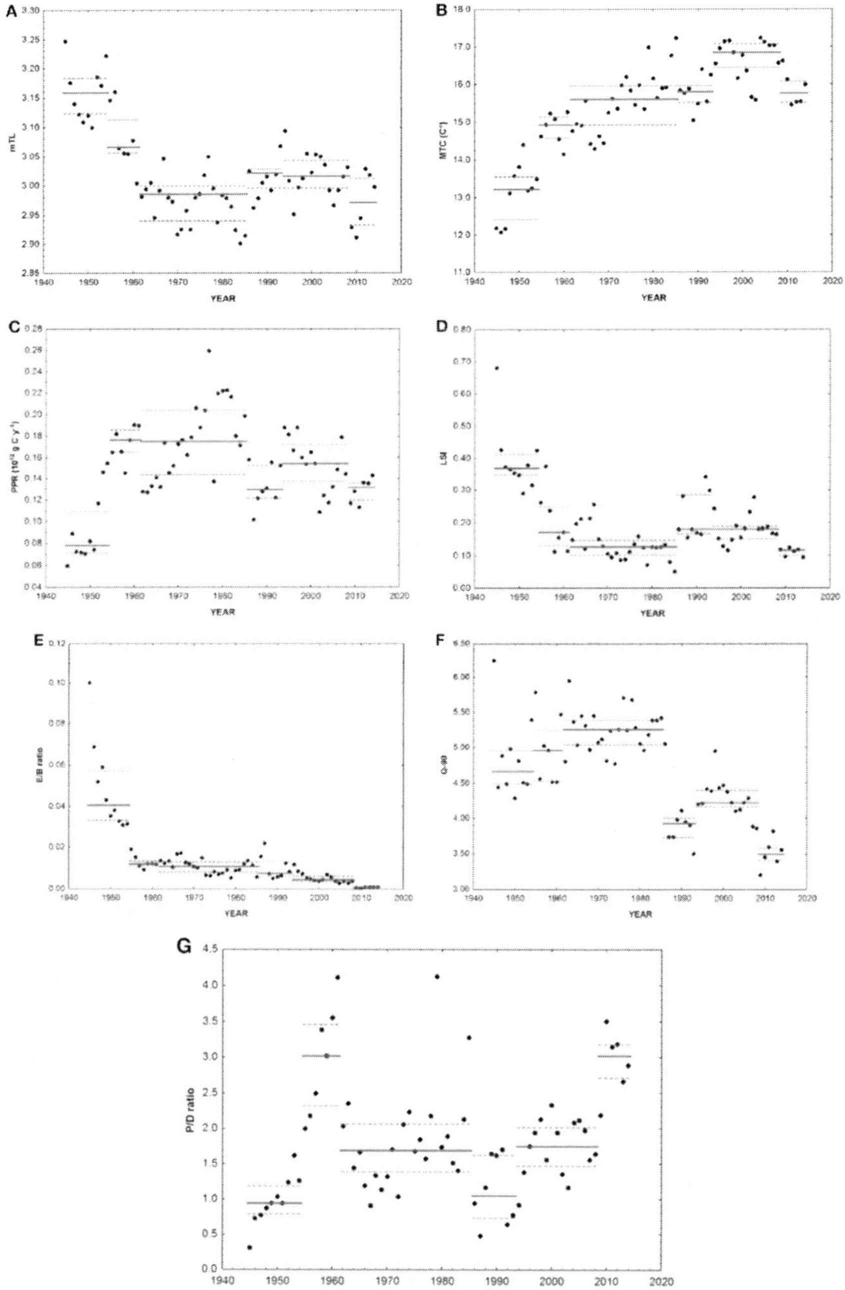

FIGURE 5. ECOLOGICAL INDICATORS CALCULATED ON THE BASIS OF LANDINGS DISAGGREGATED DATA. Black dots, data; continuous red line, median value in each period identified through the cluster analysis; dotted

red lines, interquartile range in each period. **(A)** mTL, mean trophic level; **(B)** MTC, mean temperature of the catch; **(C)** PPR, primary production required; **(D)** LSI, large species indicator; **(E)** E/B ratio, ratio between elasmobranch and bony fish in the landings; **(F)** Q-90, a variant on Kempton's Q index; **(G)** P/D ratio, ratio of small pelagic fish to demersal fish plus benthic landings.

The global BEST-test showed a moderate (ρ = 0.26–0.34) significant ($p < 0.01$) link between ecological indicators and drivers. The similarity matrices obtained with phosphorous discharge and fishing capacity correlated highest (ρ = 0.34) with the similarity matrix of ecological indicators.

Time-Series Analysis

Landings (Y) significantly increased between 1945 and 2014 ($Z = 3.36, p < 0.001$, slope = 58.9 t year^{-1}, Figure 4A). Conversely, landings rescaled over fishing capacity (LPUC) significantly declined in the whole period of study ($Z = -3.95, p < 0.001$, slope = −8.13 t GT^{-1} year^{-1}, Figure 4B). The mTL significantly declined along the whole time series ($Z = -3.38, p < 0.001$, slope = −0.002 year^{-1}, Figure 5A). MTC significantly increased from 1945 to 2014 ($Z = 7.27, p < 0.001$, slope = 0.048°C year^{-1}, Figure 5B). PPR showed no significant trend between 1945 and 2014 (Figure 5C). LSI and E/B ratio significantly declined from 1945 to 2014 ($Z = -3.21, p < 0.01$, slope = −0.002 year^{-1}; $Z = -8.67, p < 0.001$, slope = −0.0003 year^{-1}, Figures 5D,E). The same temporal trend resulted for Q-90 ($Z = -4.92, p < 0.001$, slope = −0.021 year^{-1}, Figure 5F). The P/D ratio significantly increased from 1945 to 2014 ($Z = 2.7, p < 0.01$, slope = −0.02 year^{-1}, Figure 5G).

As regards drivers, Po river mean annual discharge did not show a significant trend. However, the time-series presented a significant negative drop in mean in 2003 and a positive one in 2008 (Figure 6A). Nitrogen load significantly increased between 1945 and 2014 ($Z = 4.17, p < 0.001$, slope = 1.37 t year^{-1}). The time-series presented a significant positive drop in mean in 1972 (Figure 6B). Conversely, phosphate load did not show any significant trend, but an increase in the mean value in 1972 and a significant decrease in 1988 (Figure 6C). Neither the NAO showed any significant trend, but two negative shifts in 1962 and 2009 and a positive shift between the two in 1972 (Figure 6D). SST significantly increased in the period of study ($Z = 3.56, p < 0.001$, slope = −0.01°C year^{-1}) and showed two positive shifts in mean in 1998 and 2011 (Figure 6E). Finally, also fishing capacity significantly increased ($Z = 9.46, p < 0.001$, slope = 89.88 GT year^{-1}) without any significant shift in the mean (Figure 6F).

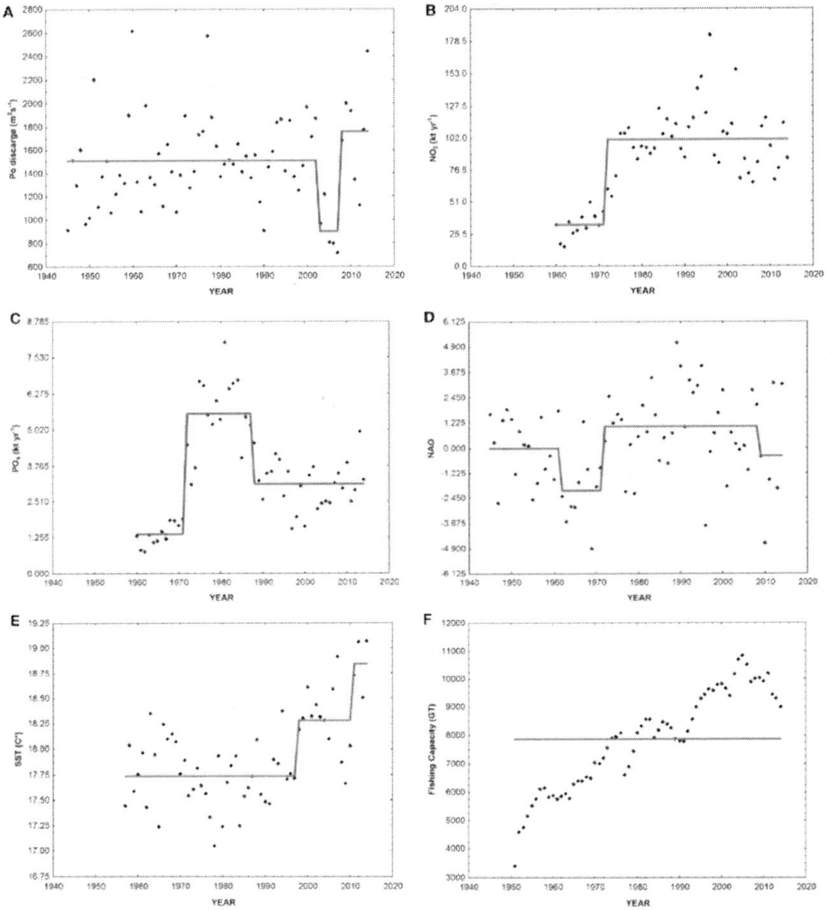

FIGURE 6. TIME-SERIES OF DRIVERS OF ECOSYSTEM CHANGE. Black diamonds, data; red line, regime shift. **(A)** Po discharge, mean annual Po river freshwater discharge; **(B)** NO_3, annual nitrogen flux from the Po river; **(C)** PO_4, annual phosporus flux from the Po river; **(D)** NAO, North Atlantic Oscillation index; **(E)** SST, sea surface temperature; **(F)** GT, total gross tonnage of vessels of the Chioggia's fishing fleet.

DISCUSSION

Six periods with significantly different community structure were identified through cluster analysis of the landings composition by functional group. The most abundant group was small pelagics (mainly European anchovy *Engraulis encrasicolus* and European pilchard *Sardina pilchardus*) in all periods, followed by cephalopods in almost all periods. Major changes in community composition between periods include skate and rays and medium pelagics decline, and bivalves dome-shaped trajectory.

The increase of bivalves' proportion in the landings from 2% in 1955–1961, to 5 and 8% in the following periods, is probably linked to the introduction in the mid-1950s of the "*rapido*" trawl, and successively in the 1970s of the hydraulic dredge, that substantially improved fishing efficiency. However, the proportion of this functional group in the landings decreased to 3% in 1994–2008, and further decreased to 2% in 2009–2014. This sharp decrease is related to the collapse of the Mediterranean scallop (*Pecten jacobaeus*) in the late 1990s (data not shown) due to overfishing. In fact, between the 1960s and the 1990s, the species showed large fluctuations determined both by the intensity of fishing effort and by mass mortalities due to hypoxic conditions that occurred episodically over wide areas of the Northern Adriatic Sea (Hall-Spencer et al., 1999). A previous study (Maurizio and Castagnolo, 1986) showed that after a 6-month period of fishing, commercially sized *P. jacobaeus* catches collapsed by 83%, indicating the high vulnerability of the species to unsustainable exploitation, and the need for the institution of selected areas closed to trawling (Hall-Spencer et al., 1999).

Moreover, a suite of ecological indicators applied to time-series of landings provided the basis for gaining useful insights on possible causes of long-term changes. The use of multiple metrics allowed to obtain a picture of how the ecosystem has changed over the past 70 years, since different metrics emphasize distinct aspects of the underlying communities, and consequently allow disentangling the role of different drivers.

Early Changes: The Decline in Size of Fished Communities

LSI, E/B ratio, and mTL showed a significant decrease between 1945 and 2014. Among the range of ecological indicators analyzed in this paper, these are directly (LSI) and indirectly (E/B ratio and mTL) related to size. Since fishing is usually size-selective, both within and among species (Jennings, 2005), these indicators are considered to be sensitive to fishing disturbance and are expected to decrease under unsustainable exploitation.

It is worth noting that LSI does not take into account the actual size-distribution of exploited populations (that requires the availability of information of the size of caught individuals), but it is based on the L_{max} of species (see Table 1). Thus, the LSI changes are attributable to changes in composition of populations rather than species' size composition. Therefore, the LSI differs substantially from the Large Fish Indicator (LFI, Greenstreet et al., 2011), but it may be useful for management purposes since it requires less detailed data than LFI for highlighting effects of fishing, and can be applied also to long-term landings data series where the size structure of catches is most often not known.

The negative trend for these indicators highlights a shift toward smaller species in the community, or anyway a decrease of proportion of larger fish. Larger fish (which usually have high trophic levels) are more vulnerable to fishing and have less capacity to sustain great rates of mortality (Dulvy and Reynolds, 2002; Jennings, 2005; Myers and Worm, 2005). This phenomenon could be further exacerbated by the possible increase of small species due to the "predation release" as their predators are depleted (Bruno and O'Connor, 2005; Jennings, 2005; Myers et al., 2007).

Therefore, fishing seems to have had significant detrimental effects on the fish community composition since the first years of the time-series. Indeed in mid-1950s, the first rearrangement of the fish community structure occurred, as emerged from the cluster analysis, with a shift from large-sized high trophic level species to small planktivorous ones. Consequently, a significant negative decrease of mTL, LSI, and E/B ratio was observed.

The decline of elasmobranchs and large-sized species started even before the 1950s in the area (i.e., in the nineteenth century; Fortibuoni et al., 2010; Raicevich and Fortibuoni, 2013), but was exacerbated afterwards probably as a consequence of the industrialization of fishery and the introduction of new highly impacting fishing gears (Ferretti et al., 2013; Barausse et al., 2014). Especially linked to the decrease of skates and rays (Barausse et al., 2014; Engelhard et al., 2015), the E/B ratio decline indicates the important role of elasmobranchs as sensitive key species for detecting early signals of fisheries disturbances (Baum et al., 2003).

However, the early collapse of large-sized species is not only a consequence of the decline of elasmobranchs. In the mid-1950s medium pelagic species reached a maximum in landings (1449 kg in 1956), and dramatically declined afterwards. Medium pelagics (mainly the Atlantic mackerel *Scomber scombrus*) represent an important group of species for the local fish market, whose decline is hardly a consequence of changes in market demand (Meneghesso et al., 2013). Thus, it can be assumed quite conservatively that the decline in catches for medium pelagics reflects a dramatic decline of populations at sea.

Overall, the decline in mTL, E/B ratio and LSI clearly highlight that there has been a long-term fishing-down food web phenomenon (*sensu* Pauly et al., 1998).

Signals of Structural Changes

The Q-90 index significantly declined between 1945 and 2014, with a significant negative median-shift in the mid-1980s, indicating a reduction in the biodiversity of landings with increasing fishing impact. The change in biodiversity of landings is concurrent to the significant decrease of total landings and LPUC, suggesting that main modification of marine communities affected local fisheries production.

In the mid-1980s, regime shifts are reported for different ecosystem components in many areas, the Mediterranean Sea, the North Sea, the Baltic Sea, and the Black Sea, possibly suggesting regional effects of a larger scale northern hemispheric pattern (Conversi et al., 2010). Barausse et al. (2011) reported that a similar shift occurred also in the Northern Adriatic fish community, including also medium-high trophic level species belonging to both the demersal and pelagic habitats. Our analysis, however, while confirming the existence of the shift, do not support the hypothesis that climatic drivers played a major effect on it, and rather point to the impact of local pressures, i.e., overexploitation and nutrient loads.

Such structural changes may have altered food-web structure and impaired its functioning and resilience. Coll et al. (2008) compared North-Central Adriatic Sea food-web structure between the 1970s and 1990s, and reported a high food-web degradation regarding overexploitation of higher trophic levels and a simplification of food-web structure (lower omnivory and higher generality). Thus, it is not surprising that LPUC showed a downwards significant shift in 1986, reaching its minimum value in 2002. Only in 2014, the index recovered to values comparable to the very beginning of the time-series, when fishing technologies were markedly less developed and efficient.

Changes in the Trophic State

The significant positive trend over time in the P/D index found in the present study may depend both from eutrophication and overexploitation of resources (Libralato et al., 2004). Indeed, eutrophication and overfishing may have similar and synergistic effects on fish communities, i.e., a decline in diversity, an initial increase in productivity of benthic/demersal and pelagic food webs, then the progressive dominance of the production system by short-lived, especially pelagic species (Caddy, 1993). However, the P/D ratio in the present case seems to point out the importance of eutrophication driven events.

Pelagic fishes are generally influenced by nutrient enrichment when it stimulates the plankton production (Caddy, 1993), while demersal fishes are influenced by the dynamics of benthic community, which generally responds negatively to the conditions of excessive enrichment as shown by de Leiva Moreno et al. (2000). These authors found a mean value of P/D equal to 3.76 in the Adriatic for the historical series 1978–1988. In the present study, the P/D index had a wide dynamic trajectory, ranging between 0.31 and 4.12 with a mean value of 1.81 (± 0.84), thus confirming the Adriatic as a mesotrophic ecosystem (de Leiva Moreno et al., 2000). The index reached some peaks that can be partly related to severe anoxic/hypoxic events. For instance, anoxias in bottom waters linked to eutrophication occurred in the periods 1955–1956 and 1972–1982 (Sangiorgi and Donders, 2004), when the P/D index reached a value of ~2. Finally, the P/D index showed an increasing trend in recent years (significant upward median-shift in 2009), probably linked to a partial recovery of small pelagic species, mainly European anchovy (Carpi et al., 2015).

PPR showed a significant positive median-shift in the mid-1950s. In the mid-1980s, PPR collapsed, mainly driven by the strong reduction of small pelagic landings. As regards anchovy, its population started declining since the late-1970s, and the reasons of this abrupt reduction have been previously ascribed to climate forcing (Santojanni et al., 2006), modified inflow of Mediterranean waters in the Adriatic Sea and associated salinity changes (Grbec et al., 2002), over-fishing, increased predation of eggs and larvae by the jellyfish *Pelagia noctiluca* and the presence of mucilage events (Regner, 1996). PPR further significantly decreased in the last period 2009–2014.

It is worth noting that during the 1970s an increase of eutrophication occurred in the basin, lasting until the mid-1980s (Giani et al., 2012). Although nitrogen loads increased up to the mid-1970s and then maintained approximately the same values up to now (Figure 6B), phosphate peaked in the mid-1980s (Figure 6C), and its following marked decline may have resulted in limiting production. In 1985, to reduce the negative effects of cultural eutrophication (e.g., anoxia events), the nutrient load delivered to the Adriatic Sea was reduced, mainly by changing the chemical composition of soap powders (banning phosphorous from the mixture) and by improving the treatment of urban sewage and farm litter products (Decree-law No. 667 of November 25th 1985—Urgent measures to limit the eutrophication of waters).

A significant decrease of phytoplankton abundance was also observed after the 1980s, along with changes in its species composition with a shift toward smaller organisms. Such changes modified also zooplankton community (Kamburska and Fonda-Umani, 2009). This trend was a consequence of a reduction of phosphorous load, being the Northern Adriatic waters phosphorous-limited (Solidoro et al., 2009), and a decline of atmospheric precipitation and the runoff in the basin (Giani et al., 2012).

Signals of Climatic Changes

MTC significantly increased between 1945 and 2014 at a decadal rate of 0.5°C, with three significant positive median-shifts in the mid-1950s, in the early-1960s, and in the mid-1990s. A significant negative median-shift was instead observed in the last period (2009–2014), even if the median value (15.8°C) was still much higher than at the one observed at the beginning of the time-series (1945–1954: 13.2°C). Thus, overall an increasing dominance in catches of warm affinity species occurred in the landings of Chioggia, coherently with the pattern observed in other Mediterranean areas (Cheung et al., 2013; Tsikliras and Stergiou, 2014; Fortibuoni et al., 2015; Tsikliras et al., 2015). Also SST significantly increased between 1957 and 2014, with a positive shift in mean in 1998. However, SST resulted not to be one of the main ecological drivers in driving community changes in the Northern Adriatic Sea, and results from this study do not allow establishing a relationship between MTC, SST in the area and NAO.

Integrating in a Coherent Framework the Community Changes

Long-term changes in the Northern Adriatic fish community resulted to be mainly related to the impacts of fisheries and nutrient dynamics, while climate had a secondary role up to now. A coherent dynamic is obtained by scaling the PPR to N:P ratio (NO_3/PO_4) to account for the changes in the limiting factor (Solidoro et al., 2009), and comparing it with the fishing capacity through time (Figure 7).

The PPR to N:P ratio increased up to the mid-1980s, showing that the exploitation development was supported by nutrient enrichment. In the mid-1980s there was an abrupt collapse after which, even for large changes in fishing capacity, PPR to N:P ratio attained much lower values (Figure 7). Notably, the mid-1980s shift is coherent with a shift between two alternative stable states, as described in Scheffer and Carpenter (2003): from the first state (line A–B in Figure 7), to a second state (line C–D–E in Figure 7). Therefore, in the mid-1980s a shift between two alternative stable states of the system likely occurred in the Northern Adriatic Sea, which is reflected also in a change in the community structure (see

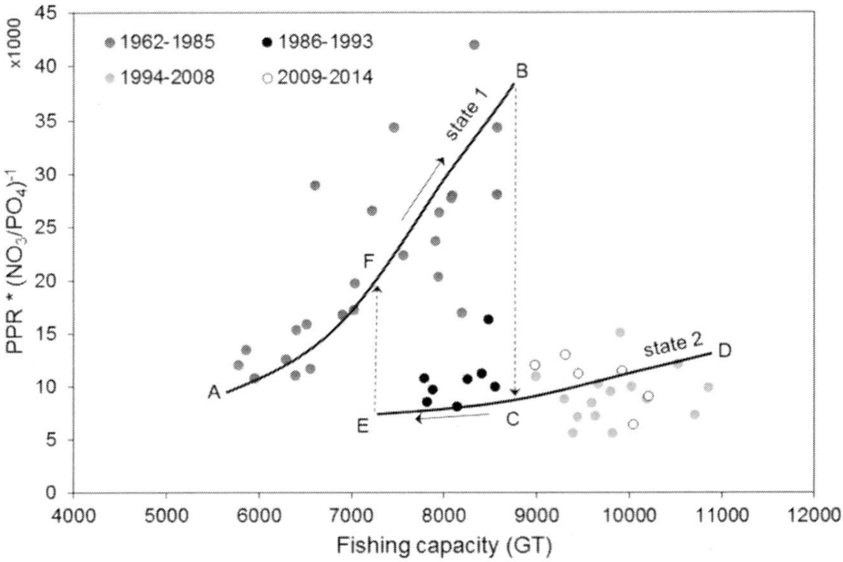

FIGURE 7. DYNAMICS OF FISHERIES PRODUCTION UNDER CHANGES OF FISHING CAPACITY AND NUTRIENT LIMITATION FROM 1960 TO 2014. PPR is scaled to N:P ratio to take into account changes in the nutrient limiting factor. Solid lines represent subjective indication of average main trajectories identifiable as two alternative states of the system (state 1: line A–B; state 2: C–D–E). The mid-1980s shift is coherent with a catastrophic shift between two alternative stable states (dashed arrow from B to C) and the presence of hysteretic behavior. For going back from the actual state (state 2, lower solid line) to a state prior to the catastrophic event of mid-1980s (state 1, upper solid line) it is necessary to considerably reduce fishing capacity (left dashed line E–F, placed subjectively).

chronological clustering; Figure 2). The shift to a new community composition and the lower trophic potential (see previous sections) thus resulted in a much lower productive capacity (here represented by PPR to N:P ratio), even for the high fishing effort exerted in the last decades of the time series (Figure 6F). Importantly, to induce a switch back to the original state, it is not sufficient to restore the conditions present before the collapse (Scheffer and Carpenter, 2003): Figure 7 shows the hysteresis (B–C–E–F) typical to alternative stable points of complex systems. From the actual situation (state 2), to reach the same PPR to N:P ratio observed before its collapse (state 1), a relevant reduction of fishing capacity is necessary (see the hypothetical line E–F).

Although from our data it is not possible to estimate the reduction in fishing capacity necessary to switch back from state 2 to state 1, Figure 7 highlights that a reduction in fishing capacity to values lower than the ones that characterized the 1980s would be necessary, together with an intervention on nutrient loads. This result points to the fact that the current critical situation of fisheries is not solely the result of mismanagement of the fisheries sector, but is due also to other environmental policies, in particular those relative to water quality. Thus, actual management needs to account for the fact that the ecosystem is now in a different state (state 2), and the reversal of the shift of the mid-1980s implies changes in fishing capacity and nutrient balance that go far beyond those of the period of the shift.

Overcoming Limitations in the Use of Landings for Ecological Analyses

This work gives support to the possibility of using landings statistics for inferring changes in marine ecosystems through the analysis of landings composition and the application of a set of ecological indicators. Despite the intrinsic limitations of fishery-dependent data (Essington et al., 2006; Hilborn, 2007), the analysis reported here highlights that opportune indicators can produce interesting and useful assessments from landings, especially if combined with local knowledge on fisheries changes. In particular, the capability to understand relationship between landings composition to community at sea is more robust when local disaggregated landings result from multi-target and multi-gear fisheries (i.e., several distinct métiers), and when changes in fishing activities (e.g., introduction of new technologies, shift from one fishing gear to another, number and tonnage of boats) are traceable.

Being aware of the poor capabilities of landings absolute quantities to assess single species abundances (Pauly et al., 2013), we used landings composition (percentage proportion of functional groups in total landings) to detect community changes. Moreover, the use of data lumped into functional groups allowed overcoming problems of changes in aggregation detail in landings statistics.

The potential is great given that landings data are the most widespread information that can be used to analyse marine ecosystem changes in the past, and probably the cheapest information that can be collected by surveying archives and statistical bulletins and that can be used also in data poor conditions.

CONCLUSIONS

The analysis of landings from the fish market of Chioggia allowed inferring information on long-term changes in the Northern Adriatic ecosystem. Most vulnerable species (i.e., elasmobranchs and large-sized species) considerably declined at the beginning of the industrialization of fishery and continued to decline in the following decades, probably because the exploitation rates were not sustainable. Indeed, fishing capacity increased enormously during the 1960s and 1970s, i.e., larger boats, higher tonnage and engine horsepower, improved fishing gears, use of high-technology equipment (Fortibuoni, 2010). Until the mid-1980s total landings continuously increased, possibly as a consequence of the modernization of fishing fleets and of the growing cultural eutrophication. However, while landings were still increasing, the LPUC was already declining.

Later, the nutrient load (in terms of phosphorus) delivered to the Adriatic Sea decreased, thus leading to a combination of high exploitation and reduced productivity, which may well-explain the collapse in landings in the following years. Indeed, from our analysis resulted that phosphorous load (a proxy of primary production) and fishing capacity (a proxy of fishing effort) were the main drivers of change among the considered explanatory variables. Conversely, climate relate variables had a smaller impact.

It is likely that long-term effects of fishing drove significant changes in fish community structure in the Northern Adriatic Sea that were partially masked or balanced by an increase in productivity in the period of high nutrient discharge. Once productivity declined, the food-web structure was already modified and probably the resilience of the system was unpaired. The ecosystem was in a "fishing status" (*sensu* Jennings and Kaiser, 1998), thus reducing its recoverability from environmental driven imbalance. As regards the role played by fishing, it is expected that over time the enhanced skipper skills, adoption of auxiliary equipment and more efficient gear and materials, replacement of old vessels by new ones and upgraded engines (Damalas et al., 2014; Engelhard, 2016), resulted in an increased catching efficiency. However, since no reliable information was available for the area we did not consider the effect of the technological creep, keeping our results rather conservative with respect to fishing impacts.

This study shows that the Northern Adriatic ecosystem and ecological drivers has dramatically changed in the last decades, and thus greater knowledge of past states is crucial to set appropriate baselines for current management. Indeed, the actual crisis faced by the Northern Adriatic Sea fishery sector may be ascribed both to long-term over-exploitation and changes in nutrient load. This evidence should be considered in fishery management, for instance by rescaling the fishing capacity according to the present status of environmental parameters (e.g., trophic conditions related to nutrient discharges). Since fisheries management in Italy (as in the whole Mediterranean) is predominantly capacity/effort-based, accounting for changes in these parameters is decisive. However, because of technological creep, measuring nominal capacity and effort in conventional terms (e.g., GT, KW, days at sea) may produce estimates far from the effective fishing mortality exerted by the fleet (Damalas et al., 2014). All this may partly explain why in Italian waters the positive impact on resources expected from fishing capacity/effort reduction was lower than expected (Cataudella and Spagnolo, 2011). Thus, research on fishing power change and fine-scale analysis of fishing effort through time are highly recommended in order to understand real changes in the capacity of fishing fleets, and their potential to exploit fish stocks (Engelhard, 2016). Moreover, the results might provide evidence on the need for considering broadly the ecosystems impacts of human interventions and management actions, since actual critical state of the fisheries have been exacerbated by regulations on water quality.

FUNDING

The authors acknowledge partial support from EU FP 7 project PERSEUS (Policy-oriented Marine Environmental Research for the Southern European Seas, http://www.perseus-net.eu/) and the Italian RITMARE Flagship Project—The Italian Research for the Sea (http://www.ritmare.it) coordinated by the Italian National Research Council and funded by the Italian Ministry of Education, University and Research within the National Research Program 2011–2013.

ACKNOWLEDGMENTS

We acknowledge Paolo Lazzari (National Institute of Oceanography and Experimental Geophysics, Italy), Virna Meccia (Max Planck Institute for Meteorology, Germany), Cristina Mazziotti and Giovanna Pirretti (Regional Environmental Protection Agency, Italy) for helping us with the collection of ecosystem drivers data (nutrients load, climatic data, Po river discharge), and Carlotta Mazzoldi (University of Padua, Italy) for providing us landings data. This study was benefited from discussions with colleagues of the ICES Working Group on the History of Fish and Fisheries. This article is based upon work from COST Action IS1403, Oceans Past Platform (OPP), supported by COST (European Cooperation in Science and Technology).

REFERENCES

1. AdriaMed (2004). *AdriaMed Seminar on Fishing Capacity: Definition, Measurement and Assessment.* AdriaMed Technical Documents, FAO-MiPAF Scientific Cooperation to Support Responsible Fisheries in the Adriatic Sea, Termoli, p. 119. GCP/RER/010/ITA/TD-13.
2. Ainsworth, C. H., and Pitcher, T. J. (2006). Modifying Kempton's species diversity index for use with ecosystem simulation models. *Ecol. Ind.* 6, 623–630. doi: 10.1016/j.ecolind.2005.08.024
3. Baeta, F., Costa, M. J., and Cabral, H. (2009). Changes in the trophic level of Portuguese landings and fish market price variation in the last decades. *Fish. Res.* 97, 216–222. doi: 10.1016/j.fishres.2009.02.006
4. Barausse, A., Correale, V., Curkovic, A., Finotto, L., Riginella, E., Visentin, E., et al. (2014). The role of fisheries and the environment in driving the decline of elasmobranchs in the northern Adriatic Sea. *ICES J. Mar. Sci.* 71, 1593–1603. doi: 10.1093/icesjms/fst222
5. Barausse, A., Michieli, A., Riginella, E., Palmeri, L., and Mazzoldi, C. (2011). Long-term changes in community composition and life-history traits in a highly exploited basin (Northern Adriatic Sea): the role of environment and anthropogenic pressures. *J. Fish Biol.* 79, 1453–1486. doi: 10.1111/j.1095-8649.2011.03139.x
6. Baum, J. K., Myers, R. A., Kehler, D. G., Worm, B., Harley, S. J., and Doherty, P. A. (2003). Collapse and conservation of shark populations in the Northwest Atlantic. *Science* 299, 389–392. doi: 10.1126/science.1079777
7. Botter, L., Giovanardi, O., and Raicevich, S. (2006). The migration of Chioggia's fishing fleet in the Adriatic from the mid-nineteenth to the early twentieth centuries. *J. Mediterr. Stud.* 16, 27–44.
8. Bruno, J. F., and O'Connor, M. I. (2005). Cascading effects of predator diversity and omnivory in a marine food web. *Ecol. Lett.* 8, 1048–1056. doi: 10.1111/j.1461-0248.2005.00808.x
9. Caddy, J. F. (1993). Toward a comparative evaluation of human impacts on fishery ecosystems of enclosed and semi-enclosed seas. *Rev. Fish. Sci.* 1, 57–95. doi: 10.1080/10641269309388535
10. Carpi, P., Santojanni, A., Donato, F., Colella, S., Keč, Č. V., Zorica, B., et al. (2015). A joint stock assessment for the anchovy stock of the northern and central Adriatic Sea: comparison of two catch-at-age models. *Sci. Mar.* 79, 57–70. doi: 10.3989/scimar.03903.29A
11. Cataudella, S., and Spagnolo, M. (2011). *The State of Italian Marine Fisheries and Aquaculture.* Rome: Ministero delle Politiche Agricole Alimentari e Forestali.
12. Cheung, W. L., Watson, R., and Pauly, D. (2013). Signature of ocean warming in global fisheries catch. *Nature* 497, 365–369. doi: 10.1038/nature12156
13. Cingolani, N., Giannetti, G., and Arneri, E. (1996). Anchovy fisheries in the Adriatic sea. *Sci. Mar.* 60, 269–277.
14. Clarke, K. R., Somerfield, P. I., and Gorley, R. N. (2008). Testing of null hypotheses in exploratory community analyses: similarity profiles and biota-environment linkage. *J. Exp. Mar. Biol. Ecol.* 366, 56–69. doi: 10.1016/j.jembe.2008.07.009
15. Clodia database (2015). *Database of Fishery Data from Chioggia, Northern Adriatic Sea.* Available online at: http://chioggia.scienze.unipd.it/Database_landing.html
16. Coll, M., Lotze, H. K., and Romanuk, T. N. (2008). Structural degradation in Mediterranean sea food webs: testing ecological hypotheses using stochastic and mass-balance modelling. *Ecosystems* 11, 939–960. doi: 10.1007/s10021-008-9171-y
17. Coll, M., Shannon, L. J., Kleisner, K. M., Juan-Jordá, M. J., Bundy, A., Akoglu, A. G., et al. (2016). Ecological indicators to capture the effects of fishing on biodiversity and conservation status of marine ecosystems. *Ecol. Ind.* 60, 947–962. doi: 10.1016/j.ecolind.2015.08.048

18. Conversi, A., Fonda Umani, S., Peluso, T., Molinero, J. C., Santojanni, A., and Edwards, M. (2010). The Mediterranean sea regime shift at the end of the 1980s, and intriguing parallelisms with other european basins. *PLoS ONE* 5:e10633. doi: 10.1371/journal.pone.0010633
19. Damalas, D., Maravelias, C. D., and Kavadas, S. (2014). Advances in fishing power: a study spanning 50 years. *Rev. Fish. Sci. Aquacult.* 22, 112–121. doi: 10.1080/10641262.2013.839620
20. de Leiva Moreno, J. I., Agostini, V. N., Caddy, J. F., and Carocci, F. (2000). Is the pelagic-demersal ratio from fishery landings a useful proxy for nutrient availability? A preliminary data exploration for the semi-enclosed seas around Europe. *ICES J. Mar. Sci.* 57, 1091–1102. doi: 10.1006/jmsc.2000.0705
21. Dulvy, N. K., and Reynolds, J. D. (2002). Predicting extinction vulnerability in skates. *Conserv. Biol.* 16, 440–450. doi: 10.1046/j.1523-1739.2002.00416.x
22. Engelhard, G. H. (2016). "On the need to study fishing power change: challenges and perspectives," in *Perspectives on Oceans Past*, eds K. S. Máñez and B. Poulsen (Dordrecht: Springer Publishers Netherlands), 89–101.
23. Engelhard, G. H., Thurstan, R. H., MacKenzie, B. R., Alleway, H. K., Bannister, R. C. A., Cardinale, M., et al. (2015). ICES meets marine historical ecology: placing the history of fish and fisheries in current policy context. *ICES J. Mar. Sci.* 73, 1386–1403. doi: 10.1093/icesjms/fsv219
24. Essington, T. E., Beaudreau, A. H., and Wiedenmann, J. (2006). Fishing through marine food webs. *Proc. Natl. Acad. Sci. U.S.A.* 103, 3171–3175. doi: 10.1073/pnas.0510964103
25. Ferretti, F., Osio, G. C., Jenkins, C. J., Rosenberg, A. A., and Lotze, H. K. (2013). Long-term change in a meso-predator community in response to prolonged and heterogeneous human impact. *Sci. Rep.* 3:1057. doi: 10.1038/srep01057
26. Fortibuoni, T. (2010). *La Pesca in Alto Adriatico dalla Caduta Della Serenissima ad oggi: un'Analisi Storica ed Ecologica*. Ph.D. thesis, University of Trieste, Trieste.
27. Fortibuoni, T., Aldighieri, F., Giovanardi, O., Pranovi, F., and Zucchetta, M. (2015). Climate impact on Italian fisheries (Mediterranean Sea). *Reg. Environ. Change* 15, 931–937. doi: 10.1007/s10113-015-0781-6
28. Fortibuoni, T., Borme, D., Franceschini, G., Giovanardi, O., and Raicevich, S. (2016). Common, rare or extirpated? Shifting baselines for common angelshark, *Squatina squatina* (Elasmobranchii: Squatinidae), in the Northern Adriatic Sea (Mediterranean Sea). *Hydrobiologia* 772, 247–259. doi: 10.1007/s10750-016-2671-4
29. Fortibuoni, T., Libralato, S., Raicevich, S., Giovanardi, O., and Solidoro, C. (2010). Coding early naturalists' accounts into long-term fish community changes in the Adriatic Sea (1800–2000). *PLoS ONE* 5:e15502. doi: 10.1371/journal.pone.0015502
30. Fortibuoni, T., Noventa, S., Rampazzo, F., Gion, C., Formalewicz, M., Berto, D., et al. (2013). Evidence of butyltin biomagnification along the northern Adriatic food-web (Mediterranean Sea) elucidated by stable isotope ratios. *Environ. Sci. Technol.* 47, 3370–3377. doi: 10.1021/es304875b
31. Froese, R., and Pauly, D. (2016). *FishBase. World Wide Web Electronic Publication.* Available online at: www.fishbase.org
32. Fu, C., Large, S., Knight, B., Richardson, A. J., Bundy, A., Reygondeau, G., et al. (2015). Relationships among fisheries exploitation, environmental conditions, and ecological indicators across a series of marine ecosystems. *J. Mar. Syst.* 148, 101–111. doi: 10.1016/j.jmarsys.2015.01.004
33. Giani, M., Djakovac, T., Degobbis, D., Cozzi, S., Solidoro, C., and Fonda Umani, S. (2012). Recent changes in the marine ecosystems of the northern Adriatic Sea. *Estuar. Coast. Shelf S.* 115, 1–13. doi: 10.1016/j.ecss.2012.08.023
34. Grbec, B., Dulcic, J., and Morovi, M. (2002). Long-term changes in landings of small pelagic fish in the eastern Adriatic—possible influence of climate oscillations over the Northern Hemisphere. *Clim. Res.* 20, 241–252. doi: 10.3354/cr020241

35. Greenstreet, S. P. R., Rogers, S. I., Rice, J. C., Piet, G. J., Guirey, E. J., Fraser, H. M., et al. (2011). Development of the EcoQO for fish communities in the North Sea. *ICES J. Mar. Sci.* 68, 1–11. doi: 10.1093/icesjms/fsq156
36. Hall-Spencer, J. M., Froglia, C., Atkinson, R. J. A., and Moore, P. G. (1999). The impact of Rapido trawling for scallops, *Pecten jacobaeus* (L.), on the benthos of the Gulf of Venice. *ICES J. Mar. Sci.* 56, 111–124. doi: 10.1006/jmsc.1998.0424
37. Hammer, Ø., Harper, D. A. T., and Ryan, P. D. (2001). PAST: Paleontological Statistics Software package for education and data analysis. *Palaeontol. Electron.* 4, 1–9.
38. Hilborn, R. (2007). Reinterpreting the state of fisheries and their management. *Ecosystems* 10, 1362–1369. doi: 10.1007/s10021-007-9100-5
39. Hopkins, T. S., Artegiani, A., Kinder, C., and Pariante, R. (1999). "A discussion of the northern Adriatic circulation and flushing as determined from the ELNA hydrography," in *The Adriatic Sea. Ecosystem Research Report 32*, eds T. S. Hopkins, A. Artegiani, G. Cauwet, D. Degobbis, and A. Malej (Brussels: European Commission), 85–106.
40. Hurrell, J. (2016). *The Climate Data Guide: Hurrell North Atlantic Oscillation (NAO) Index (Station-Based).* Available online at: https://climatedataguide .ucar.edu/ climate-data/hurrell-north-atlantic-oscillation-nao-index-station-based
41. Jackson, J. B. C., Kirby, M. X., Berger, W. H., Bjorndal, K. A., Botsford, L. W., Bourque, B. J., et al. (2001). Historical overfishing and the recent collapse of coastal ecosystems. *Science* 293, 629–637. doi: 10.1126/science.1059199
42. Jennings, S. (2005). Indicators to support an ecosystem approach to fisheries. *Fish Fish.* 6, 212–232. doi: 10.1111/j.1467-2979.2005.00189.x
43. Jennings, S., and Kaiser, M. J. (1998). The effects of fishing on marine ecosystems. *Adv. Mar. Biol.* 34, 201–352. doi: 10.1016/S0065-2881(08)60212-6
44. Kaiser, M. J., Ramsay, K., Richardson, C. A., Spence, F. E., and Brand, A. R. (2000). Chronic fishing disturbance has changed shelf sea benthic community structure. *J. Anim. Ecol.* 69, 494–503. doi: 10.1046/j.1365-2656.2000.00412.x
45. Kamburska, L., and Fonda-Umani, S. (2009). From seasonal to decadal inter-annual variability of mesozooplankton biomass in the northern Adriatic Sea (Gulf of Trieste). *J. Mar. Syst.* 78, 490–504. doi: 10.1016/j.jmarsys.2008.12.007
46. Kempton, R. A., and Taylor, L. R. (1976). Models and statistics for species diversity. *Nature* 262, 818–820. doi: 10.1038/262818a0
47. Kleisner, K., Zeller, D., Froese, R., and Pauly, D. (2013). Using global catch data for inferences on the world's marine fisheries. *Fish Fish.* 14, 293–311. doi: 10.1111/j.1467-2979.2012.00469.x
48. Legendre, P., and Legendre, L. (1998). *Numerical Ecology, 2nd Edn.* Amsterdam: Elsevier.
49. Libralato, S., Pranovi, F., Raicevich, S., Da Ponte, F., Giovanardi, O., Pastres, R., et al. (2004). Ecological stages of the Venice Lagoon analysed using landing time series data. *J. Mar. Syst.* 51, 331–344. doi: 10.1016/j.jmarsys.2004.05.020
50. Lotze, H. K., Coll, M., and Dunne, J. A. (2011). Historical changes in marine resources, food-web structure and ecosystem functioning in the Adriatic Sea, Mediterranean. *Ecosystems* 14, 198–222. doi: 10.1007/s10021-010-9404-8
51. Lotze, H. K., Lenihan, H. S., Bourque, B. J., Bradbury, R. H., Cooke, R. G., Kay, M. C., et al. (2006). Depletion, degradation, and recovery potential of estuaries and coastal seas. *Science* 312, 1806–1808. doi: 10.1126/science.1128035
52. Ludwig, W., Dumonta, E., Meybeck, M., and Heussner, S. (2009). River discharges of water and nutrients to the Mediterranean and Black Sea: major drivers for ecosystem changes during past and future decades? *Prog. Oceanogr.* 80, 199–217. doi: 10.1016/j.pocean.2009.02.001
53. Maurizio, C., and Castagnolo, L. (1986). Variazioni di taglia in un banco commerciale di *Pecten jacobaeus* L. e *Chlamys opercularis* L. al largo di Chioggia. Un anno di studi. *Nova Thalassia* 8, 635–640.

54. Mazzoldi, C., Sambo, A., and Riginella, E. (2014). The Clodia database: a long time series of fishery data from the Adriatic Sea. *Sci. Data.* 1:140018. doi: 10.1038/sdata.2014.18
55. Meneghesso, C., Riginella, E., La Mesa, M., Donato, F., and Mazzoldi, C. (2013). Life-history traits and population decline of the Atlantic mackerel *Scomber scombrus* in the Adriatic Sea. *J. Fish Biol.* 83, 1249–1267. doi: 10.1111/jfb.12223
56. Mion, M., Piras, C., Fortibuoni, T., Celić, I., Franceschini, G., Giovanardi, O., et al. (2015). Collection and validation of self-sampled e-logbook data in a Mediterranean demersal trawl fishery. *Reg. Stud. Mar. Sci.* 2, 76–86. doi: 10.1016/j.rsma.2015.08.009
57. Mozetič, P., Solidoro, C., Cossarini, G., Socal, G., Precali, R., Francé, J., et al. (2009). Recent trends towards oligotrophication of the Northern Adriatic: evidence from chlorophyll a time series. *Estuar. Coasts* 33, 362–375. doi: 10.1007/s12237-009-9191-7
58. Munyandorero, J., and Guenter, C. B. (2010). A six-decade portrait of florida marine fisheries via landings-based trophodynamic indicators. *N. Am. J. Fish. Manage.* 30, 259–280. doi: 10.1577/M08-202.1
59. Myers, R. A., Baum, J. K., Shepherd, T. D., Powers, S. P., and Peterson, C. H. (2007). Cascading effects of the loss of apex predatory sharks from a coastal ocean. *Science* 315, 1846–1850. doi: 10.1126/science.1138657
60. Myers, R. A., and Worm, B. (2005). Extinction, survival or recovery of large predatory fishes. *Philos. Trans. R. Soc. B* 360, 13–20. doi: 10.1098/rstb.2004.1573
61. Palomares, M. L. D., and Pauly, D. (2016). *SeaLifeBase.* World Wide Web Electronic Publication. Available online at: http://www.sealifebase.org
62. Pauly, D. (1995). Anecdotes and the shifting baseline syndrome of fisheries. *Trends Ecol. Evol.* 10:430. doi: 10.1016/S0169-5347(00)89171-5
63. Pauly, D., and Christensen, V. (1995). Primary production required to sustain global fisheries. *Nature* 374, 255–257. doi: 10.1038/374255a0
64. Pauly, D., Christensen, V., Dalsgaard, J., Froese, R., and Torres, F. Jr. (1998). Fishing down marine food webs. *Science* 279, 860–863. doi: 10.1126/science.279.5352.860
65. Pauly, D., Hilborn, R., and Branch, T. (2013). Fisheries: does catch reflect abundance? *Nature* 494, 303–306. doi: 10.1038/494303a
66. Pauly, D., and Watson, R. (2005). Background and interpretation of the 'Marine Trophic Index' as a measure of biodiversity. *Philos. Trans. R. Soc. B* 360, 415–423. doi: 10.1098/rstb.2004.1597
67. Piet, G., and Pranovi, F. (2005). *A Review of the Indicators for Ecosystem Structure and Functioning.* Project No. 513754. INDECO - Development of Indicators of Environmental Performance of the Common Fisheries Policy.
68. Pinnegar, J. K., Jennings, S., O'Brien, C. M., and Polunin, N. V. C. (2002). Long-term changes in the trophic level of the Celtic Sea fish community and fish market price distribution. *J. Appl. Ecol.* 39, 377–390. doi: 10.1046/j.1365-2664.2002.00723.x
69. Piroddi, C., Gristina, M., Zylich, K., Greer, K., Ulman, A., Zeller, D., et al. (2015). Reconstruction of Italy's marine fisheries removals and fishing capacity, 1950–2010. *Fish. Res.* 172, 137–147. doi: 10.1016/j.fishres.2015.06.028
70. Pranovi, F., Raicevich, S., Franceschini, G., Torricelli, P., and Giovanardi, O. (2001). Discard analysis and damage to non-target species in the "rapido" trawl fishery. *Mar. Biol.* 139, 863–875. doi: 10.1007/s002270100646
71. Raicevich, S., and Fortibuoni, T. (2013). "Assessing neoextirpations in the Adriatic Sea: an historical ecology approach" in *Marine Extinctions - Patterns and Processes*, ed F. Briand (Monaco: CIESM Publisher), 97–111.
72. Regner, S. (1996). Effects of environmental changes on early stages and reproduction of anchovy in the Adriatic Sea. *Sci. Mar.* 60, 167–177.
73. Rodionov, S. N. (2004). A sequential algorithm for testing climate regime shifts. *Geophys. Res. Lett.* 31, L09204. doi: 10.1029/2004GL019448
74. Rodionov, S. N. (2006). Use of prewhitening in climate regime shift detection. *Geophys. Res. Lett.* 33, L12707. doi: 10.1029/2006GL025904

75. Romanelli, M., Cordisco, C. A., and Giovanardi, O. (2009). The long-term decline of the *Chamelea gallina* L. (Bivalvia: Veneridae) clam fishery in the Adriatic Sea: is a synthesis possible? *Acta Adriat.* 50, 171–205.
76. Rosenberg, A. A., Bolster, W. J., Alexander, K. E., Leavenworth, W. B., Cooper, A. B., and McKenzie, M. G. (2005). The history of ocean resources: modeling cod biomass using historical records. *Front. Ecol. Environ.* 3, 84–90. doi: 10.2307/3868514
77. Russo, A., Rabitti, S., and Bastianini, M. (2002). Decadal climatic anomalies in the Northern Adriatic Sea inferred from a new oceanographic data set. *Mar. Ecol.* 23, 340–351. doi: 10.1111/j.1439-0485.2002.tb00032.x
78. Sàenz-Arroyo, A., Roberts, C. M., Torre, J., Carino-Olvera, M., and Enriquez-Andrade, R. R. (2005). Rapidly shifting environmental baselines among fishers of the Gulf of California. *Proc. R. Soc. Lond. B Bio.* 272, 1957–1962. doi: 10.1098/rspb.2005.3175
79. PubMed Abstract | CrossRef Full Text | Google Scholar
80. Salmi, T., Määttä, A., Anttila, P., Ruoho-Airola, T., and Amnell, T. (2002). *Detecting Trends of Annual Values of Atmospheric Pollutants by the Mann-Kendall Test and Sen's Slope Estimates: the Excel Template Application MAKESENS*. Publications of Air Quality.
81. Sangiorgi, F., and Donders, T. H. (2004). Reconstructing 150 years of eutrophication in the north-western Adriatic Sea (Italy) using dinoflagellate cysts, pollen and spores. *Estuar. Coast. Shelf S.* 60, 69–79. doi: 10.1016/j.ecss.2003.12.001
82. Santojanni, A., Arneri, E., Bernardini, V., Cingolani, N., Di Marco, M., and Russo, A. (2006). Effects of environmental variables on recruitment of anchovy in the Adriatic Sea. *Clim. Res.* 31, 181–193. doi: 10.3354/cr031181
83. Scheffer, M., and Carpenter, S. R. (2003). Catastrophic regime shifts in ecosystems: linking theory to observation. *Trends Ecol. Evol.* 18, 648–656. doi: 10.1016/j.tree.2003.09.002
84. Shephard, S., Fung, T., Houle, J. E., Farnsworth, K. D., Reid, D. G., and Rossberg, A. G. (2012). Size-selective fishing drives species composition in the Celtic Sea. *ICES J. Mar. Sci.* 69, 223–234. doi: 10.1093/icesjms/fsr200
85. Shin, Y.-J., Bundy, A., Shannon, L. J., Simier, M., Coll, M., Fulton, E. A., et al. (2010). Can simple be useful and reliable? Using ecological indicators to represent and compare the states of marine ecosystems. *ICES J. Mar. Sci.* 67, 717–731. doi: 10.1093/icesjms/fsp287
86. Solidoro, C., Bastianini, M., Bandelj, V., Codermatz, R., Cossarini, G., Melaku Canu, D., et al. (2009). Current state, scales of variability, and trends of biogeochemical properties in the northern Adriatic Sea. *J. Geophys. Res.* 114, C07S91. doi: 10.1029/2008JC004838
87. Stergiou, K. I., and Karpouzi, V. S. (2002). Feeding habits and trophic levels of Mediterranean fish. *Rev. Fish Biol. Fish.* 11, 217–254. doi: 10.1023/A:1020556722822
88. Swartz, W., Sala, E., Tracey, S., Watson, R., and Pauly, D. (2010). The spatial expansion and ecological footprint of fisheries (1950 to Present). *PLoS ONE* 5:e15143. doi: 10.1371/journal.pone.0015143
89. Tsikliras, A. C., Peristeraki, P., Tserpes, G., and Stergiou, K. I. (2015). Mean temperature of the catch (MTC) in the Greek Seas based on landings and survey data. *Front. Mar. Sci* 2:23. doi: 10.3389/fmars.2015.00023
90. Tsikliras, A. C., and Stergiou, K. I. (2014). Mean temperature of the catch increases quickly in the Mediterranean Sea. *Mar. Ecol. Prog. Ser.* 515, 281–284. doi: 10.3354/meps11005
91. Van Beveren, E., Fromentin, J.-M., Rouyer, T., Bonhommeau, S., Brosset, P., and Saraux, C. (2016). The fisheries history of small pelagics in the Northern Mediterranean. *ICES J. Mar. Sci.* 73, 1474–1484. doi: 10.1093/icesjms/fsw023
92. Zavatarelli, M., Raicich, F., Bregant, D., Russo, A., and Artegiani, A. (1998). Climatological biogeochemical characteristics of the Adriatic Sea. *J. Mar. Syst.* 18, 227–263. doi: 10.1016/S0924-7963(98)00014-1

Index